GAS ENGINE MANUAL

Edwin P. Anderson
Revised by Ted Pipe

THEODORE AUDEL & CO.
a division of
HOWARD W. SAMS & CO., INC.
4300 West 62nd Street
Indianapolis, Indiana 46268

SECOND EDITION

THIRD PRINTING—1981

Copyright © 1962, 1965 and 1977 by Howard W. Sams Co., Inc., Indianapolis, Indiana 46268. Printed in the United States of America.

All rights reserved. Reproduction or use, without express permission, of editorial or pictorial content, in any manner, is prohibited. No patent liability is assumed with respect to the use of the information contained herein. While every precaution has been taken in the preparation of this book, the publisher assumes no responsibility for errors or omissions. Neither is any liability assumed for damages resulting from the use of the information contained herein.

International Standard Book Number: 0-672-23245-6
Library of Congress Catalog Card Number: 76-45883

FOREWORD

The purpose of this book is to serve as a helpful guide to mechanics and students whose work deals with the operation, maintenance and repairs of modern gas engines of various types and sizes.

Although the gas engine principle is not new, it is only in recent years that its service as a prime mover has multiplied to include almost every field of human activity, from that of motor transportation, to the running of various appliances around the home. Other uses include outboard motors, marine and aircraft engines.

The book explains the operating principles of various types of gas engines, it then goes on to illustrate the function of the various engine parts and necessary accessories, such as carburetors, fuel ignition methods, cooling and lubricating systems, etc.

It also deals with troubleshooting and modern service operations, including engine tune-up and emission control procedures. The various ignition system items that affect engine performance are fully listed and illustrated.

Contents

Chapter 1

GAS ENGINE FUNDAMENTALS 9
  Operating principles

Chapter 2

CLASSIFICATION OF ENGINES 22
  Cooling—valve arrangement—multicylinder engines—cylinder arrangement—firing order

Chapter 3

GAS ENGINE PARTS 31
  Stationary parts—moving parts

Chapter 4

PISTONS 37
  Piston requirements—piston material—piston slap—constant clearance pistons—collapse of skirt—cam grinding—T-slot pistons—piston temperature—piston clearances

Chapter 5

PISTON RINGS 47
  Compression rings—oil rings—miscellaneous rings

Chapter 6

CONNECTING RODS AND WRIST PINS 59
  Connecting rod

Chapter 7

CRANKSHAFTS 65
Construction—crankshaft balance—crankshaft throw arrangement—built-up and single piece crankshafts—main bearings

Chapter 8

ENGINE FLYWHEEL 73
Torsional vibration—classes of vibration dampers

Chapter 9

VALVES AND VALVE GEARS 77
Valve seats—valve stem guides—camshafts and cams—how to design a cam—valve operating mechanism

Chapter 10

VALVE TIMING 87
How valves are timed—how to find the dead center—determining correct position of the camshaft

Chapter 11

LUBRICATING SYSTEMS 93
Combination lubricating systems—oil pumps—oil filter—oil gauge—engine oils

Chapter 12

COOLING SYSTEMS 105
Water circulation systems—variable-speed fan—a variable pitch flex fan—engine water jacket—radiators—water circulating pump—temperature control—radiator pressure cap—antifreeze solutions—water and oil circulating systems—air cooling systems—air and oil circulating systems

Chapter 13

FUEL SYSTEMS 119
Carburetor fuel system—purpose and types of fuel pumps—fuel filter—air cleaner—intake manifold—exhaust manifold—muffler—fuel tanks—fuel gauges—a fuel-injection system—gasoline—gas/oil mixture for two-stroke cycle engines

Chapter 14

CARBURETORS AND FUEL-INJECTION COMPONENTS 137
Air/fuel ratio—carburetor operating principles and types—fuel flow circuits—fuel injectors and associated components—engine speed governors

Chapter 15

EMISSION CONTROL SYSTEMS 169
Classification of controls—summary—catalytic converters

Chapter 16

FRICTION CLUTCHES 201
Clutch principles

Chapter 17

HORSEPOWER MEASUREMENT 207
Prony brake—rope brake—dynamometer—indicated horsepower—SAE horsepower—efficiency—thermal efficiency—mechanical efficiency—volumetric efficiency—piston displacement—super-chargers

Chapter 18

FUNDAMENTAL ELECTRICITY 219
Ohm's law—kinds of current—magnetism—electromagnetic induction—cells—cell circuits—primary induction coils—secondary induction coils

Chapter 19

IGNITION SYSTEMS 229
Electro-mechanical battery ignition—electronic battery ignition—magneto ignition.

Chapter 20

ELECTRICAL SYSTEM 253
Storage battery—generating system—a-c alternator system—starting system

Chapter 21

SPARK PLUGS 279

Chapter 22
TROUBLESHOOTING 285
    Service diagnosis—high oil consumption—engine noises

Chapter 23
ENGINE TUNE-UP 305
    Minor engine tune-up—reasons why spark plugs fail—major engine tune-up

Chapter 24
CYLINDER BLOCK SERVICE 319
    Reconditioning cylinder bores—removing carbon—scoring of cylinders

Chapter 25
PISTON AND PISTON RINGS SERVICE 325
    Expansion of pistons—removing pistons from cylinders—fitting pistons—piston ring service

Chapter 26
CONNECTING RODS AND CRANKSHAFT SERVICE 335
    Removing piston and connecting rod—fitting connecting rod bearings—crankshaft service—checking bearing clearance—clearance measurement

Chapter 27
VALVES AND VALVE GEAR SERVICE 341
    Reconditioning valves and seats—valve guides—refacing valves—refacing valve seats—reaming valve guides—valve spring—camshaft service

Chapter 28
CARBURETOR AND FUEL INJECTION 349
    Servicing the carburetor—cleaning carburetor parts—carburetion—service diagnosis—fuel-injection system service

Chapter 29
ELECTRICAL SYSTEM SERVICE 361
    Dwell—condenser—relationship of coil to condenser—spark timing—electronic ignition service—magneto ignition service—generator servicing—generator regulator service—alternator servicing—starting motor service—testing starting motor parts

Chapter 30
EMISSION CONTROLS SERVICES 383
    Fuel-evaporation emission controls.

INDEX 388

CHAPTER 1

Gas Engine Fundamentals

The gas engine is an internal combustion machine which derives its power from the heat generated when a compressed air-fuel mixture is ignited within its cylinders.

The fuel most commonly used in internal combustion engines as the source of power is gasoline. Other fuels include benzol, alcohol, fuel oil, butane, propane and natural gas. Any of these fuels may be used efficiently in the cylinder of an internal combustion engine.

OPERATING PRINCIPLES

The process by means of which the engine produces power is based on a fundamental law of physics which states that gas will expand upon application of heat. If the gas be confined, however, with no outlet for expansion, then the pressure of the gas will be increased when heat is applied as the result of igniting the compressed gas in an internal combustion engine.

In an engine, this pressure acts against the head of a piston causing it to move away from the combustion chamber. The piston being connected to the crankshaft by the connecting rod, converts this linear or straight line force into rotary motion and supplies power to a crankshaft and associated flywheel.

The air-fuel mixture is admitted to the engine intermittently, and the amount supplied at each admission is known as the *charge*. The combustion of each charge takes place under pressure attained by *compression* as a result of the upward movement of the piston after the charge is admitted and all valves closed.

The effect produced by igniting the mixture after compression is commonly called an *explosion* which is simply a quick burning or rapid combustion of the mixture.

This sudden explosion causes a high degree of heat within the combustion chamber, resulting in considerable initial pressure, and gives the piston an impulse, which decreases in intensity while the piston advances to make the power stroke by reason of the expansion of the gases. The products of combustion are finally exhausted from the cylinder.

The term "cycle" as applied to an engine is defined as a series of events which are repeated in regular order, and which constitute the principle of operation. Expressed briefly, the cycle of a gas engine embraces:

1. The admission of a fresh charge of gas and air into the cylinder.
2. Compression and igniting the explosive mixture.
3. Expansion of the ignited charge and absorption of its energy.
4. Expulsion of the burned gases.

These four events are called: *Admission (intake), compression, expansion (power) and exhaust.*

In operation of a gas engine the number of strokes required to complete the cycle varies with the type of engine. Thus, in a two-stroke cycle engine, the cycle of events is completed in two strokes of the piston. On such engines, each cylinder delivers a power stroke at every revolution of the crankshaft.

In the four-stroke cycle engine, on the other hand, the cycle is extended through four strokes, two downward and two upward, thus, there is only one power stroke for every two revolutions of the crankshaft.

Two-Stroke Cycle Engines

Two-stroke cycle engines, may be of the two port, or three port type with valve, or a variation of these. Valves may be of the rotary, poppet or reed type.

In two-stroke cycle engines, the crankcase is used as a receiver for the air-fuel mixture before it enters the cylinders through passages connected with port openings in the cylinder walls. The ports are covered and uncovered by the action of the piston as it moves up and down within the cylinder. See Fig. 1.

The air-fuel mixture enters the crankcase through the fuel admission port opening, equipped with a vacuum controlled poppet valve. The air-fuel ratio is controlled by carburetion action, the carburetor being connected between the fuel tank and the port opening in the crankcase.

Gas Engine Fundamentals

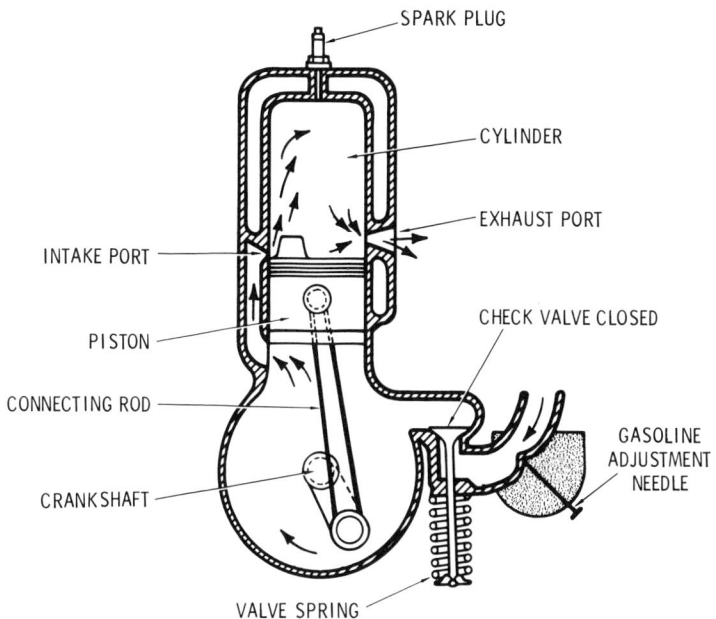

Fig. 1. The principal parts of a two-stroke cycle gas engine.

It is in this manner that the admission or intake port supplies the combustible mixture to the crankcase, and the exhaust port discharges the burnt gases into the exhaust pipe and muffler.

The basic principle of a complete operating cycle from the moment the mixture is ignited by the spark plug, will be as follows: With reference to Fig. 2, showing the operating features, it will be noted that at this point, the piston is almost at the end of its upward stroke and the crankshaft throw is about to pass over center.

A charge of gasoline and air have been pulled into the crankcase by the vacuum created during the upward movement of the piston. This mixture remains trapped in the crankcase because both the intake and exhaust ports are covered by the piston skirt. The fresh charge of gasoline and air are being compressed by the upward movement of the piston. The ignition timing mechanism causes the spark plug to fire at about this point.

In Fig. 3, the piston is part way down on the power stroke. The exploding gasoline expands and forces the piston downwards which causes the crankshaft throw to deliver power to pull the load. As the

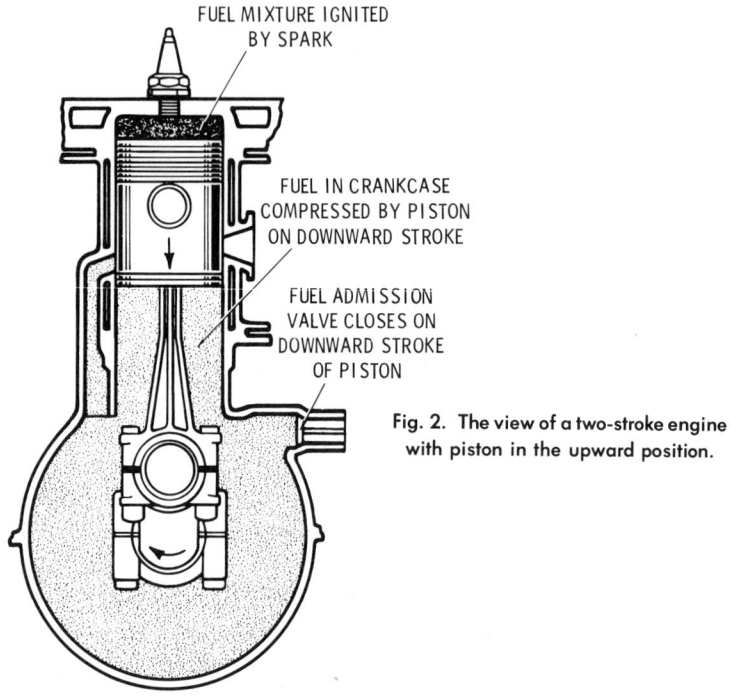

Fig. 2. The view of a two-stroke engine with piston in the upward position.

piston travels down the valve the reeds are forced onto their seats, preventing the mixture of air and gasoline trapped in the crankcase from escaping back into the carburetor. This mixture is slightly compressed by the downward travel of the piston.

In Fig. 4, the piston has reached the end of its downward movement or *power stroke*. The compressed crankcase mixture is now permitted to enter the intake port. As this fresh material speeds into the cylinder chamber it assists to expel the exploded materials through the exhaust port.

On the upward stroke of the piston, Fig. 5, a fresh charge of gasoline oil and air are taken into the crankcase (fuel admission valve open). As the piston on its upward movement reaches the top of the cylinder, the spark plug will again ignite the fuel charge. The downward movement of the piston results in a constant turning of the crankshaft and consequent supply of power to the load. The foregoing cycle must be repeated several times a second to maintain operation and produce usable power.

Gas Engine Fundamentals

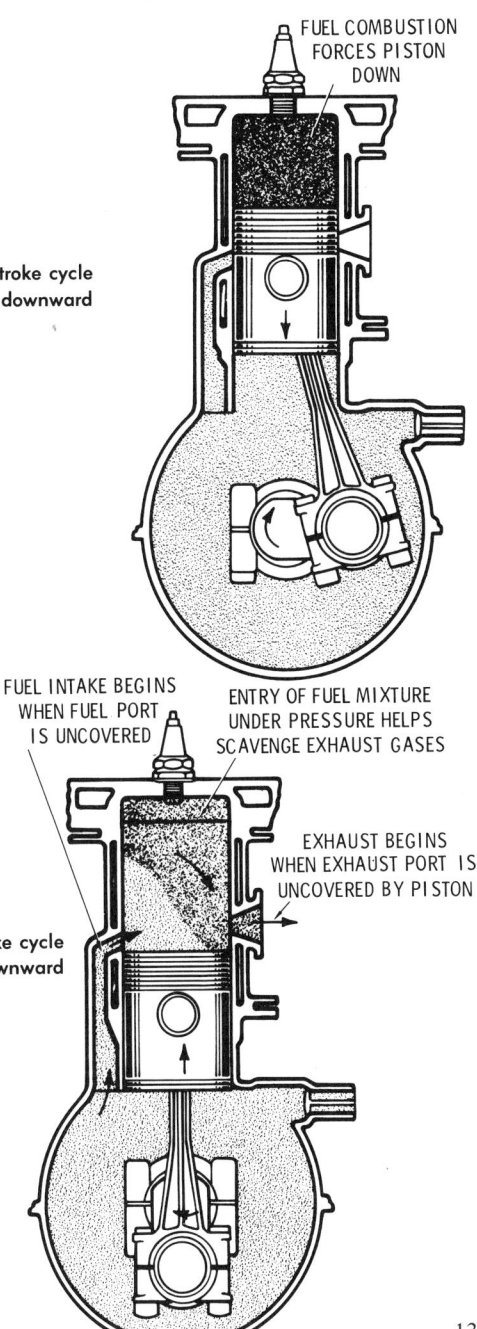

Fig. 3. The view of a two-stroke cycle engine with the piston on its downward movement.

Fig. 4. The view of a two-stroke cycle engine with the piston in the downward position.

13

GAS ENGINE MANUAL

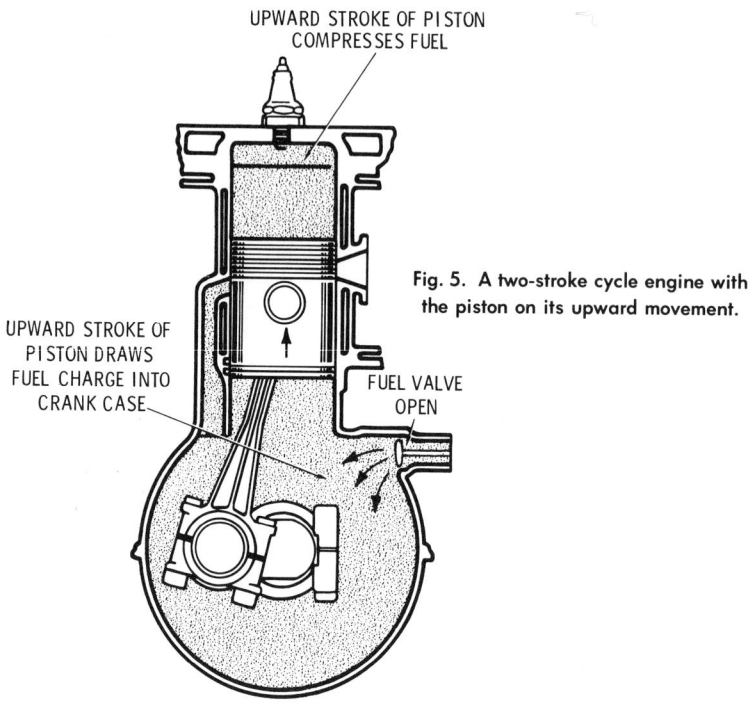

Fig. 5. A two-stroke cycle engine with the piston on its upward movement.

Four-Stroke Cycle Engine

Gas engines that need four strokes of the piston, two up and two down, to complete the cycle are known as *four-stroke cycle* engines. Thus, it requires two complete revolutions of the crankshaft to complete the cycle.

Engines of this type operate in the same manner as that of the well known automobile engine. As noted in Figs. 6 thru 9, which illustrate the four strokes forming the complete cycle, the inlet valve controls the fuel admission into the combustion chamber, and the exhaust valve as the name implies, controls the expansion of the burnt gases from the cylinder. These valves are controlled by an operating mechanism from the engine shaft. The other parts of the engine are substantially the same as on the two-stroke engine.

The fundamental difference between the operations of the four-stroke cycle and the two-stroke cycle engine is that in a four-stroke cycle single acting engine all operations take place separately within the cylinder

GAS ENGINE FUNDAMENTALS

above the piston, whereas in a two-stroke cycle engine the operations take place on both sides of the piston.

The four strokes comprising the working cycle are termed as: (1) *admission (intake)*; (2) *compression*; (3) *expansion (power)* and (4) *exhaust*. The foregoing events are produced substantially as follows:

The admission stroke Fig. 6 is the first step in the cycle and commences with the piston in its topmost position. As the piston moves downward a vacuum is created in the cylinder above the piston, permitting the air-fuel mixture to be drawn into the cylinder from the carburetor through the open fuel admission valve. When the piston reaches the bottom of its stroke the fuel admission valve closes, sealing the fuel mixture in the cylinder. The piston has now reached the bottom dead center. Note position of cams.

The compression stroke is the next step in the cycle. As noted in Fig. 7, at the beginning of the stroke the piston is at the bottom dead center position and both valves are closed and remain closed during the entire stroke. As the piston moves upward the air-fuel charge is compressed to as little as one-tenth of its original volume in the combustion chamber above the piston. The crankshaft at this point has made one complete revolution. This completes the compression stroke.

The third step in the cycle is the *power* or *expansion stroke*, and the only one that contributes to the engine output. As noted in Fig. 8, the piston is at its topmost or dead center position. The compressed fuel mixture at this time is ignited by an electric spark, generated at the spark plug gap. The heat of combustion causes the burning gases to expand, forcing the piston to move downward and produces mechanical energy to turn the crankshaft. Both the fuel admission and exhaust valve remains closed during the power stroke. The crankshaft at this point of the cycle has made one and one-half complete revolution.

The final step in the cycle is termed the *exhaust stroke*, since its function is to exhaust the burned gases in preparation for the commencement of another cycle. During the upward movement of the piston, from the lower to the upper dead center, Fig. 9, the exhaust valve gradually opens to expel the burnt gases from all parts of the cylinder. When the piston again reaches its maximum upward position the exhaust valve closes, and the cylinder is then ready for another cycle.

GAS ENGINE MANUAL

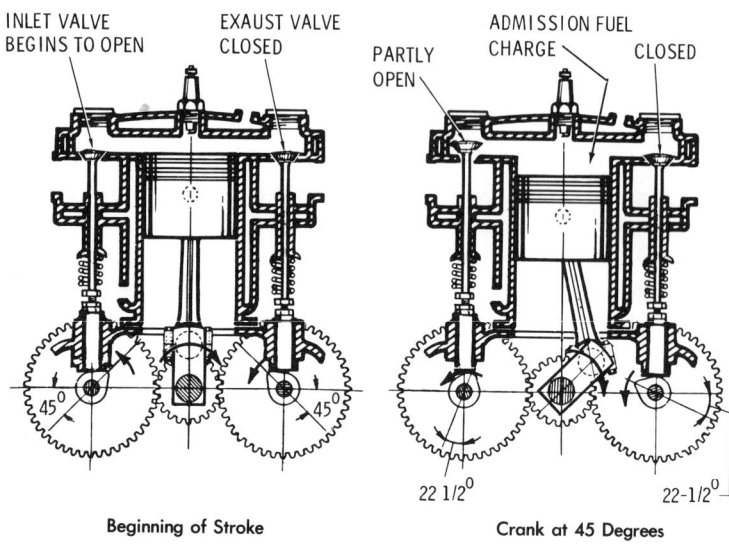

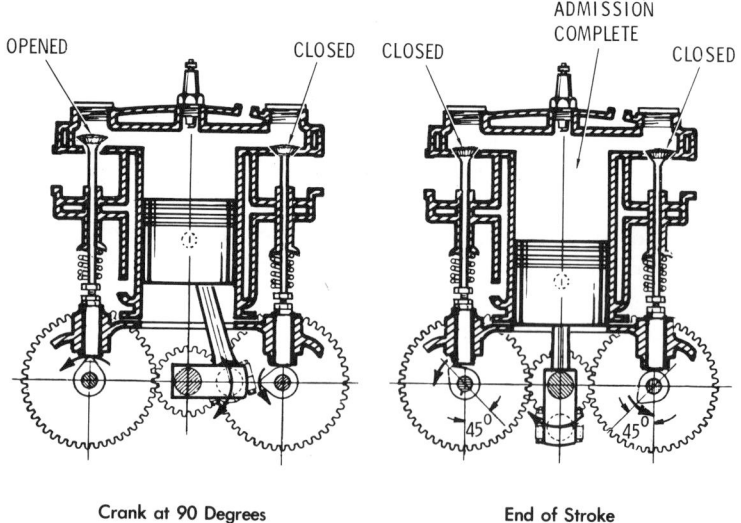

Fig. 6. Illustrating the admission stroke in a four-stroke engine.

Gas Engine Fundamentals

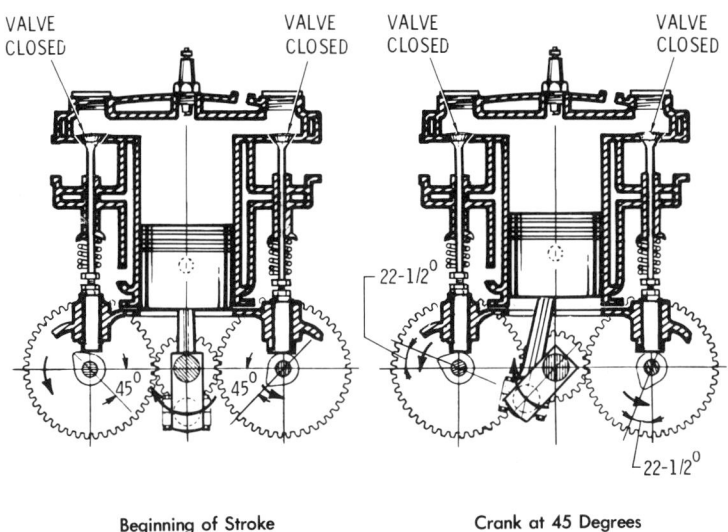

Beginning of Stroke

Crank at 45 Degrees

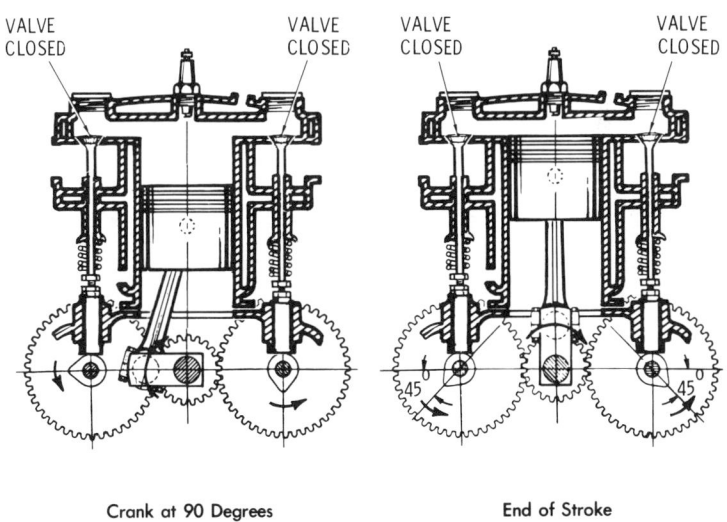

Crank at 90 Degrees

End of Stroke

Fig. 7. Illustrating the compression stroke in a four-stroke engine.

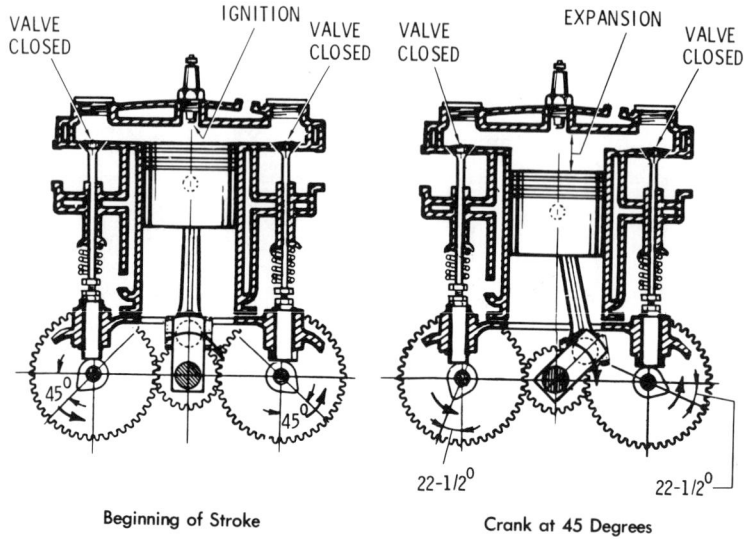
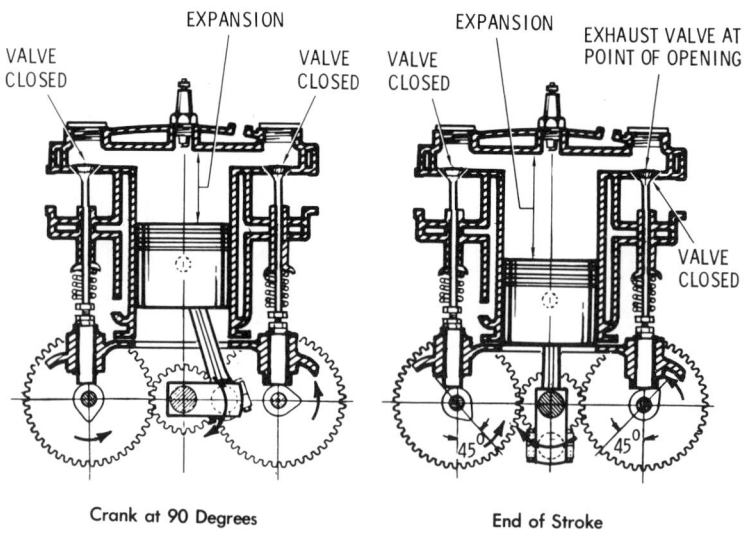

Fig. 8. Illustrating the power stroke.

Gas Engine Fundamentals

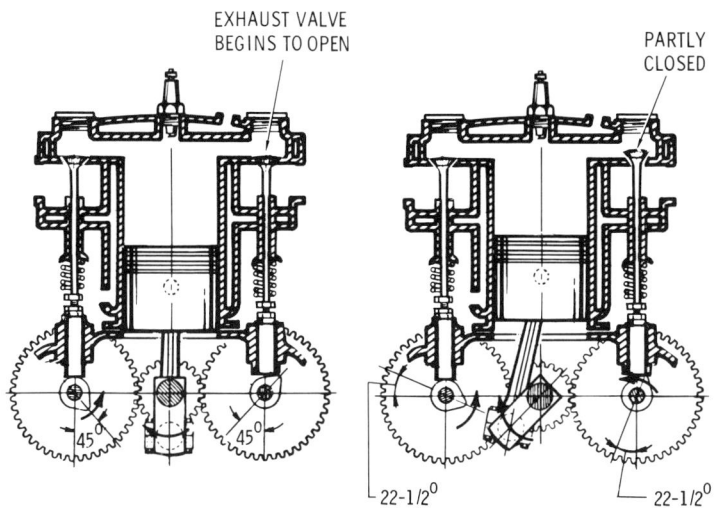

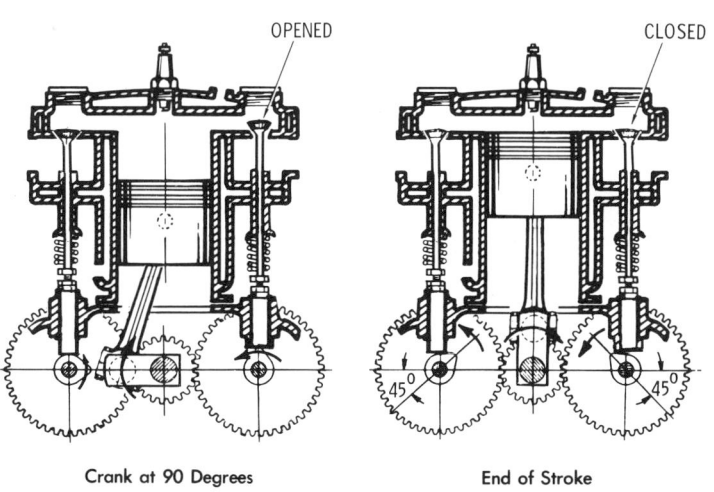

Fig. 9. Illustrating the exhaust stroke.

It will be noted that during the step-by-step description of the four-stroke cycle, the crankshaft made two complete revolutions and the piston made four strokes. The same sequence of *admission, compression, expansion* and *exhaust* must always occur in the same order and must be repeated several times a second to operate the engine and produce useful power.

Valve Overlap

In the foregoing series of illustrations it was assumed that the valves begin to open when the piston passes its dead-center, but this, however, is not the case in the highly developed high speed engines. To compensate for the time required by the air or gas to flow through the manifold, the valves are timed to overlap a certain amount, the amount of overlap depending upon the design and speed of the engine.

In theory, the admission valve should open at the exact time the piston starts down in the cylinder and should close at the instant the piston starts upward in the cylinder on the compression stroke. Both the admission and exhaust valves should remain closed during compression and expansion, and the exhaust valve would then open at the end of the power stroke and close at the end of the exhaust stroke. Such an arrangement may be satisfactory for a slow speed engine, but with increased speed a certain amount of valve overlap is necessary to obtain better distribution of the charge.

Power Output

In our discussion of the fundamental principles of operation, it will be noted that the two-stroke cycle engine delivers twice as many power impulses per cycle to the crankshaft as the four-stroke cycle type. Theoretically then, the two-stroke cycle engine would have twice the power output as a four-stroke cycle engine of the same size. This, however, is not true, because of the waste in fuel and power when some of the incoming fuel mixture combines with the exhaust gases and goes out with them. The volumetric efficiency of the two-stroke cycle engine is thus reduced to a considerable extent.

Comparison of Engines

As previously noted, the two-stroke cycle engine has the advantage of extreme simplicity, owing to the absence of valves or other moving parts which would be likely to require adjustment and care.

Gas Engine Fundamentals

This valveless feature of the two-stroke cycle type, while providing simplicity will, at the same time, give rise to certain irregularities in the action of the engine.

The action of the gas in the cylinder is somewhat uncertain, since it is hardly to be expected that the inflow of gas will continue exactly long enough to fill the cylinder. Because of this it is entirely possible that either some of the exhaust gases may not have time to escape or that some of the fresh charge may pass over and out through the exhaust port.

There are also more disadvantages which may be termed structural. Intake and exhaust ports of a two-stroke cycle engine are cut into the cylinder wall instead of in the top of the combustion chamber as in a four-stroke cycle engine, and while the working parts of the two-stroke cycle type are simple they are entirely enclosed and thus are not easily examined and serviced.

As the crankcase must be gas tight, any leakage around the crankshaft bearings causes a pressure loss with consequent loss of power. Also any leakage around the piston will allow the partially burnt gases to pass down and deteriorate the quality of the fresh gas in the crankcase.

The four-stroke cycle engine, although more complicated, is more dependable in its action, as the behavior of the gas is mechanically controlled. Owing to the mechanical regulation there is less chance of fuel waste, and the economy is therefore somewhat greater than those of the two-stroke cycle type. Since no enclosed crankcase is necessary in the four-stroke cycle engine, the working parts are more easily accessible for adjustment and servicing.

As a general conclusion it may be stated that for small light engines which receive little attention and where economy is not of great importance, the two-stroke cycle engine is preferred. For single-cylinder engine, the two–stroke cycle type is preferable because the vibration is less than in the single-cylinder engine of the four-stroke cycle type.

For engines of larger size, where increased reliability and fuel economy is an important factor, the four-stroke cycle engine is preferable.

CHAPTER 2

Classification Of Engines

Gas engines may be classified according to cycle arrangement whether of the two-stroke or four-stroke cycle, type of cooling employed or valve and cylinder arrangement. They all operate on the internal combustion principle, and the application of basic principle of construction to particular needs or system of manufacture has caused certain designs to be recognized as conventional.

COOLING

Engines may also be classified as to whether they are air or liquid cooled. All engines are cooled by air to some extent, but air cooled engines are those in which air is the only external cooling medium. Lubricating oil and fuel assist somewhat in cooling all engines, but there must be an additional external means of dissipating the heat absorbed by the engine during the power stroke.

Air Cooled

Air cooled engines are used extensively on outboard motor boats, lawn mowers, motor scooters, air-craft, etc. This type of engine is employed where there must be an economy of space and weight, since it does not require a radiator, water jacket, or pump to circulate the coolant.

In air cooled engines, the cylinders are cooled by conduction of heat to metal fins on the outside of the cylinder wall and head. To accentuate the cooling, air is circulated between the fins. Also when possible, the engine is installed so that it is exposed to the air stream of the vehicle or appliance

CLASSIFICATION OF ENGINES

and has baffles to direct the air over the fins. If the engine cannot be mounted in the air stream, a fan is customarily employed to force the air through the baffles.

Liquid Cooled

Liquid cooled engines are those that require a water jacket to hold the coolant around the valve ports, combustion chambers and cylinders; a radiator to dissipate the heat from the coolant to the surrounding air and a pump to circulate the coolant through the engine. Liquid cooled engines also require a fan to draw air through the radiator for proper dissipation of heat.

VALVE ARRANGEMENT

Engines may be classified according to the position of the admission and exhaust valves, that is, whether they are located in the cylinder block or in the cylinder head. Various arrangements have been used, but the most common are the **L-head, T-head** and **I-head**. The latter designation is used because the shape of the combustion chamber resembles the form of the letter identifying it.

L-Head

The L-head as shown in Fig. 1, has both valves on the same side of the cylinder. The valve operating mechanism is located directly below the valves, and one camshaft serves both the admission and exhaust valves.

T-Head

The T-head type of valve arrangement Fig. 1, has one or more admission valves on one side with the exhaust valve or valves on the opposite side of the cylinder. This type of arrangement has been largely supplemented by the L-head, the disadvantage being that two complete valve operating mechanisms were required.

I-Head

Engines employing the I-head construction, Fig. 1, are commonly termed "valve-in-head" or "overhead" valve engines, because both the admission and exhaust valves are mounted in the cylinder head above the cylinder. This arrangement requires a lifter, a push rod, and a rocker arm above the cylinder to reverse the direction of valve movement, but only one camshaft is required for both valves, as noted in Fig. 2.

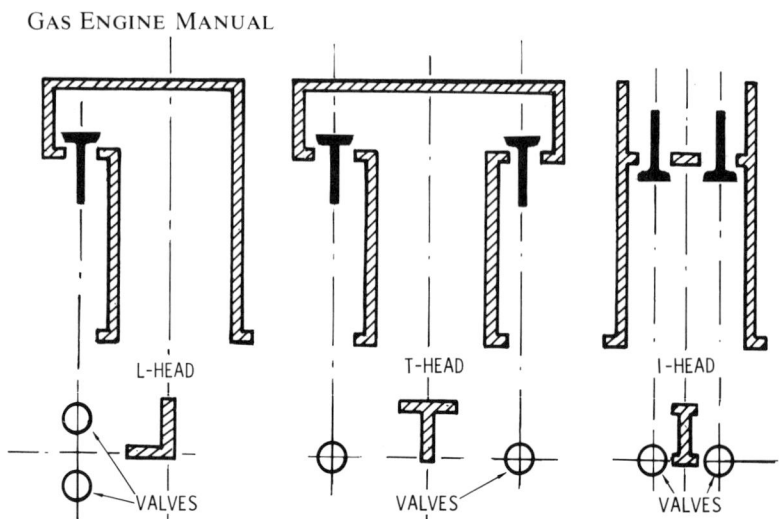

Fig. 1. Valve arrangements for various type engines.

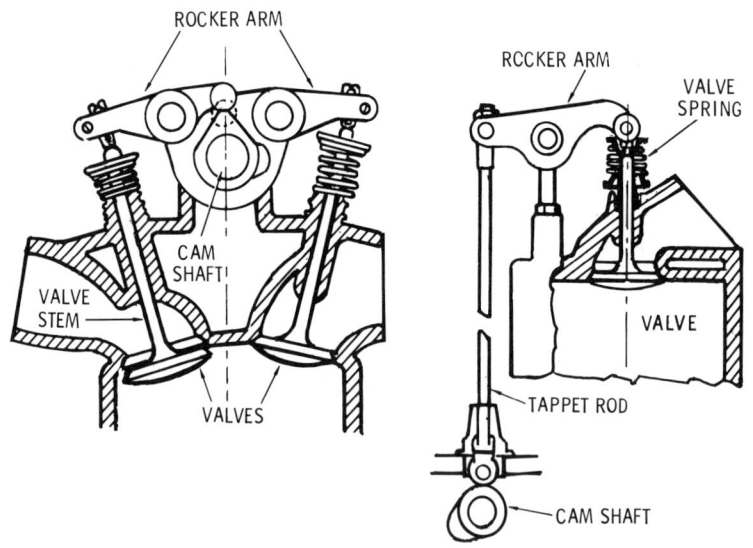

Fig. 2. Overhead camshaft with rocker arms, and overhead valves operated by tappet rods from the camshaft.

Each of the foregoing valve arrangements have certain advantages and disadvantages, although the I-head or "valve-in-head" type is favored because it provides the most efficient distribution of the charge in and out of the cylinder in addition to being the best arrangement for obtaining maximum power.

MULTICYLINDER ENGINES

The number of cylinders required in a particular engine depends upon its size and application. Due to the fact that the four-stroke cycle gas engine delivers only one power stroke in two revolutions of the crankshaft, a one cylinder engine would be very objectionable in automotive service. Hence modern practice in automotive and similar applications is to divide the power between several cylinders.

In early automotive applications, one cylinder, two cylinder and some three cylinder power plants were used, but the many advantages of a larger number of cylinders soon led to their adoption throughout the industry.

Amongst the disadvantages of one or two cylinder engines are:
1. The uneven turning effect.
2. Excessive vibration.
3. Large cylinders required.
4. Tendency to stall.

In order to overcome these defects, additional cylinders were provided. Thus, for a time the four cylinder engine reigned supreme; then came engines with six, eight, twelve and sixteen cylinders.

It is usually impracticable, however, to employ more than eight cylinders in automotive application on account of the greater length of the power plant and the much stronger and heavier crankshaft required. When the number of cylinders are increased above six, the solution for the best arrangement is found in two sets of cylinders inclined at an angle, thus providing an engine of the same length but increased power. This engine is called the V-type. In this type of engine the angle between the cylinders usually varies from 90° to 45°. The advantages of multiple cylinders are: decreased vibration, greater flexibility, overlapping power strokes, lighter reciprocating parts and higher speeds.

To clearly understand the necessity for multicylinder engines where larger amounts of power are required, the comparison of a four cylinder engine with a one cylinder engine of equal power will be considered.

Example.—Assume the four cylinder engine has a 3 in. diameter cylinder. Its area is 7.07 sq. ins. and if the pressure of combustion is 300 lbs. then the sudden piston load will be 7.07 × 300 or 2121 lbs.

The piston of a single cylinder engine of equal horsepower must consequently be four times the area of the four cylinder engine or $7.07 \times 4 = 28.28$ sq. ins. The load absorbed by the piston at ignition would in this latter case be 28.28×300 or $8,484$ lbs.

Now, since the piston in the single cylinder engine will be four times larger in area for the same power and pressure per square inch, it follows that the parts of the single cylinder engine must be much heavier than those of the four cylinder engine. Also, a single cylinder engine would easily stall if a sudden load were applied owing to the intermittent nature of the torque. This defect has been partly eliminated by the addition of cylinders. The use of the greater number of cylinders has resulted in a more flexible engine. In this connection it should be noted, however, that the foregoing advantages are obtained at an additional gasoline consumption and cost with respect to maintenance and repair.

Because of constantly increasing gasoline taxes, the present tendency toward a more economical engine has resulted in the development of automotive engines with fewer cylinders and less power per engine. Thus, at the present time four and six cylinder engines are again gaining favor, resulting in smaller and more compact engines and cars.

CYLINDER ARRANGEMENT

Gas engine cylinders also vary in their arrangement. Cylinder arrangement in liquid cooled engines is usually "in line" or "V-type," whereas in air cooled engines it is generally radial or horizontally opposed.

In-Line

The vertical in-line cylinder arrangement is one of the most commonly used types. In a design of this sort all the cylinders are cast or assembled in a straight line above a common crankshaft which is mounted immediately below the cylinders. A variation of this design is the inverted in-line type. See Fig. 3.

V-Type

In the V-type engine (Fig. 3) two "banks" of in-line cylinders are mounted in a V-shape above a common crankshaft. This type is designated by the number of degrees contained in the angle between the banks of cylinders. In automotive use the angle of the V is usually 90° for eight

CLASSIFICATION OF ENGINES

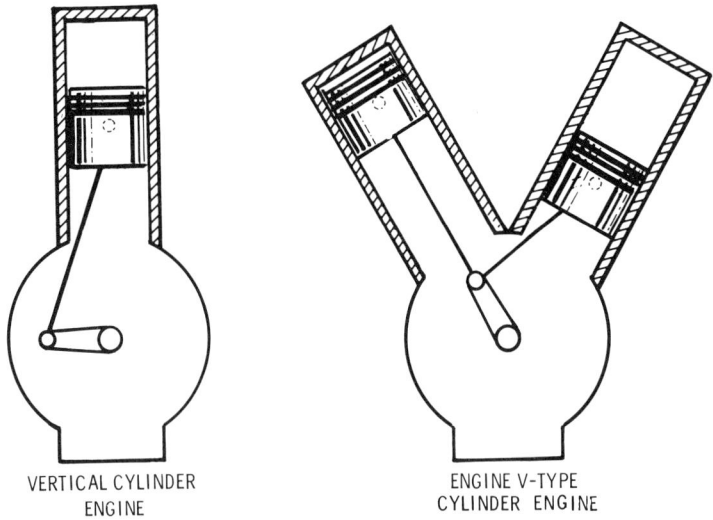

VERTICAL CYLINDER ENGINE   ENGINE V-TYPE CYLINDER ENGINE

Fig. 3. Illustrating the cylinder arrangement in vertical (in line) and V-type engines.

cylinder engines; 75, 60 or 45° for 12 cylinder engines, etc. Crankshafts for V-type engines generally have only half as many throws as there are cylinders, since two connecting rods, one for each bank, generally are connected to each throw.

Horizontally Opposed

The horizontally opposed engine (Fig. 4) has its cylinders placed horizontally in two rows 180 degrees apart with the crankshaft mounted in the center. This type of cylinder arrangement is used primarily for conservation of space, and is found in passenger busses, and similar applications, where the height of the engine would prevent its normal placement.

Radial

The radial engine, as used on aircraft, has cylinders placed in a circle around the crankshaft. In this design the crankshaft has only one throw, and one piston is connected to this throw by a "master" rod. The connecting rods from the other pistons are fastened to the master rod, making the power flow first to the master rod and then to the crankshaft. The result is the same as if each rod were connected directly to the

GAS ENGINE MANUAL

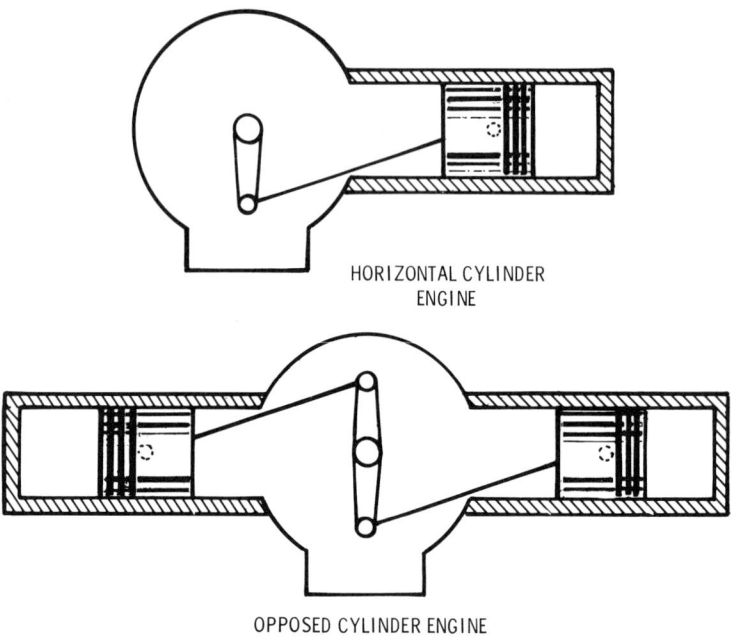

Fig. 4. Horizontal and horizontally-opposed cylinder arrangement.

crankshaft. High powered radial engines may have two rows of cylinders in which each row operates on its own throw of the crankshaft.

FIRING ORDER

There are two possible firing orders or methods of timing for four cylinder engines, as 1-2-4-3 and 1-3-4-2, neither of which has any appreciable advantage over the other. See Fig. 5.

With six cylinder engines, there are six possible firing orders, and if the cylinders are cast in pairs, and the exhaust passages of the cylinders in each pair are "siamesed," or come close to the cylinders, it is found advantageous to choose one of the following orders, in which the two cylinders of any pair never fire in direct succession, as 1-5-3-6-2-4, 1-4-2-6-3-5 or 1-3-2-6-3-5.

The order in which the cylinders will fire depends upon the sequence of cranks and cams and ignition hookup. The construction being such that

28

CLASSIFICATION OF ENGINES

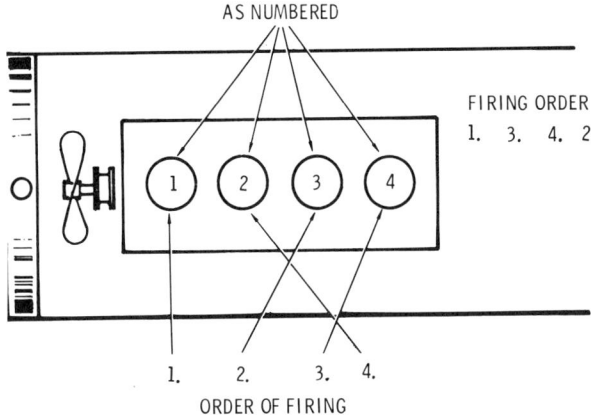

Fig. 5. Typical firing order of a four-cylinder engine.

they will not fire in consecutive order from front to rear of engine, but in some other order selected by the engine designer, the object being to reduce vibration as much as possible.

CHAPTER 3

Gas Engine Parts

In the preceding chapters, the gas engine has been treated with respect to its principles of operation; the explanations being accompanied by elementary diagrams. Drawings of this kind, however, do not show the construction details but only the principles upon which the engine operates.

In this connection it should be noted that due to the various services to which a gas engine is being put the designs will vary greatly as will the component parts; the operating principles, however, will in each instance be similar.

The various parts which make up the engine proper may be classed as:
1. **The stationary parts:**
    a. Cylinders
    b. Crankcase
    c. Oil pan
2. **The moving parts:**
    a. Pistons
    b. Connecting rods
    c. Crankshaft
    d. Flywheel
    e. Valves
    f. Valve gear

In addition to these parts, various necessary accessories are attached to the engine as shown in Fig. 1:
1. **Fuel system**
2. **Ignition system**

Gas Engine Manual

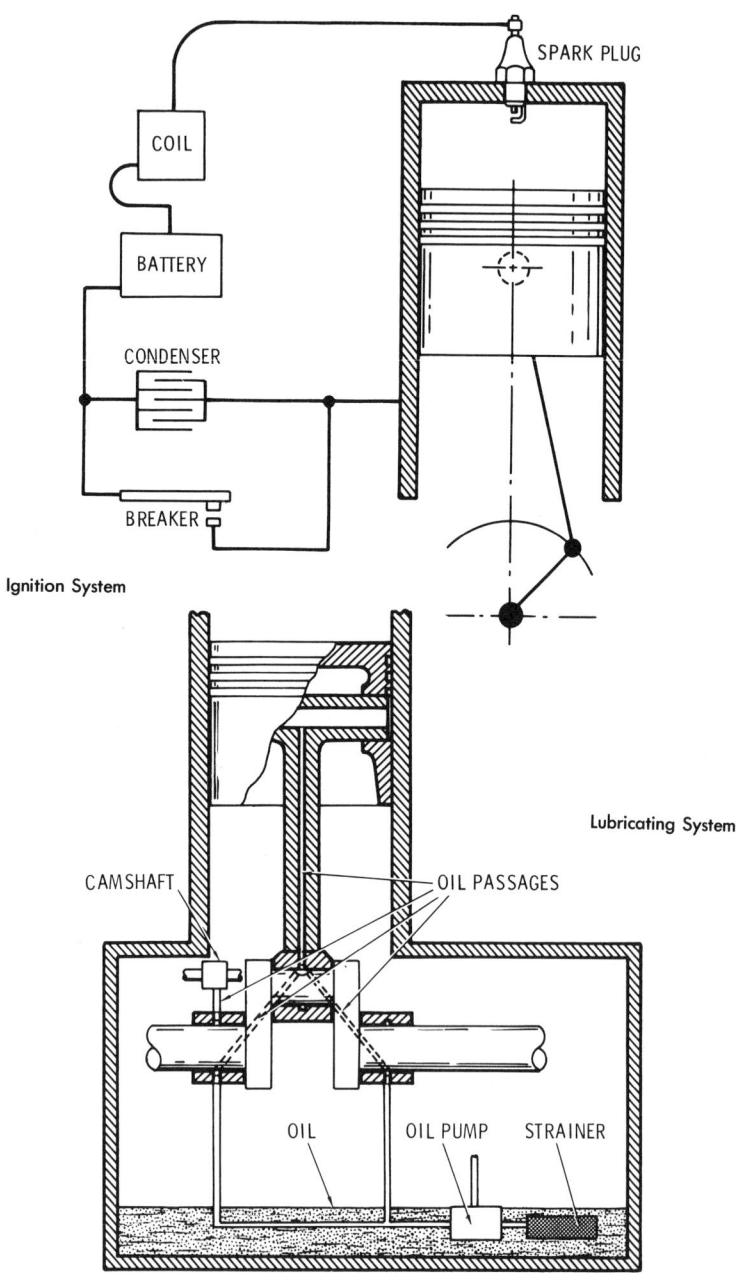

Fig. 1. Necessary

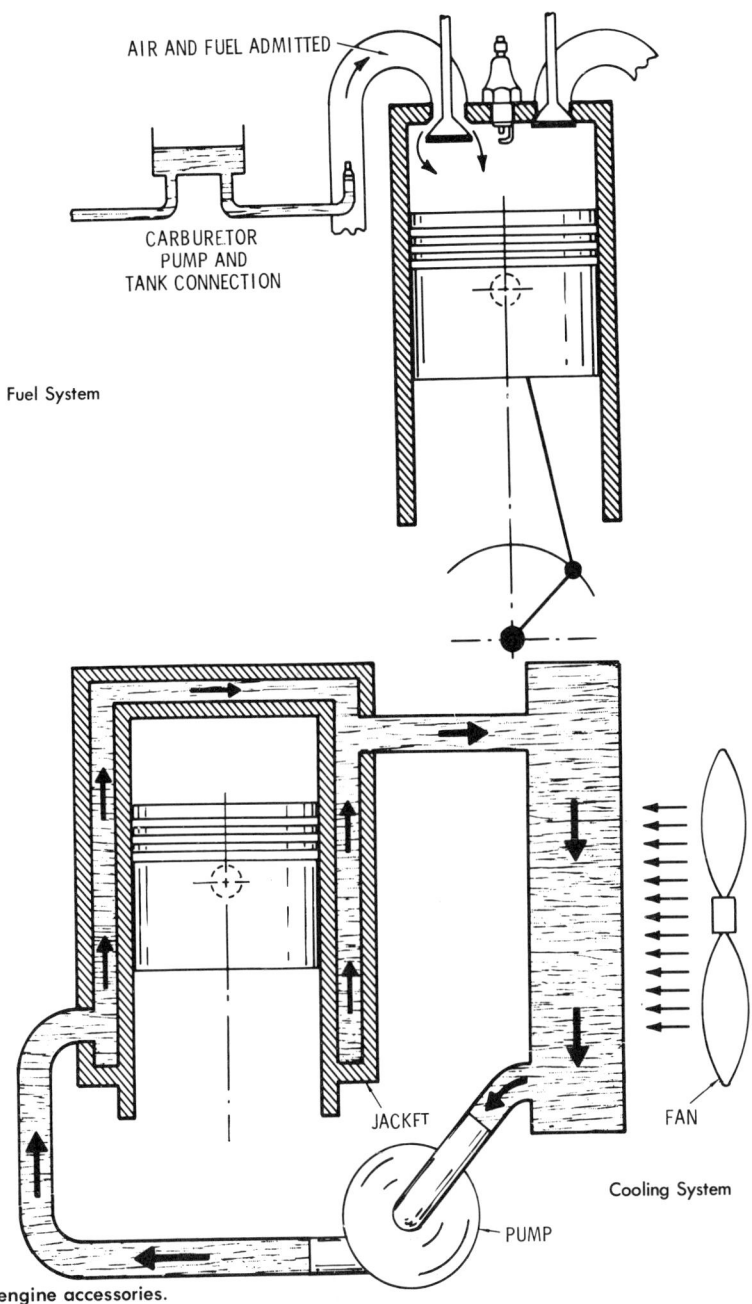

engine accessories.

3. Lubrication system
4. Cooling system

STATIONARY PARTS

In the make up of a gas engine the main stationary parts are: (1) the cylinder casting, (2) the crankcase (3) the cylinder head and (4) the oil pan.

Combined Cylinder and Crankcase Casting

In late engine construction, the practice has been in most cases to combine these two parts in one casting, that is; the cylinder and crankcase are cast "in-block" or in one piece called the cylinder block. The advantage of the in-block method are so numerous that it has become almost universal. Casting cylinders in-block produces more compact, shorter, and more rigid construction at less cost than casting cylinders singly or in pairs. It also simplifies the assembly, and provides for a simplified enclosure of the valve operating mechanism.

In the in-block design, the crankcase is usually extended a short distance below the center line of the crankshaft. This construction is what is known as the "split" type of crankcase, that is, the lower part enclosing the bottom of the crankcase or oil pan is separated and bolted to the lower flange of the crankcase as distinguished from the "Barrel Type" in which the lower part is integral.

Cylinder Head

The cylinder head is a detachable portion of the engine, fastened securely to the cylinder block. It contains passages which match those of the cylinder block and allow the coolant to circulate in the head. All cylinder heads are formerly made of cast iron, but numerous engines are now being built with cylinder heads of cast aluminum alloy because of its better conduction of heat, thus providing improved engine cooling.

Oil Pan

The lower member of the split-type crankcase or oil pan consists of a pressed steel stamping which is attached to the lower flange of the crankcase by a number of cap screws or small bolts. Horizontal baffles are placed across the pan at proper points to reduce oil splashing.

The oil pan usually has a depression forming an oil sump in which is located a drain plug at the lowest point which permits complete oil

GAS ENGINE PARTS

drainage. In engines using the splash system of lubrication, there are two pans, the lower or oil pan proper which carries the oil and the upper pan or trough into which scoops in the end of the connecting rods dip and scoop up oil supplied from the lower pan.

Attached to the cylinder block assembly are various auxiliary devices such as fuel, ignition, lubrication and cooling systems, but owing to the space required to properly explain these devices they are presented in separate chapters.

MOVING PARTS

The various moving parts which act to transform the energy contained in the incoming fuel mixture finally into useful power, may be classed as:

1. **The reciprocating parts:**
    a. Piston
    b. Connecting rod
    c. Valve gear
2. **The rotating parts:**
    a. Crankshaft
    b. Camshaft
    c. Auxiliary shafts, for pump operation, etc.

The degree of perfection in operation of the engine depends largely upon the proper proportion and efficient arrangements of these parts, and is an index of the engineering ability of the designer of the engine.

CHAPTER 4

Pistons

The piston is one of the most important parts of the engine mechanism. For the gas engine, it consists essentially of a long cylindrical casting, closed at the top and open at the bottom end, having attached at an intermediate point a wrist pin which transmits the thrust due to the burning gases, to the connecting rod and thence to the crankshaft.

A piston performs the following duties:
1. It transforms heat energy into mechanical energy.
2. It transmits the lateral thrust, due to the angularity of the connecting rod, to the cylinder walls.
3. It carries off some of the excessive heat generated in the combustion chamber.

Pistons employed in early engines were crude affairs, as compared with late designs. The improved designs are a result of the requirements becoming more and more severe to meet the needs of increased piston speeds and higher compression pressures.

PISTON REQUIREMENTS

The design of the piston is controlled to some extent by the design of other parts of the engine, and this accounts for the fact that all engines do not use the same kind of piston. One problem with respect to piston design is to produce a piston that will run with the minimum amount of clearance. See Fig. 1.

PISTON MATERIALS

The materials commonly used in the manufacture of pistons are:
 (a) Cast iron.

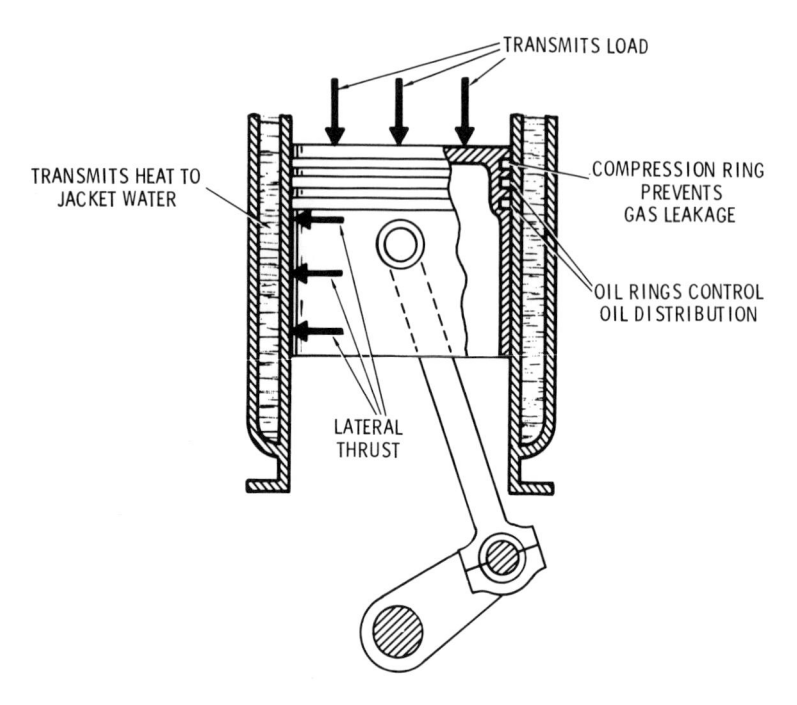

Fig. 1. Typical water-cooled cylinder assembly.

(b) Semisteel.
(c) Aluminum alloy.

Cast Iron Pistons

The trouble with cast iron for pistons is its weight, being nearly three times as heavy as aluminum alloy. Moreover it is not as good a heat conductor as aluminum alloy.

Formerly, before engine speeds and compression became so great as now, the above properties of cast iron were not so objectionable. However, a heavy piston running at several thousand strokes per minute has considerable inertia, that is, great force must be applied to start and stop the piston at the ends of the stroke resulting in excessive vibration and greater stresses on the bearings.

To reduce vibration and heavy stresses, attempts were made to make the piston as light as possible by the rib construction. The ribs extending across the head and down the skirt were depended on almost entirely for strength.

PISTONS

An advantage of cast-iron pistons is that they have a rate of expansion the same as that of the cylinder and therefore can be fitted with minimum clearance.

Semisteel Pistons

The tendency in the construction of semisteel pistons is to make them very light, in fact some designs weigh little more than aluminum alloy pistons. Owing to the very thin walls of these pistons, they are slightly flexible and will shape themselves to the cylinder wall in which they are working.

Aluminum Alloy Pistons

A desirable characteristic of the aluminum alloy piston is its lightness. In modern high speed engines, the pistons have to travel up and down, that is, reciprocate from a few hundred to thousands times a minute.

A weight moving up and down at such high speed requires a very great force to stop it and a very great force to start it because of its *inertia*.

This force due to inertia is transmitted by the connecting rod to the bearings and it is this force which pounds against the bearings and produces vibration. It is accordingly important that this force be reduced to a minimum and that was the reason for the introduction of aluminum alloy pistons, since aluminum is very light. See Fig. 2.

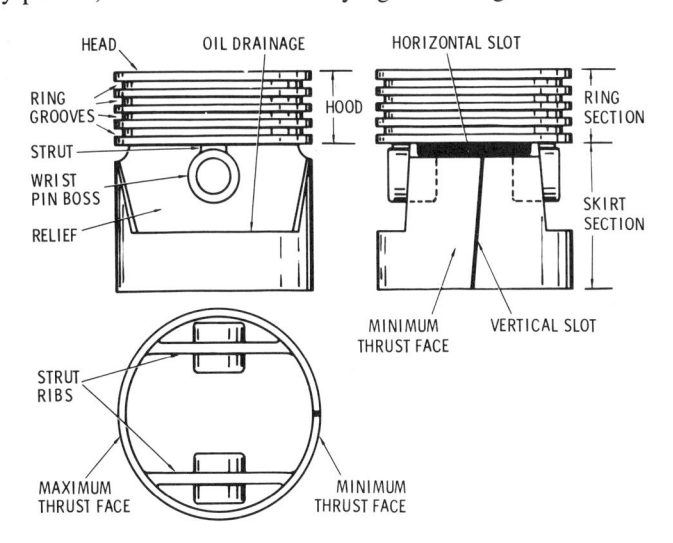

Fig. 2. Showing the various parts of the piston.

To reduce the weight of cast-iron pistons they must be made very thin. While this might give the desired lightness, a thin piston offers so much resistance to heat flow that the piston would become unduly hot. Quicker acceleration is possible with aluminum alloy pistons because of the reduction in weight.

Another advantage is that aluminum is a much better conductor of heat than iron, hence the temperature of the head of the piston is lower than when made of cast iron.

A disadvantage is that aluminum alloy expands much more than cast iron with rise of temperature, as a result early aluminum alloy pistons were made with considerable clearance so that they would not be too large at the working temperature. This resulted in piston slap.

PISTON SLAP

Piston slap is caused by the piston moving from one side of the cylinder to the other. It is a result of the forces acting on the piston under influence of the angularity of the connecting rod as illustrated in Fig. 3.

During the complete cycle, the piston travels from one side of the cylinder to the other. All of these changes, however, are rather gradual

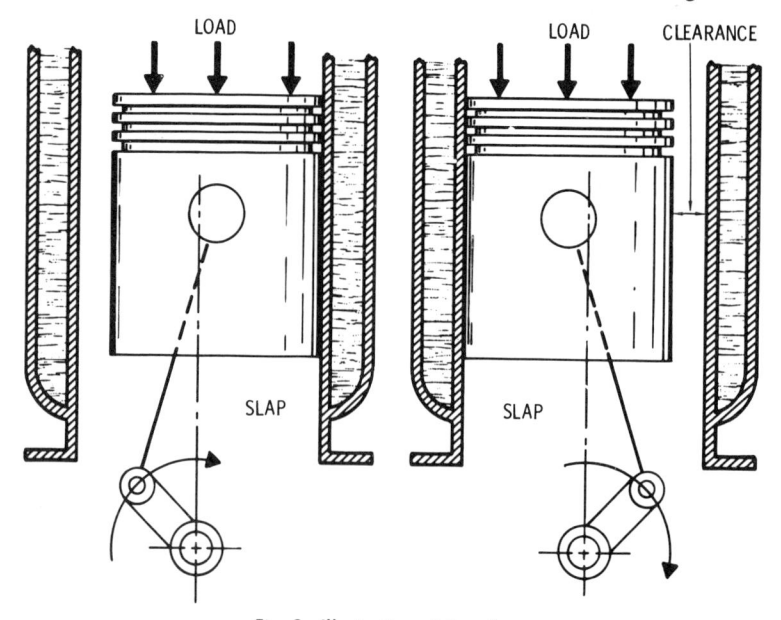

Fig. 3. Illustrating piston slap.

except the one at the beginning of the power stroke, which produces the noise known as *piston slap*.

It is not possible to prevent the piston moving from side to side in the cylinder, but the noise due to slapping can be reduced by reducing the clearance. Since aluminum alloy expands when heated at a much greater rate than cast iron, this was not possible in the early solid skirt aluminum alloy pistons, but designers overcame this in an improved construction known as the constant clearance piston.

CONSTANT CLEARANCE PISTONS

It was found that most of the heat which enters the piston leaves through the rings, and that the temperature of the skirt is not high enough to account for all the clearance which was formerly necessary to give to aluminum alloy pistons. The skirt expansion was therefore due to two causes:
1. The expansion caused by the rise in temperature of the skirt.
2. The distortion of the skirt caused by the expansion of the much hotter piston head.

By means of slots in the skirt it has been possible to avoid the two effects just mentioned, thus producing a piston that will work with a greatly reduced clearance.

By cutting a vertical slot as shown in Fig. 4, constant clearance is obtained, with the result that when heated, the skirt expands circumferentially instead of radially, which causes the diameter of the skirt to remain constant.

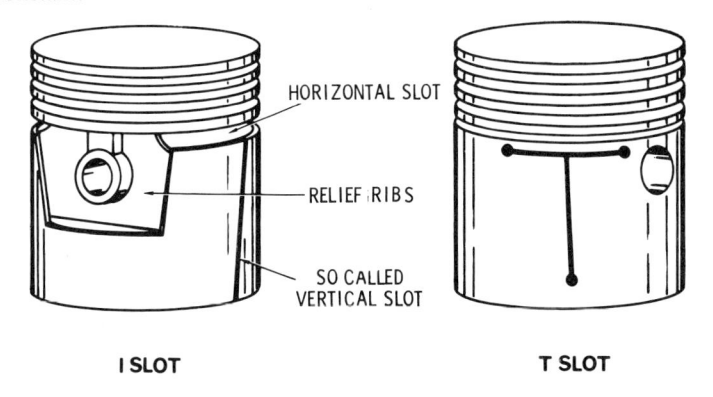

Fig. 4. Piston slots as a means to compensate for skirt expansion due to temperature variations.

The horizontal slot at the top of the skirts partially isolates the skirt from the much hotter ring section and thus avoids distortion by the expansion of the head.

By means of ribs connecting the head and wrist pin bosses, the load on the piston head is transferred directly to the wrist pin, thus relieving the skirt of the load.

COLLAPSE OF SKIRT

As pistons wear, or if they become overheated, the skirt is liable to collapse or become smaller in diameter. This condition will cause piston slap in the cylinder, and high oil consumption due to the increased clearance caused by the collapse of the skirt.

To remedy this, piston expanders are sometimes used. There are a great number of these devices on the market. They vary in design but commonly consist of a spring steel strut that is compressed and installed in such a manner that it exerts internal pressure on the split piston skirt.

CAM GRINDING

Many aluminum alloy pistons are *cam ground*, that is, they are purposely machined with the skirts oval. This oval shaping of the piston commonly referred to as "cam grinding" is intended to compensate for expansion. The reason for this is that the skirts will be slightly oval when cold in such fashion that the thrust faces will have the greater diameter, but the skirt will become more nearly round when the piston expands at operating temperature. Fig. 5.

T-SLOT PISTONS

This type of piston shown in Fig. 6 combines the features of solid and split skirts, that is, the vertical slot is not extended all the way to the bottom of the skirt. The term "T-slot" is derived from the resemblance of the two slots to the letter T.

As noted in Fig. 6, the horizontal slot is placed in the lower groove or near the top of the skirt according to the results aimed at by the designer. Drilled holes are placed at the ends of the slots to prevent possible splitting due to stresses set up by the expansion.

PISTONS

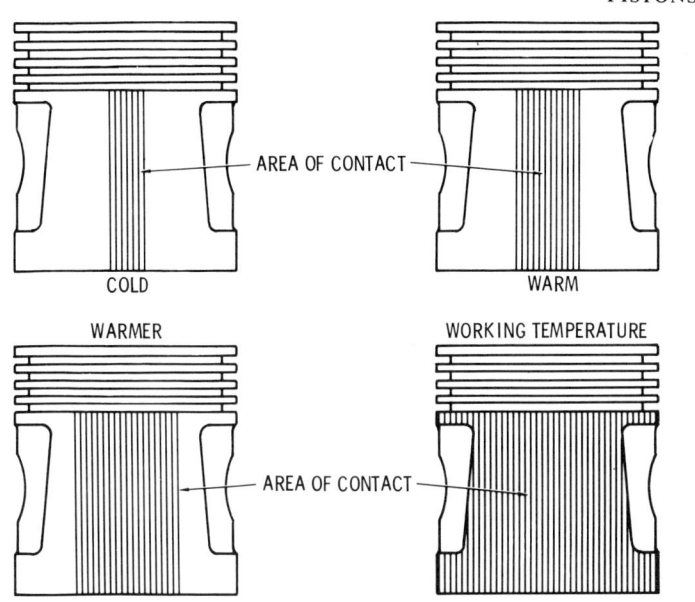

Fig. 5. Variable contact of the piston skirt face with the cylinder walls due to temperature changes progressively shown for cam ground piston.

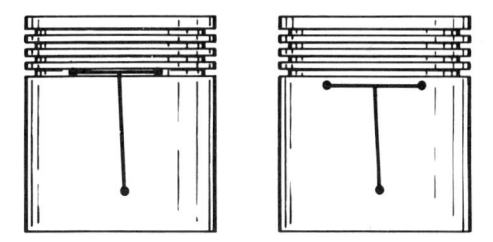

Fig. 6. Two types of T-slot pistons.

The maximum thrust face is complete with the piston hood and is not slotted. The expansion characteristics of this piston differ somewhat from others.

PISTON TEMPERATURES

There is a great variation in the temperature of different parts of a piston. The hottest part is at the top and the coolest part at the bottom of the skirt. See Fig. 7.

The intense heat of combustion naturally heats the head more than other parts. In construction, as a result of this, there must be ample

GAS ENGINE MANUAL

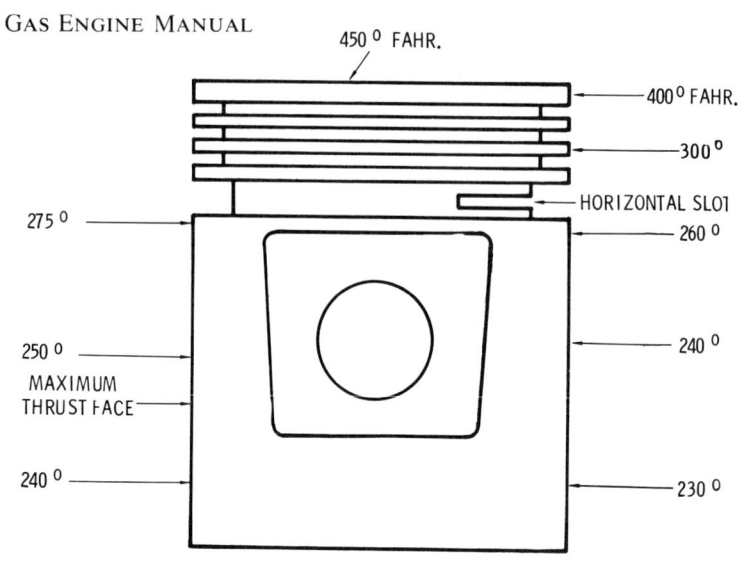

Fig. 7. Working temperatures of different parts of a piston.

thickness of metal in the head to freely transmit the heat as well as withstand the great pressure to which it is subjected by the combustion of the fuel charge. The head may be flat, concave, convex or a great variety of shapes to promote turbulence or control combustion. See Fig. 8.

It must be understood that in any particular engine the temperatures of the piston are always varying, depending upon many running conditions, such as load (amount of throttle opening) speed, efficiency of the circulating water, ambient temperature, etc.

PISTON CLEARANCES

In automotive and small gasoline engines the end clearance should be at least .004 in. per inch of cylinder diameter. An exception to this rule may be noted in regard to motorcycle engines, particularly those used in races. In this case, it is recommended that the end clearance be increased to at least .008 in. per inch of cylinder diameter.

Assuming the width of the ring groove to be the exact fractional size, the minimum side clearance is found by subtracting the maximum face width of the ring from the exact fractional groove width.

In all cases it is advisable to install the top ring with slightly more side clearance that the other rings. Thus, on automotive and small gasoline engines, a minimum side clearance of .001 in. is recommended for all

PISTONS

rings except the top, and for the top ring, a clearance of .002 in. is consistent with good practice.

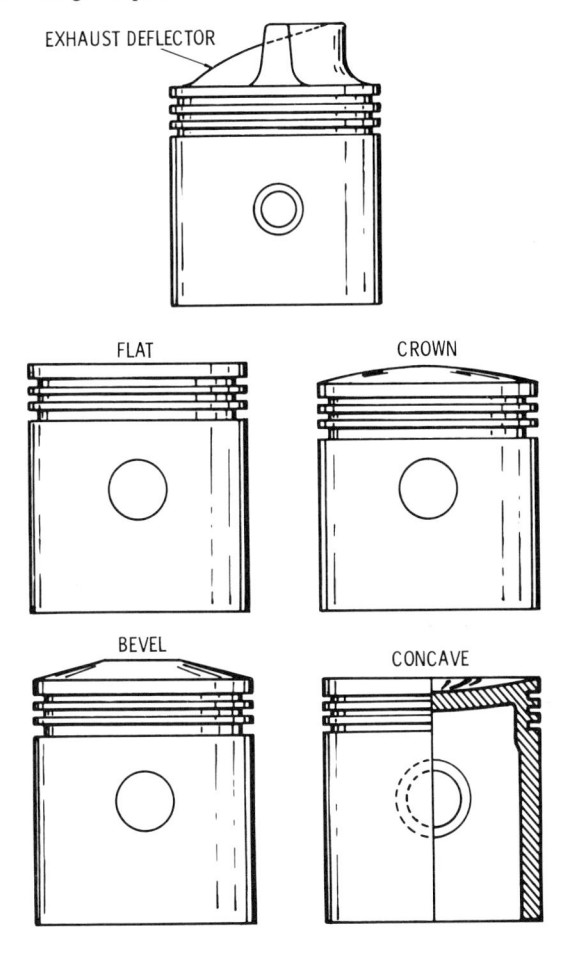

Fig. 8. Various shapes of pistons.

CHAPTER 5

Piston Rings

The function of piston rings is to provide a seal between the piston and cylinder, permitting the gases to be compressed in the cylinder. The piston rings therefore, must contact the cylinder walls evenly and should fit snugly in the ring grooves to prevent the compressed gases from escaping past the rings.

In the gas engine, the piston ring appears to be a very simple thing, but due to the exacting requirement of its use, it has taken years of experience, research and testing to develop the successful technique of manufacture.

Piston rings have been designed in a great number of variations to serve the present day requirements, whereas originally there were only simple split rings made of cast iron. See Fig. 1.

Modern piston rings are made of steel as well as cast iron and are often made in multiple sections instead of in a single piece. Their designs are often quite complicated, they are heat treated in various ways and plated with other metals. They are, furthermore, made in two distinct classifications, namely:

1. Compression rings.
2. Oil control rings.

COMPRESSION RINGS

Perhaps no part of a gas engine mechanism has received more attention with respect to preventing leakage of the compressed charge into the crankcase and leakage of oil into the combustion chamber, than has been given to piston rings whose function it is to control both compression and oil leakage.

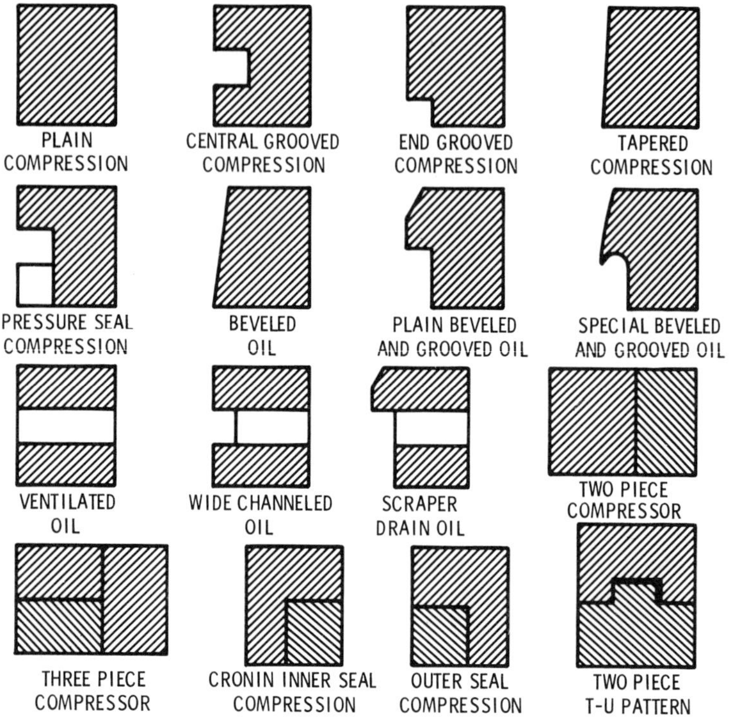

Fig. 1. Various piston ring sections.

(Top row) PLAIN COMPRESSION; CENTRAL GROOVED COMPRESSION; END GROOVED COMPRESSION; TAPERED COMPRESSION.

(Second row) PRESSURE SEAL COMPRESSION; BEVELED OIL; PLAIN BEVELED AND GROOVED OIL; SPECIAL BEVELED AND GROOVED OIL.

(Third row) VENTILATED OIL; WIDE CHANNELED OIL; SCRAPER DRAIN OIL; TWO PIECE COMPRESSOR.

(Fourth row) THREE PIECE COMPRESSOR; CRONIN INNER SEAL COMPRESSION; OUTER SEAL COMPRESSION; TWO PIECE T-U PATTERN.

The first function of rings is to prevent compression loss, that is, to prevent any loss of pressure of the compressed charge by leakage or "blow-by" as it is usually called. The provision against this is the compression rings. They are of two general types, namely:

1. Plain,
2. Grooved.

The plain ring is a one function ring and the grooved ring a double function ring. The plain compression ring is shown in Fig. 2, illustrating the various cuttings of the joint, as straight, angle and step joints. The joint to be used is determined principally by the proportions of the ring.

The straight joint is suitable for diameters up to 8 ins.; the angle joint for large diameters and the step joint is suitable for extremely narrow widths. The plain compression ring depends upon its ability to prevent loss of compression by its tight or precision fit all around its circumference.

PISTONS RINGS

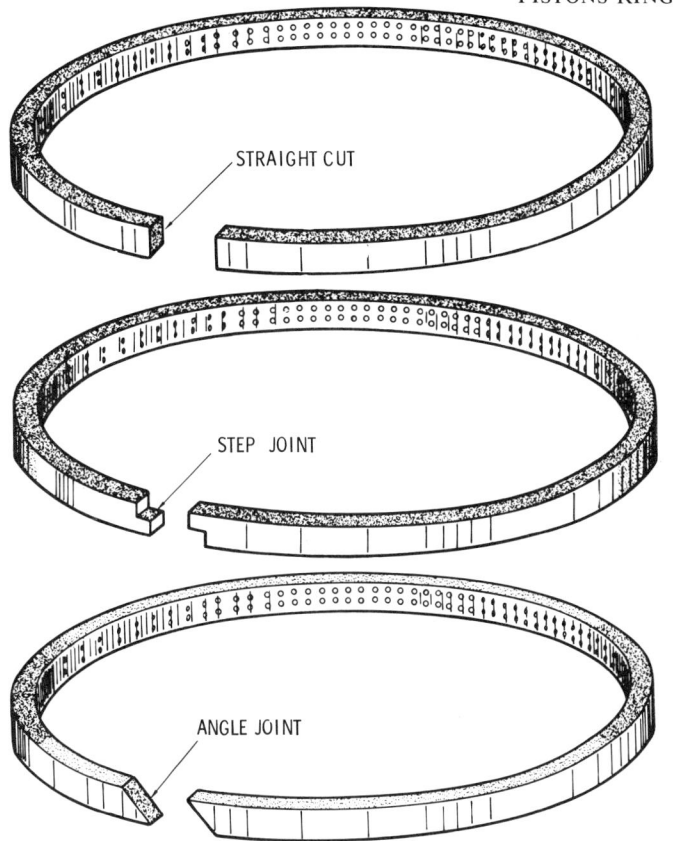

Fig. 2. Plain compression piston rings.

As an auxiliary to the plain compression ring is the grooved or oil seal ring. It is a good heat ring for the top piston groove; or preferably in the second groove in combination with the plain compression ring to seal compression, that is to say, a final stop against compression leakage.

In construction, as shown in Fig. 3, a centralized groove on the face of the ring holds the proper amount of oil, maintaining an oil film between the ring and the cylinder wall. In operation even if the groove should fill with carbon (which it naturally does) the ring continues to function as intended because the carbon is oil saturated and cannot harden. This ring is sometimes called the scraper oil ring, but the author prefers to consider it as the grooved compression ring in order to differentiate or draw the line more clearly between compression rings and oil control rings.

GAS ENGINE MANUAL

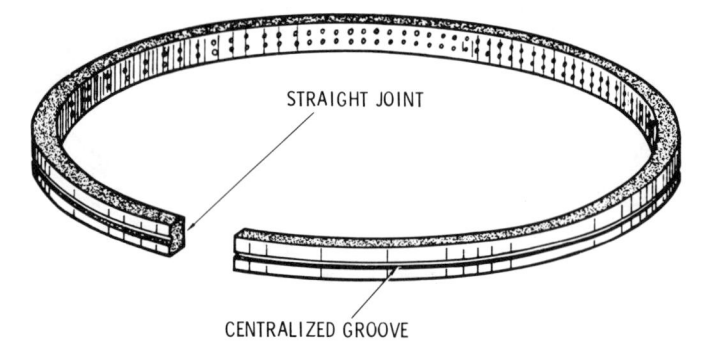

Fig. 3. Centralized grooved compression ring, sometimes called the oil seal compression ring.

There are many types of compression rings in both the plain and grooved classes, such as
1. Lower end grooved.
2. Tapered.
3. Step seal.
4. Pressure seal.

Fig. 4 shows an end grooved compression ring.

In construction it has a continuous groove around the lower outer edge. The object of this groove is to collect a small amount of oil which helps lubrication and forms an added "hydraulic" seal to prevent the escape of the compressed gases from the combustion chamber. This type ring is used chiefly as an oil regulating compression ring and should be installed in the lower compression groove on automobile and diesel pistons.

The tapered compression ring, as shown in Fig. 5 differs from other plain compression rings in that it has a 2° taper on its outside or acting face.

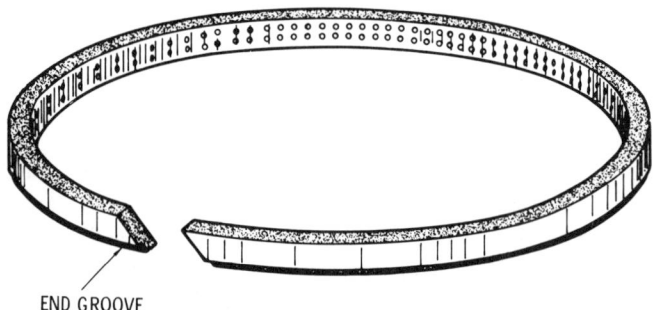

Fig. 4. End groove compression ring.

PISTONS RINGS

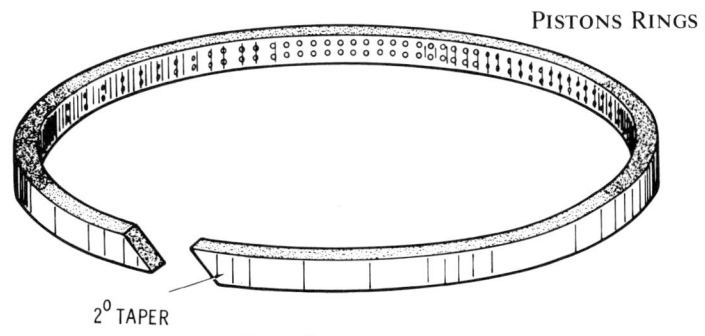

Fig. 5. Tapered compression ring. (2° TAPER)

This angle is so small that it cannot be discerned by the naked eye and therefore the acting face from which the taper starts outwardly and which should be turned toward the piston head is stamped "UP" as an aid to correct installation. This ring is used mainly as an oil control compression ring to keep oil down while the rings seal in. Most frequently it is used in air craft engines to form an oil seal during the initial "run in" period. It also has its use in elliptical or distorted cylinders.

The step seal ring, as shown in Fig. 6, is a one piece hammered concentric ring. In construction, one leg of the joint is triangular in section and the other leg is pentagonal.

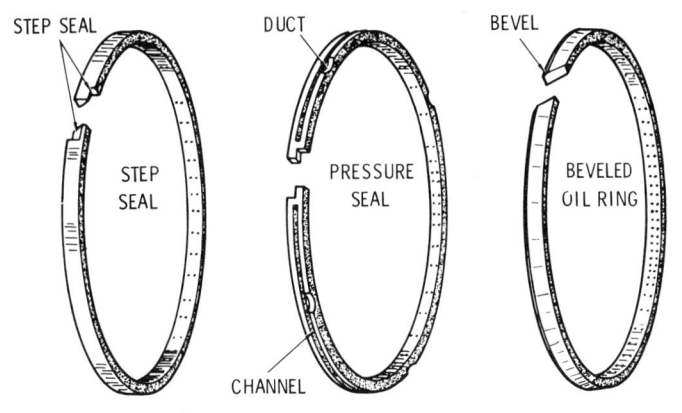

Fig. 6. Various type piston rings.

It is a good replacement ring in worn cylinders where the cylinder is non-circular and taper is present. This ring when properly installed combines the strength of the single piece ring with the sealing qualities of the multipiece ring.

Gas Engine Manual

It is difficult to always classify rings as compression or oil control rings as many are a combination of both classes, that is, double function.

For instance, what is called a pressure seal ring, as shown in Fig. 6, should be (according to the manufacturers) installed in the second groove from the head of the piston with a plain compression ring in the first groove. It acts as a fire check ring in the groove and oil rings in the lower grooves.

In gas, oil and diesel engines, the pressure seal ring is used to advantage to replace one or more of the compression rings; however, it is advisable to use plain compression rings in the first one or two grooves as fire check rings.

OIL RINGS

The rapid increase in compression pressures and piston speeds have taxed the ingenuity of manufacturers of piston rings to cope with the situation and has resulted in the introduction of a multiplicity of rings — both compression and oil types, due to research and experimentation on the part of the manufacturers. On account of this, piston rings are now available to meet the very severe conditions imposed on them.

With respect to oil control rings, numerous types have been developed as listed in the classification. Fig. 6 shows a beveled type oil ring.

This is a one piece concentric hammered ring having the upper part of the circumferential surface beveled and presenting only a narrow active face or bearing surface to the cylinder walls. In operation on the up stroke this ring rides over the oil and presses it to a thin film on the cylinder walls. On the down stroke, the ring scrapes the excess oil back toward the crankcase. See Fig. 7.

CONCENTRIC RING   ECCENTRIC RING

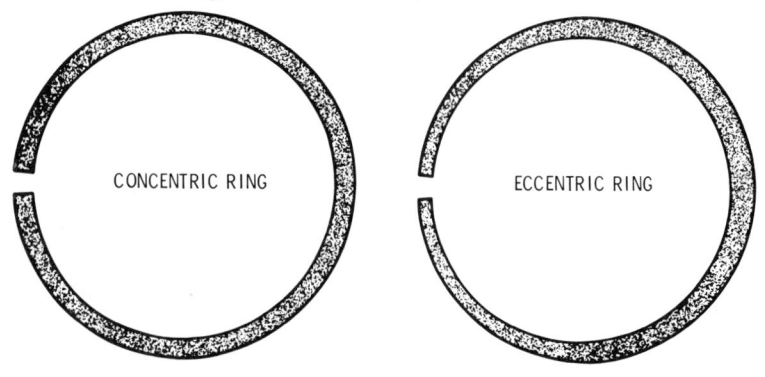

Fig. 7. Concentric and eccentric piston rings.

PISTONS RINGS

In installation, the beveled ring should be installed in the groove farthest from the head of the piston, with the beveled side up or toward the piston head. On a piston containing one or more grooves suitably relieved for oil drainage, the beveled oil ring should be installed in the groove or grooves so relieved.

Oil drainage can be provided in several ways. One method is to undercut the skirt of the piston below the oil ring groove, allowing sufficient clearance for the oil to drain back to the crankcase.

Another method is to machine an oil collecting groove on the piston around the lower, outer edge of the oil ring groove and to drill a series of holes at an angle through the piston wall from this oil collecting groove. This type of oil ring is used largely in two-cycle engines and in aeronautical engines. It can be used wherever a scraping ring is required.

Fig. 8 shows a plain beveled and grooved oil ring. The groove is rectangular and the bevel is not so pronounced as the special beveled type. The scraping edge of the groove in this ring remains always exactly in the center of the ring, and in installations subjected to much wear, this ring is recommended instead of the special beveled and grooved oil ring.

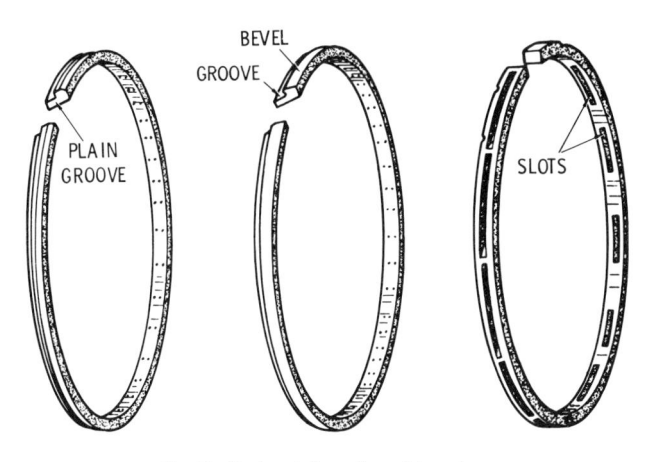

Fig. 8. Horizontal engine piston rings.

A special beveled and grooved oil ring is shown in Fig. 8. *In construction*, it is a one piece concentric hammered ring with the upper edge gradually tapering to a narrow cylinder contacting surface which forms a scraping edge for an undercut corner groove opening downward and outward.

In operation, the ring rides over the oil with the least possible scraping action on the up stroke of the piston and to scrape the oil into the groove in the ring on the down stroke of the piston. On the next up stroke of the piston, oil that is trapped in the groove is again delivered to the cylinder walls for lubrication.

Some form of oil relief should be provided on the piston in order to drain the excess oil back to the crankcase. This oil relief may be in the form of holes drilled at an angle through the piston groove, or in case the ring groove is near the bottom of the skirt of the piston, the diameter of the skirt below the ring groove, may be reduced so as to permit the free drainage of the oil past this section of the piston.

This ring should be installed in the groove farthest from the head of the piston with the beveled side up or toward the piston head. If one or more grooves on the piston is suitably relieved for this oil ring, then install in grooves so relieved. This type of ring is used in both horizontal and vertical single acting oil engines and air compressors, particularly in the larger sizes.

On rings for horizontal machines, the cylinder contacting surface of each ring is broader than on rings for vertical machines. Fig. 8 shows a ventilated oil ring. *In construction*, it is a one piece, concentric, hammered ring having elongated slots passing through the body of the ring, with cylinder contacting lands between all slots. The slots are closely spaced and of ample width to insure maximum drainage. The narrow lands between the slots represent solid wear-resistant bearing surfaces which insure long life to the rings.

In operation, when the ring moves rapidly over the surface of the cylinder, considerable oil pressure is built up ahead of the ring. This pressure builds up to such an extent that oil passes the cylinder contacting face of the ring and the pressure is then suddenly relieved as the slot moves over the oil so that the surface tension of the oil causes a wave of oil to rise up into the slots. The greater portion of this wave is then cut off by the following edge of the ring. Thus, excessive oil is removed from the cylinder by a method similar to that used in an ordinary vacuum cleaner.

In application, the ventilated oil ring is used extensively in normal or high speed vertical engines of all designs and should be installed in the piston groove farthest from the head of the piston. It is frequently advantageous to install two ventilated oil rings in the two lowest grooves when the piston contains a total of four or more grooves.

The wide channeled oil ring, as shown in Fig. 9, is a one-piece concentric hammered piston ring having a series of extra wide slots

PISTONS RINGS

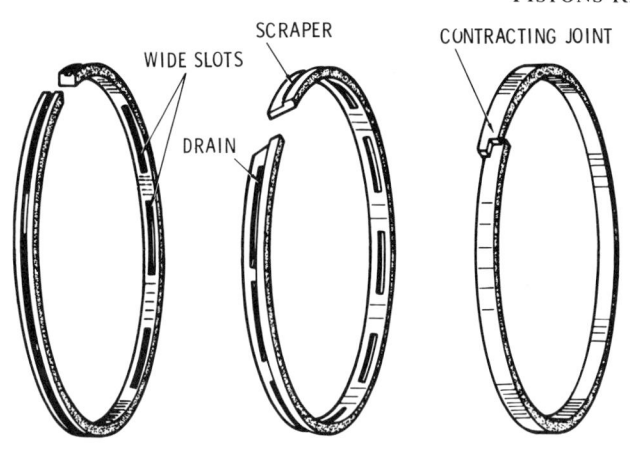

Fig. 9. Wide channel piston rings.

opening into two channels. One of these channels passes across the point and extends for a distance of 90° on each side of the joint. The other channel extends for approximately 180° around the ring on the side opposite the joint and the two channels are separated one from the other by narrow dams. *In operation*, it removes the excess oil from the cylinder wall by a method similar in principle to that employed in a vacuum cleaner.

The wide channeled ring has a smaller bearing surface and accordingly exerts a higher unit pressure than the ventilated type ring, thus rendering it more effective as an oil control ring. *In operation*, by having two separate channels in the ring in place of one continuous channel, oil that is collected in these two channels must drain back through the ring instead of flowing around the ring and collecting in one place. Thus, not only is ample drainage taken care of by preventing too great an accumulation of oil in one spot, but lubrication is also improved because of a better distribution of oil.

The excess oil is drained off from the channels through the slots to the rear of the ring and thence to the interior of the piston through oil relief holes. These holes should be drilled in the back of the piston groove, perpendicular to the axis of the piston, and they should be as large and numerous as the strength of the piston will permit. This ring is suitable for vertical high speed engines and compressors. It should be installed in the groove farthest from the piston head.

Fig. 9 shows a scraper drain oil ring. *In construction*, it is a one piece concentric hammered ring having the upper, outer edge beveled and a corner groove formed on the lower, outer edge with oil passages extending from the upper corner of the groove to the back of the ring.

In operation, the ring rides over the oil on the cylinder wall on the up stroke of the piston and scrapes oil into the groove in the ring on the down stroke of the piston. The oil thus collected passes straight through the oil passages in the ring and then through holes in the piston.

It is essential to have oil relief holes in the piston in the back of the groove in which the scraper drain oil ring is to be used. The greater the drainage area in the piston grooves, the greater the amount of oil that will be removed. This ring is suitable for the groove farthest from the head of the piston unless the piston contain four or more ring grooves with one on the skirt and then it is advisable to install two rings of this type, one in the groove on the skirt and one in the groove just above the piston pin.

MISCELLANEOUS RINGS

In addition to the rings already described there are numerous other types having features designed to meet some special conditions. Fig. 9 shows a contracting ring.

In construction, this ring is hammered on its outer circumference so that it will contract and bear on its inner circumference. It can be beveled so as to act as a wiper ring which will ride over the oil on one stroke of the piston and scrape oil on the reverse stroke.

The contracting ring is designed primarily for installation in a specially constructed groove in the cylinder, so as to wipe oil from the skirt of a long trunk piston in a manner similar to that in which the beveled oil ring, when installed on the piston, removes excess oil from the cylinder.

A two piece compressor ring designed to prevent undue cylinder wall pressure is shown in Fig. 10. The inner ring provides the necessary tension and effectually seals the joint of the outer ring. When subjected to high pressure that may build up behind the ring, causing the ring to expand, there is a definite restraining action obtained due to friction between the cooperating surfaces of the two rings which prevents undue cylinder wall pressure.

A three piece compressor ring is shown in Fig. 10. Two types of ring known as inner seal and outer seal and especially suitable for diesel engines are shown in Fig.11. These are special compression rings intended to be effective in preventing leakage in both new and worn cylinders.

PISTONS RINGS

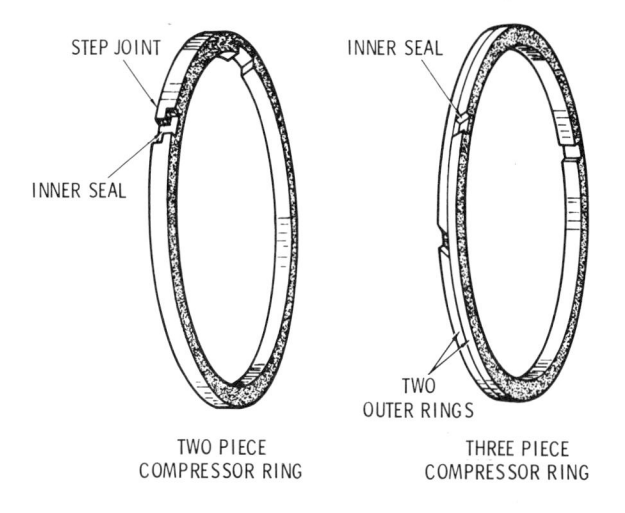

Fig. 10. Two and three piece compressor rings.

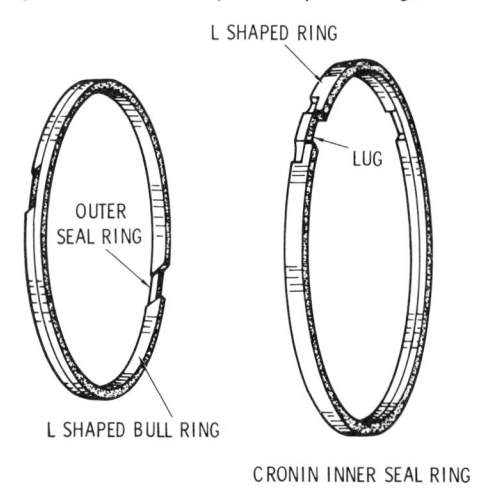

Fig. 11. Inner and outer seal ring.

Another ring well adapted to diesel engines is shown in Fig. 12. It is a double ring and known as the T-U ring because of the cross sectional shape of the two component rings. Each ring contains an overlapping joint intended to prevent leakage to the rear of the groove and when assembled to give efficient joint and groove seal.

GAS ENGINE MANUAL

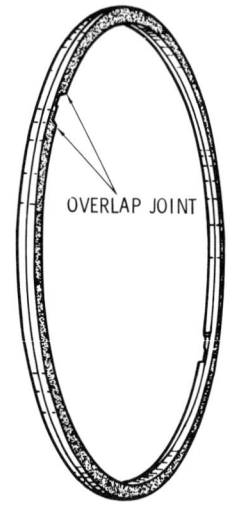

Fig. 12. T-U ring used in diesel engines.

The object sought in the segmental type rings (Fig. 13) is to avoid the necessity of springing the rings when inserting in the grooves. This is important in the case of thick heavy rings. They are usually made in three interlocking segments as shown and when assembled in the groove form a continuous seal tight ring.

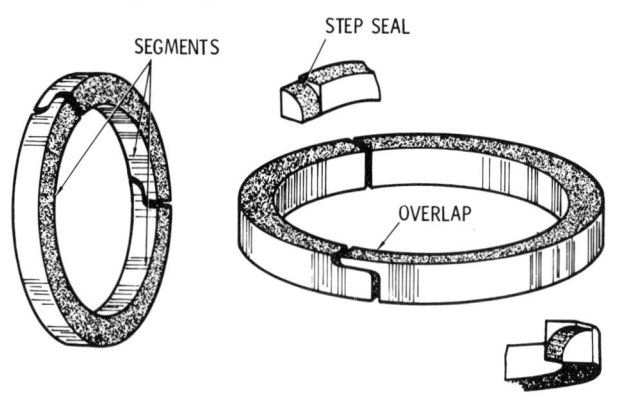

Fig. 13. Segmental rings.

CHAPTER 6

Connecting Rods and Wrist Pins

Piston pins also known as "wrist pins" or "gudgeon" pins serve as a connection between the upper end of the connecting rod and the piston. Piston pins are made of alloy steel with a precision finish and are case-hardened and sometimes chromium-plated to increase their wearing qualities. It forms a pivot connecting one end of the connecting rod to the piston which permits lateral oscillating motion of the rod with reciprocating motion of the piston.

There are three methods of fastening a piston pin to the piston and connecting rod:

1. An *anchored* or *fixed* pin is attached to the piston by a set screw running through one of the bosses; the connecting rod oscillating on the pin as shown in Fig. 1.
2. A *semifloating* pin is anchored to the connecting rod and turns in the piston pin bosses, Figs. 1 and 2.
3. A *full-floating* pin is free to rotate in the connecting rod and in the bosses but is prevented from working out against the sides of the cylinder by plugs or snap-ring locks, Fig. 1.

As will be noted, piston pins are of tubular construction which gives them a maximum of strength with a minimum of weight. They are lubricated by splash from the crankcase or by pressure through passes bored in the connecting rods.

The *full-floating* pin, is not anchored to either the piston or connecting rod and is free to turn in the connecting rod end or in the piston bosses. It is a plain pin and is sometimes called by that name. This pin is generally used with aluminum alloy pistons.

GAS ENGINE MANUAL

HOLE FOR SET SCREW

HALF SLOT TO ENGAGE WITH CLAMP

SET SCREW

SEMIFLOATING

FULL FLOATING

HOLE IN PIN FOR SET SCREW

BRONZE BUSHINGS IN PISTON BOSSES

SNAP RING LOCKS PREVENT CONTACT WITH CYLINDER WALLS

SET SCREW ANCHORS PIN TO PISTON BOSS

HALF SLOT ANCHORS PIN TO CONNECTING ROD END

PIN FREE TO TURN IN PISTON BOSSES AND IN CONNECTING ROD END

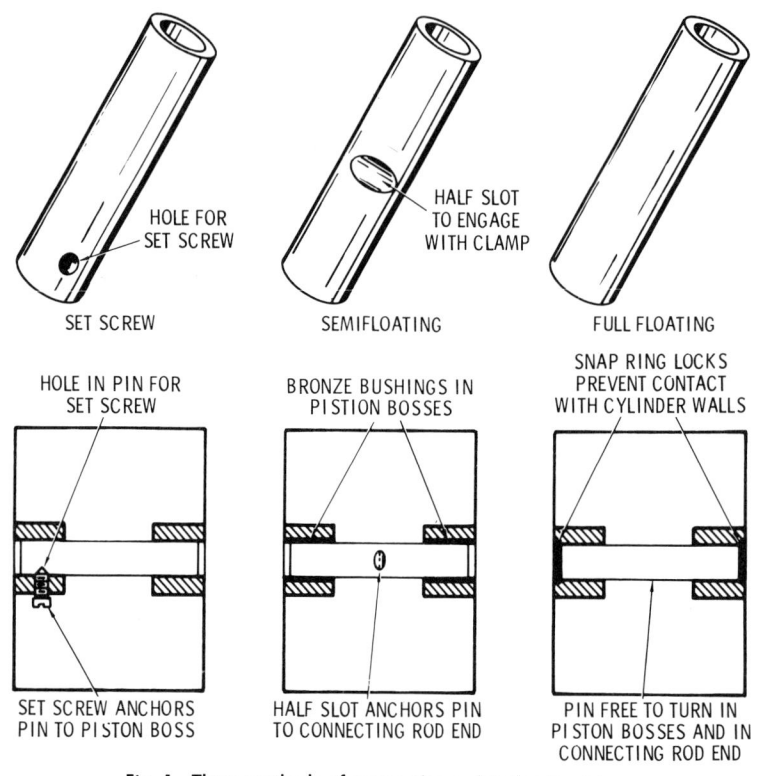

Fig. 1. Three methods of connecting wrist pins to pistons.

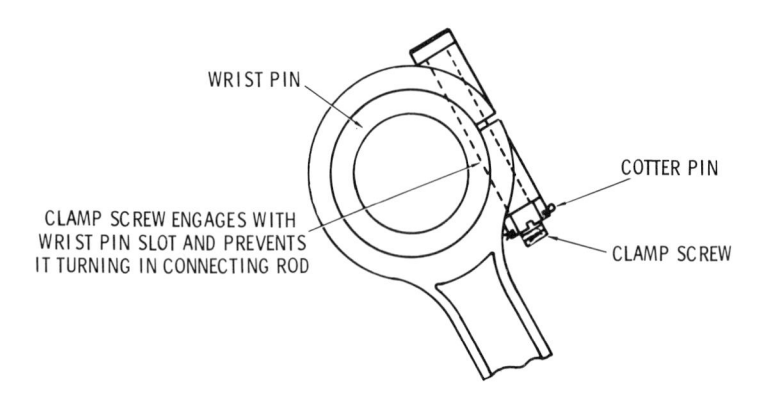

WRIST PIN

COTTER PIN

CLAMP SCREW ENGAGES WITH WRIST PIN SLOT AND PREVENTS IT TURNING IN CONNECTING ROD

CLAMP SCREW

Fig. 2. Semifloating wrist pin connection.

CONNECTING RODS AND WRIST PINS

Since the full-floating pin is free to rotate or move sidewise some form of locking device is necessary to prevent the pin working to one side and rubbing against the cylinder walls with resulting damage to the walls. Numerous devices are used, a typical one consists of a snap ring fitted into recesses cut in the bosses at the ends. Fig. 3 illustrates numerous types of wrist pin construction.

CONNECTING ROD

The connecting rod is a connecting link between the piston and the crankshaft. It serves to transform the reciprocating motion of the piston into rotary motion at the crankshaft.

Construction

Connecting rods must be light and yet strong enough to transmit the thrust of the piston. Automotive connecting rods are drop forged from a steel alloy capable of withstanding heavy loads without deflection, that is, without bending or twisting. They are usually made in the form of an I-beam for lightness with maximum strength. Holes at the upper and lower ends are machined to permit accurate fitting of bearings. It is very important that these holes be parallel.

The upper end of the connecting rod is connected to the piston by the piston pin. If the piston be locked in the piston pin bosses or if it floats in both piston and connecting rod, the upper hole of the connecting rod will have a solid bearing or bushing of bronze or similar material. As the lower end of the connecting rod revolves with the crankshaft, the upper end is forced to rotate on the piston pin. Although this movement is not great, the bushing is necessary because the temperature and unit pressures exerted are high. If the piston pin be semifloating, that is, anchored to the connecting rod a bushing is not required. See Fig. 4.

The lower hole of the connecting rod is split to permit it to be clamped around the crankshaft. The bottom part or cap, is made of the same material as the rod and is attached by two or more connecting rod bolts. The surface which bears on the crankshaft is generally a bearing material in the form of a separate split shell although in a few cases, it may be spun or die-cast in the inside of the rod during manufacture.

The two parts of the separate bearing are positioned in the rod and cap by dowel pins, projections, or short brass screws. The shell may be all bronze or of babbit metal face spun or die-cast on a backing of bronze or steel. Roller-type bearings are also frequently employed, particularly on

GAS ENGINE MANUAL

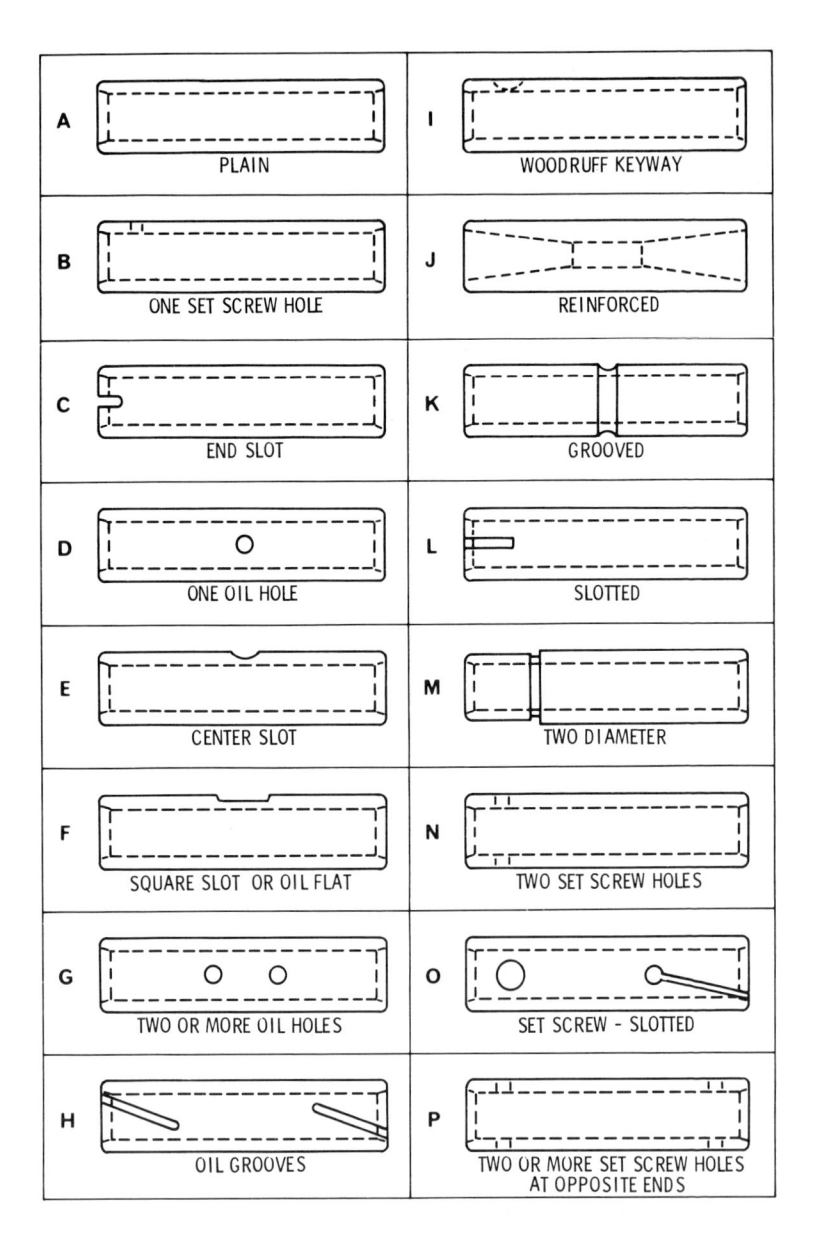

A	PLAIN	I	WOODRUFF KEYWAY
B	ONE SET SCREW HOLE	J	REINFORCED
C	END SLOT	K	GROOVED
D	ONE OIL HOLE	L	SLOTTED
E	CENTER SLOT	M	TWO DIAMETER
F	SQUARE SLOT OR OIL FLAT	N	TWO SET SCREW HOLES
G	TWO OR MORE OIL HOLES	O	SET SCREW - SLOTTED
H	OIL GROOVES	P	TWO OR MORE SET SCREW HOLES AT OPPOSITE ENDS

Fig. 3. Various standard types of wrist pins.

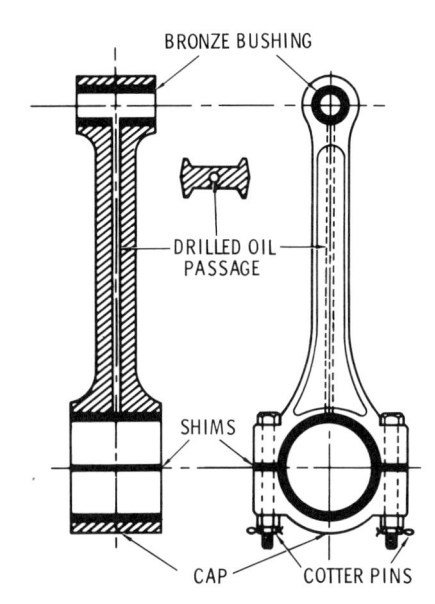

Fig. 4. Connecting rod construction.

high speed engines. Split bearings may be of the precision or semiprecision type.

The precision type is accurately finished to fit the crank pin and does not require further machining during installation. It is positioned by projections on the shell which match reliefs in the rod and cap. The projections prevent the bearings from moving sideways, but they permit rotary movement after the bearing cap is removed, thus making it possible to replace the bearing without removing the connecting rod from the engine. The semiprecision type is usually fastened securely to the rod and cap. Prior to installation, it is machined to the proper inside diameter with the cap and rod bolted together. See Fig. 5.

Aluminum Alloy Connecting Rods

In an aluminum rod the sectional area is bigger than in steel rods and the designer has an excellent opportunity to shape it for the best rigidity in both planes, having only the consideration of clearance between the edges of the cylinder wall and rod as the limiting factors against this design.

Gas Engine Manual

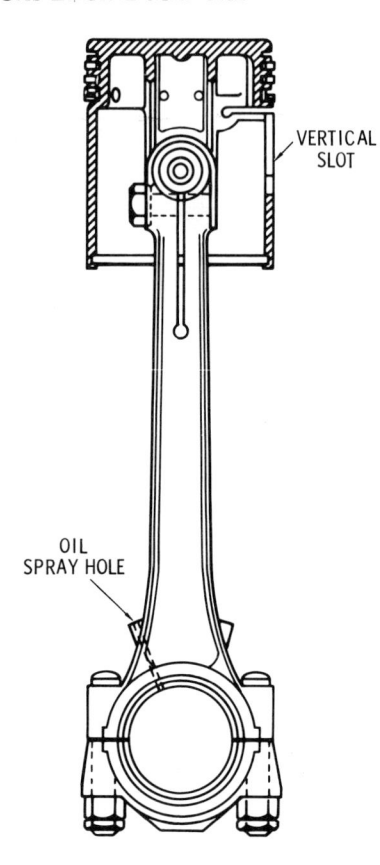

VERTICAL SLOT

OIL SPRAY HOLE

Fig. 5. Piston and connecting rod assembled.

CHAPTER 7

Crankshafts

The function of a crankshaft is to convert the reciprocating motion of the piston and connecting rod into rotary motion, and to transmit the resulting torque to the flywheel and clutch.

In its simplest form, it consists essentially of a shaft with one or more throws along its length as noted in Fig. 1. The arangement of the throws is determined by the desired firing order of the engine cylinders. The desired firing order is regulated by the relationship of the camshaft and crankshaft.

Crankshafts must be correctly proportioned to carry the great loads imposed upon them while turning at high speed, also to resist the shocks due to the nature of the gas engine cycle.

CONSTRUCTION

Crankshafts are generally made of forged steel, or cast from an alloy of steel and nickel. The rough casting or forging is then machined to exact specifications. After machining, the nonbearing surfaces are plated with a light coating of copper. When the plating process is completed, the whole crankshaft is placed in a carburizing oven or an electro-induction furnace, where surfaces of the crankshaft not coated with copper become alloyed with the carbon, producing a thin, hard surface or bearing area.

This process is known as "case-hardening". The crankshaft is completed by grinding the case-hardening surfaces.

CRANKSHAFT BALANCE

Owing to the high rotative speed for which modern engines are designed, most manufacturers provide counterweights for the shaft. These are usually forged integrally with the crankshaft. See Fig. 2.

GAS ENGINE MANUAL

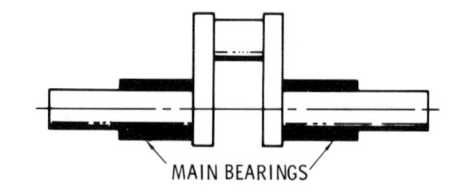

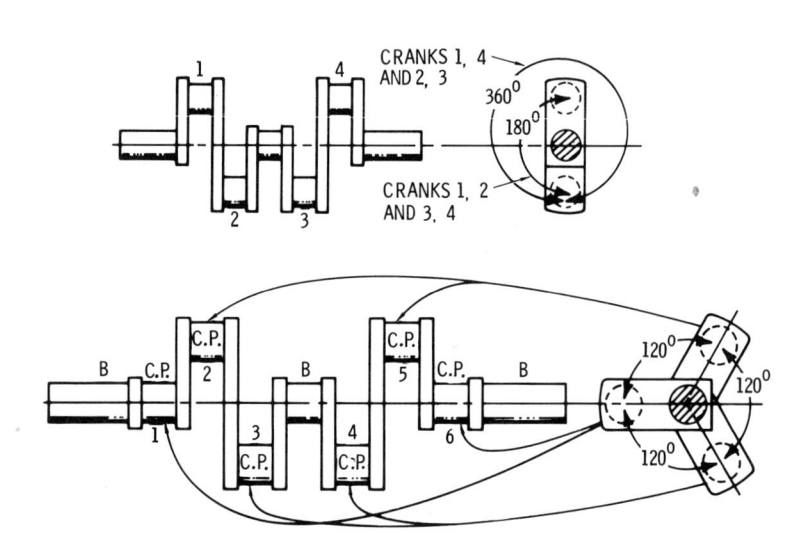

Fig. 1. Showing one, four, and six throw crankshaft.

The functions of a counterweight are:
1. To balance the weights of the piston, connecting rod, crank arms and crank pin, so that the assembly of moving parts will be in static equilibrium for all points of the stroke.
2. To provide an opposing centrifugal force to counteract the oppositely directed centrifugal force due to the connecting rod, crank arms and crank pin.
3. To counteract the inertia loads due to the moving parts during the accelerating and retarding portions of their travel and in this way avoid a considerable amount of vibration which would otherwise occur. Fig. 2A shows a shaft without counterweights.

The tendency of this arrangement is for the unbalanced weight of the piston, rod, crank arms and pin to move downward to the lower dead center. Fig. 2B shows the effect of a counterweight for static balance.

CRANKSHAFTS

It must be evident that if enough metal is put into the counterweight so that its tendency to rotate to the bottom center is the same as that due to the weight of the piston, connecting rod, etc. the assembly will be in equilibrium for any crank position, that is, there will be no tendency to rotate, and the assembly will be in static equilibrium.

CRANKSHAFT THROW ARRANGEMENT

While engine crankshafts are similar in design, the number of throws and their angular arangement depends upon the number of cylinders to be operated and their firing order.

A single cylinder engine will have one throw on the crankshaft as shown in Fig. 2C. A two cylinder opposed engine with one cylinder on each side of the crankshaft (four-stroke cycle) will have two throws spaced 180 degrees apart. An arrangement for alternate firing cylinders is shown in Fig. 3.

An identical crankshaft throw arrangement may be used for a two-stroke cycle opposed engine having two cylinders firing at the same time.

Crankshafts four cylinder engines, Fig. 4-B and C, have either three or five bearings. The four throws are in one plane, the throws for cylinders No. 2 and 3 being advanced 180 degrees over cylinders No. 1 and 4.

Crankshafts for six cylinder engines may have either three, four or seven bearings. The throws for the connecting rod bearings are forged in three planes 120 degrees apart, with two throws in each plane. As noted in Fig. 1B and C, throws No. 1 and 6 are in the first plane, throws No. 2 and 5 are in the second plane, and No. 3 and 4 in the third plane.

An eight cylinder V-type engine may have two identical four throw arrangements positioned end to end, with one set advanced 90 degrees over the other. This is known as a 4-4 shaft, Fig. 5A.

In the other design, known as a 2-4-2 shaft Fig. 5B, a set of four throws is positioned between two sets of two throws each. The end cylinders are advanced 90 degrees over the center group.

BUILT-UP AND SINGLE PIECE CRANKSHAFTS

The built-up type of crankshaft has each of its parts made separately and then jointed strongly together as by keyways and force fits, as in Fig. 6. Quite often built-up shafts are fitted with ball bearings as they are the only kind of shaft that can be fitted with ball bearings if they have more than two bearings.

Gas Engine Manual

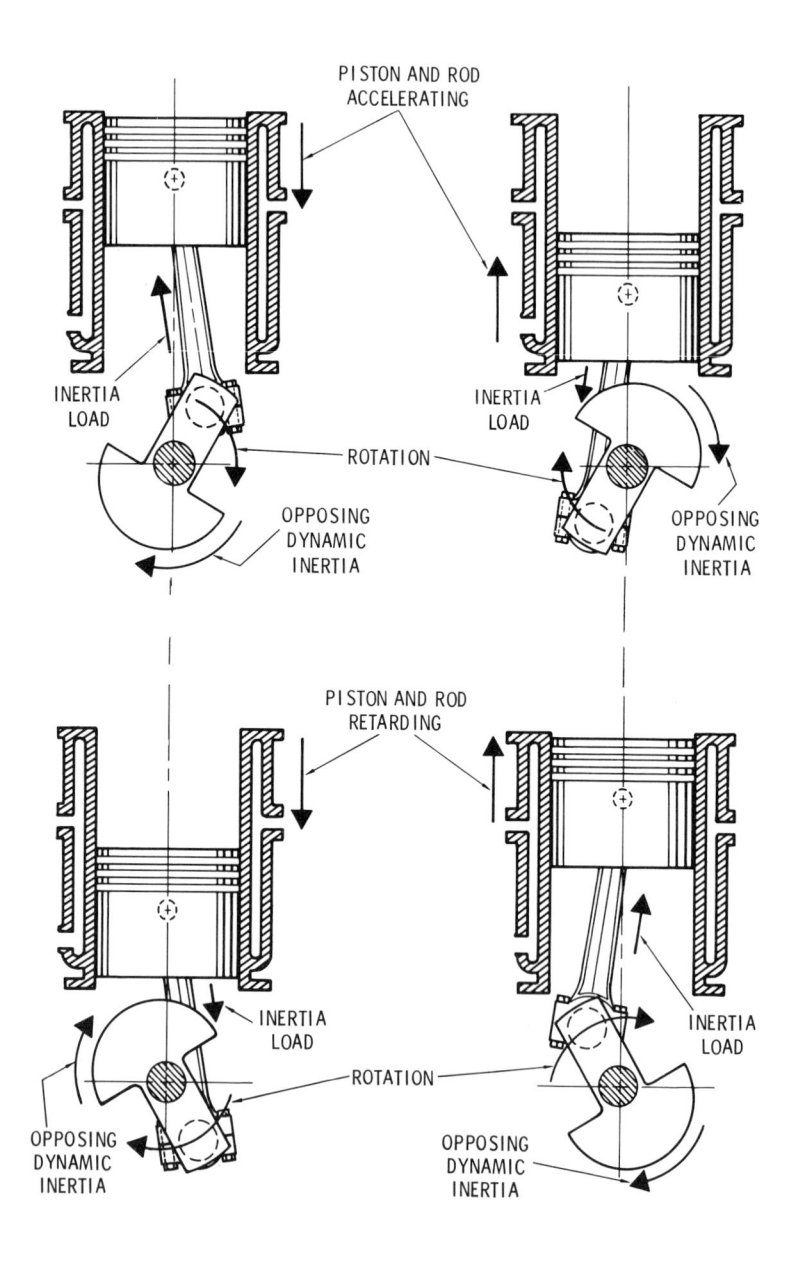

Fig. 2. Counterweights forged

CRANKSHAFTS

Such shafts are sometimes used on motorcycles where the two crank arms in the shape of discs may perform the functions as a flywheel and also act as counterweights.

MAIN BEARINGS

The crankshaft of an engine rotates in the main bearings. These bearings are located at both ends and at a few intermediate points along the crankshaft. The number of bearings will depend upon the type of engine and number of cylinders, although a crankshaft for a given number of cylinders may be constructed for different numbers of bearings.

A one or two cylinder engine will have two main bearings, one front and one rear; the rear main bearing always being fitted adjacent to the

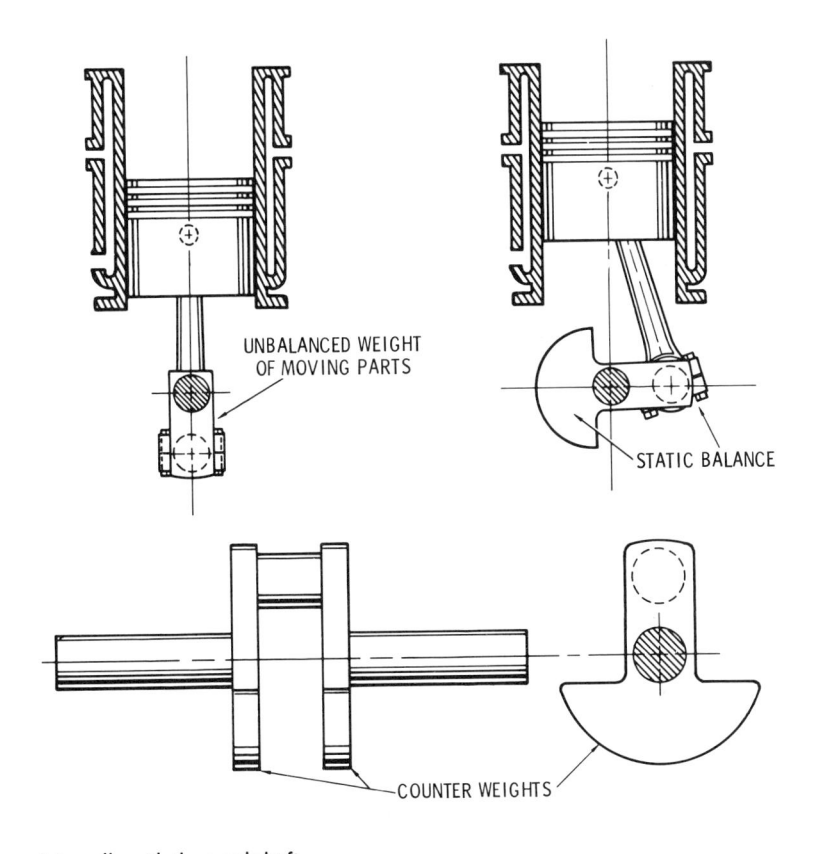

UNBALANCED WEIGHT OF MOVING PARTS

STATIC BALANCE

COUNTER WEIGHTS

integrally with the crankshaft.

GAS ENGINE MANUAL

flywheel. A four cylinder engine normally has three bearings, one at the front, one between the cylinders and one at the rear.

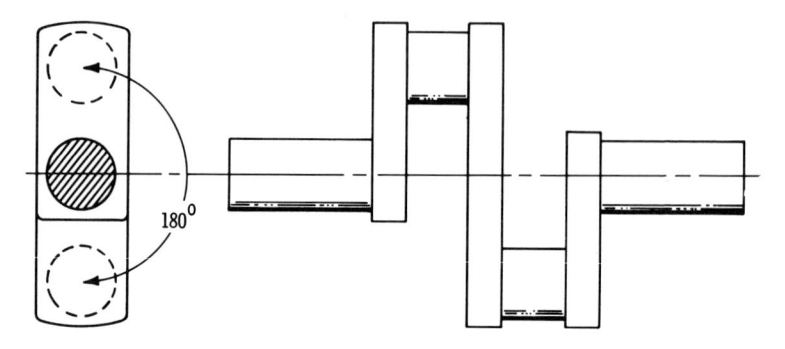

Fig. 3. Illustrating a typical two-throw or 180 degree crankshaft.

The prevailing practice is to fit the bearings into the webs or ribs and ends of the crankcase. The bearings with this construction are upside

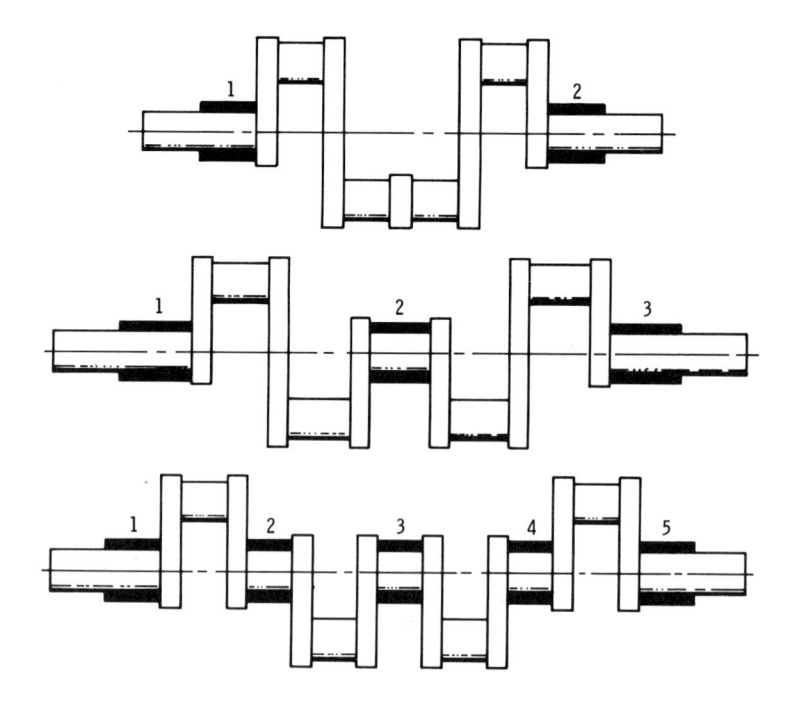

Fig. 4. Four-throw crankshafts illustrating two, three and five bearings.

CRANKSHAFTS

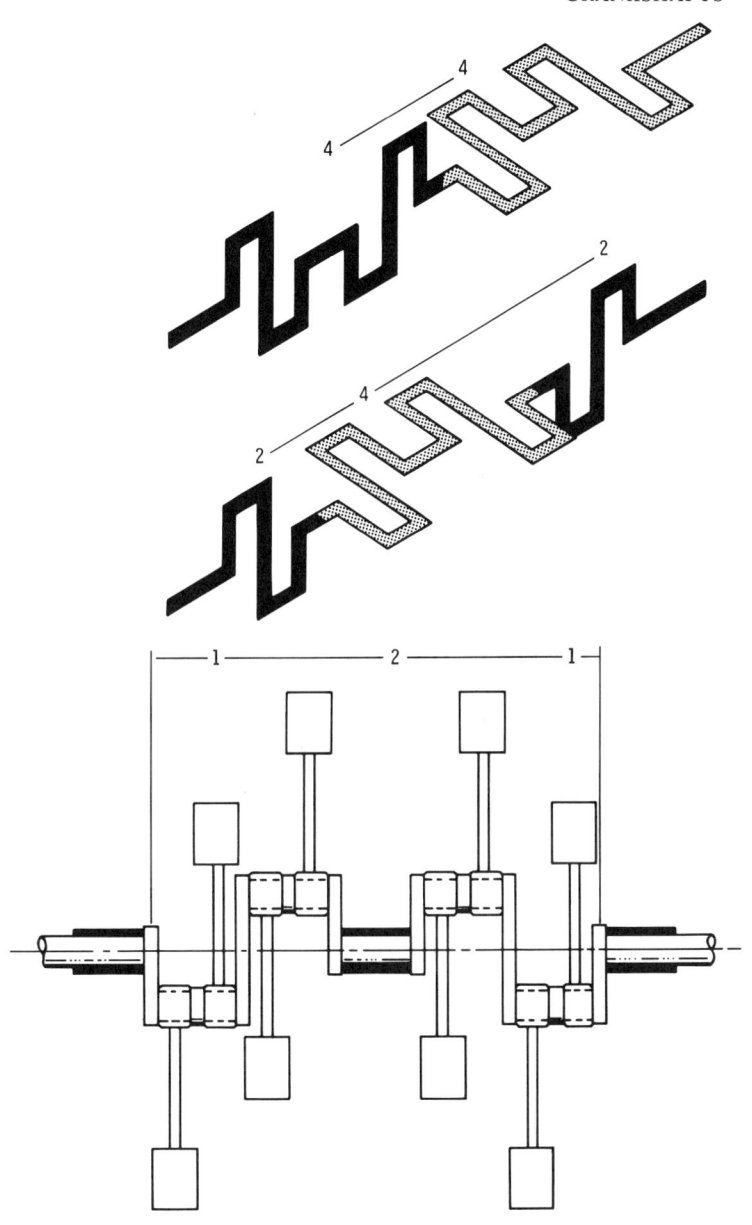

Fig. 5. Two crankshaft arrangements for a straight eight engine illustrating the 4-4 and 2-4-2 sequence.

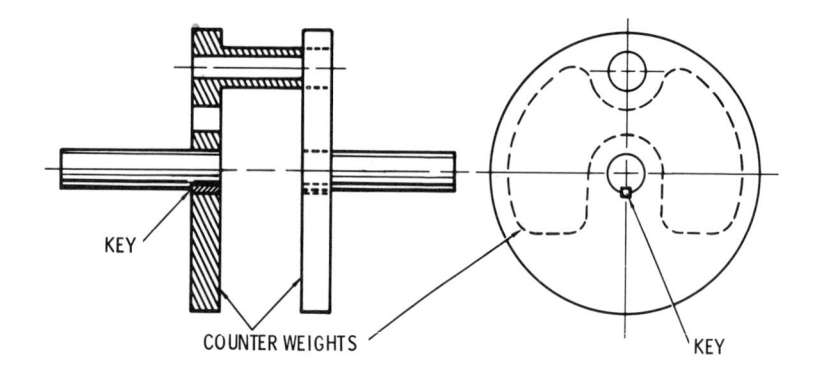

Fig. 6. General construction of a disc built-up counterweighted shaft.

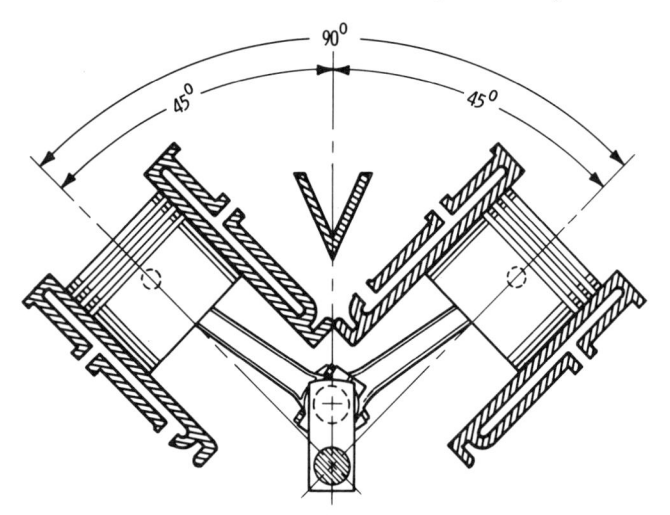

Fig. 7. Angular arrangement of cylinders on a V-8 engine.

down so to speak, but are accessible by removing the oil pan. The practice of manufacturers differ with respect to the number of bearings to be used.

A little consideration will show that a shaft for a given number of cylinders may be constructed for different numbers of bearings.

For example, Fig. 4 shows how a 1-2-1 shaft for a four cylinder engine could be designed for two, three or five bearings. The arrangement shown in Fig. 4A would require a shaft of very liberal diameter to resist the bending action and would not be as good as the three bearing shaft shown in Fig. 4B.

CHAPTER 8

Engine Flywheel

The purpose of the flywheel is to secure momentum which is necessary to keep the crankshaft turning when it is not receiving power impulses from the pistons.

The flywheel will thus permit the engine to idle smoothly through those parts of the cycle when power is not being produced and to keep the engine turning at a nearly uniform rate of rotation.

It is self evident from the foregoing that the heavier the engine flywheel is, the smoother the engine will idle. An excessively heavy flywheel, on the other hand, will cause the engine to accelerate and decelerate slowly because of its inertia. It is for these reasons that heavy duty engines are built with large inertia flywheels, and small high speed engines have light flywheels. On engines equipped for electric starting, the flywheel rim carries a gear teeth ring, either integral with the flywheel or shrunk on, which meshes with the starter driving gear for cranking the engine at starting.

Flywheels employed on small two- and four-cycle engines are usually made of cast or forged aluminum alloy, the hub forming an integral part of the flywheel shell, or a steel hub is attached to the shell. On some engines of this type, the flywheel encloses the magnets, and has a permanent magnet attached to its inner surface which functions as a part of the ignition system. On air cooled engines, the flywheel usually has fins on its outer surfaces which acts as a fan, promoting circulation of air over the cylinder. See Fig. 1.

TORSIONAL VIBRATION

Torsional vibration is a twisting rotation, usually noticeable in in-line six and eight cylinder engines with long crankshafts. It is caused by the reciprocating movement of the piston acting on the rotating crankshaft.

GAS ENGINE MANUAL

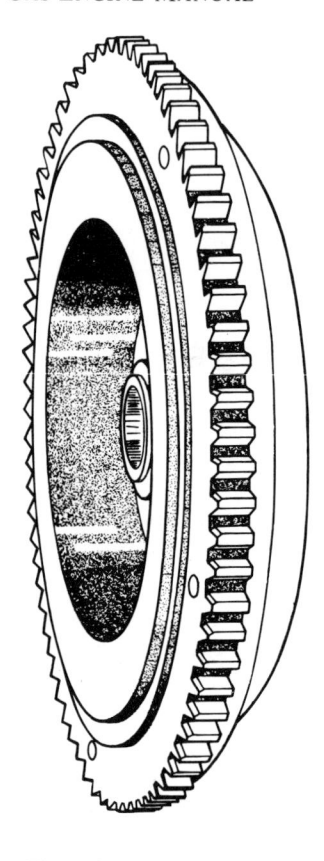

Fig. 1. Typical gear type flywheel used on small and medium size engines.

Thus, for example, when the front cylinder is fired, it would turn the crankshaft very rapidly, but the inertia of the flywheel would tend to prevent this rapid increase in speed at the rear of the crankshaft.

The result is a "winding" or "twisting" action in the crankshaft. As the force exerted by the front cylinder decreases, this twisting action will cease. Although this twisting action is very small, it is nevertheless large enough to set up torsional vibration in the engine.

CLASSES OF VIBRATION DAMPERS

Vibration dampers may be divided into two classes, as:
1. The slipping flywheel type,
2. The harmonic balancer.

ENGINE FLYWHEEL

The slipping flywheel type is a device developed to overcome torsional vibration. It consists of a small secondary flywheel which runs loosely on the front of a crankshaft, as illustrated in Fig. 2. A friction clutch, having a slipping torque, tends to make the flywheel rotate with the crankshaft. Normally, torsional vibration would result, but the dampers flywheel slips, preventing the winding up. (See Fig. 3.)

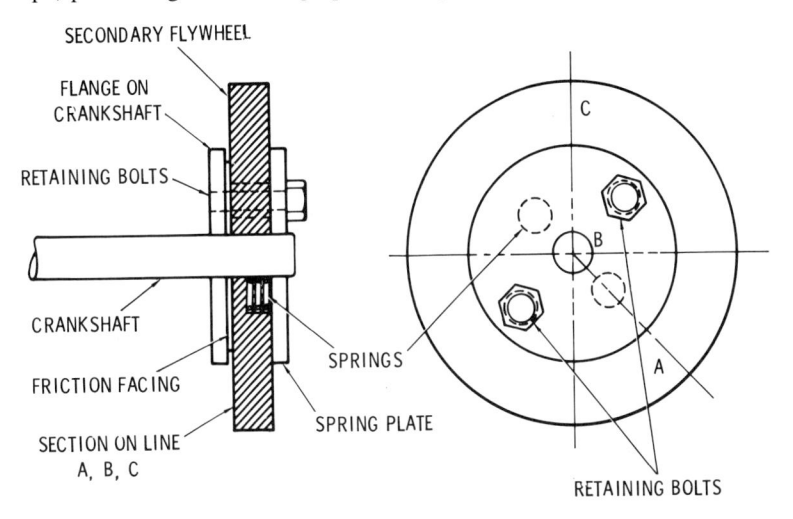

Fig. 2. Essential parts of the friction type torsional vibration damper.

When the speed of rotation of the crankshaft and main flywheel is steady, the clutch grips the flywheel and both flywheel and crankshaft rotate together. When the speed of rotation of the crankshaft is reduced, as it is by the intake stroke, the crankshaft will tend to lag behind the main flywheel and thus cause torsional vibration. In this case, the flywheel again slips, and the friction disks in the clutch will exert a force on the crankshaft and assist in bringing it up to speed.

The harmonic balancer has a flywheel or inertia weight mounted by means of leaf springs on the fan drive pulley. When the crankshaft begins to vibrate torsionally, the weight also will vibrate but, because of its inertia and type of mounting, its motion will be out of phase with the crankshaft and will in this manner reduce the intensity of vibration.

Evidently at one instant there will be a maximum effort on the cranks tending to rotate the shaft and at another instant a minimum effort. The result is the shaft tends to follow these variations with an increase or decrease in speed of rotation.

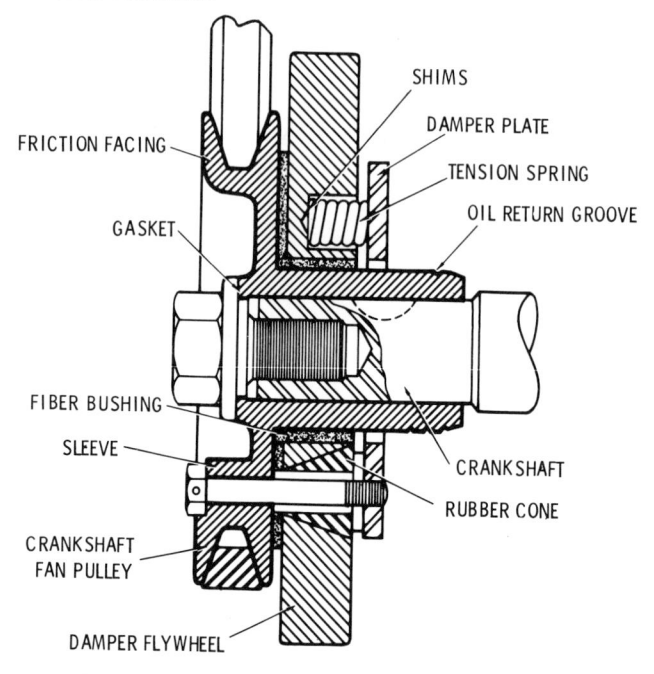

Fig. 3. Combined rubber and friction torsional damper.

If it were not for the considerable amount of energy stored up in the main flywheel in the form of dynamic inertia the operation of the engine would be so jerky as not to be practical.

The main bearings are often channeled for oil distribution and may be lubricated with crankcase oil by pressure through drilled passages or by splash. To prevent loss of engine lubricating oil, oil seals are placed at the main bearings where the the crankshaft extends through the crankcase.

CHAPTER 9

Valves and Valve Gears

Every cylinder of any four-stroke cycle engine must have at least one admission and one exhaust valve, the admission valve to permit the mixture to enter the cylinder, and the exhaust valve to allow the burned gases to escape.

The type of valves usually employed in internal combustion engines are called "poppet" or "mushroom" valve, the word "poppet" being derived from its action (it pops in and out) and the word "mushroom" from its general appearance or shape.

With reference to Fig. 1, the poppet valve consists essentially of a disc, around a circumference of which is a face which provides a ground joint seal with the valve seat. Projecting from the center is a stem which holds the valve in place centrally and transmits the movement of the valve gear to the valve.

The valves are usually made in one piece from special alloy steel. The admission valves are ordinarily made of chromium-nickel alloy and the exhaust valves of silichrome alloy because of the extremely high temperature they must withstand.

In some engines, particularly the air cooled types, the exhaust valve contains sodium in a sealed cavity extending from the head through the stem from where it is conducted to the valve guide, thus aiding in cooling.

VALVE SEATS

The matched circular surface upon which the valve face rests and which is a part of the opening leading into the combustion chamber of the cylinder is termed the *valve seat*.

There are at least two such openings or ports in each cylinder, to which are connected the admission and exhaust manifolds. Since exhaust valve

GAS ENGINE MANUAL

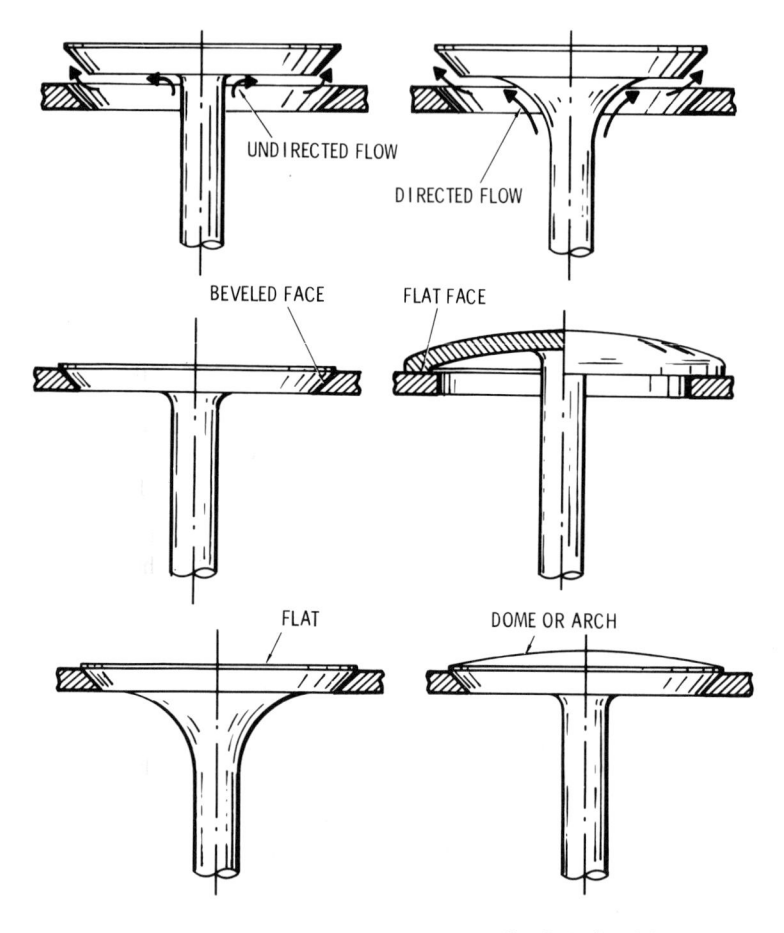

Fig. 1. Valve detail showing

seats are subjected to intense heat, valve grindings and re-seating are usually necessary from time to time to renew the sealing surfaces.

VALVE STEM GUIDES

The function of the valve stem guide, as the name implies, is to guide the valve stem during its reciprocating motion, actuated by the cam on the camshaft. The guides may be an integral part of the cylinder block or cylinder head, depending upon the type of valves used, or they may be removable sleeves which can be replaced when worn, Removable valve

VALVES AND VALVE GEARS

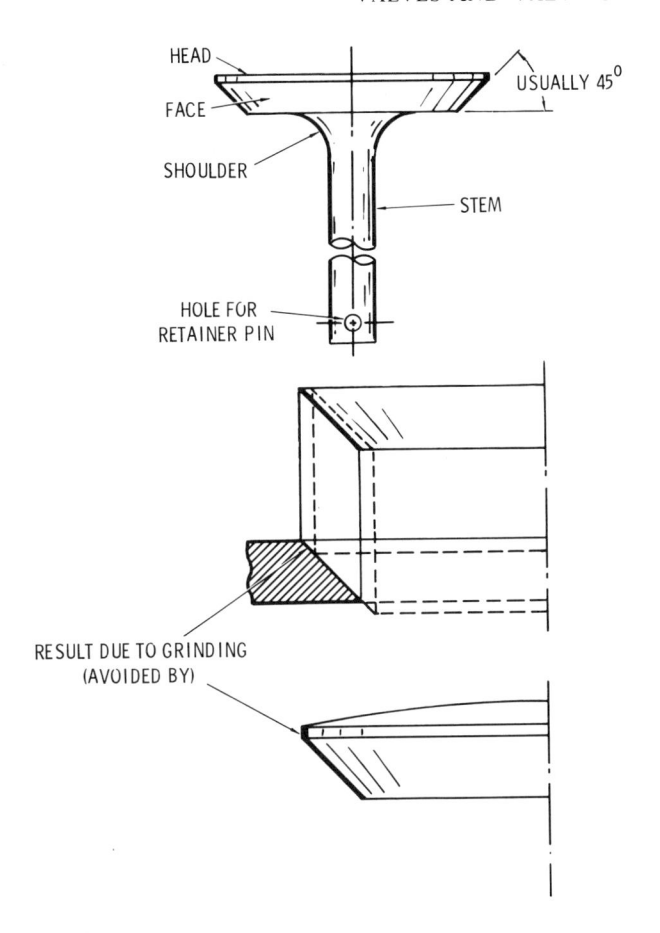

various parts and shapes.

guides are usually made of cast iron, although bronze has been used in some engines because of better wearing characteristics.

The valve stems are ground to fit the guides in which they operate and the reamed hole in the guide must be aligned and square with the valve seat to insure proper seating of the valve face.

CAMSHAFTS AND CAMS

The camshaft in its simplest form is a straight shaft on which eccentric lobes or cams are forged as an integral part. In multiple cylinder engines there are ordinarily as many cams as there are valves to be operated. A one cylinder four-stroke cycle poppet valve engine would have two cams, one for the admission and one for the exhaust valve.

In order for the engine to operate properly the cams must have the proper location on the shaft and must be designed to lift the valves at precisely the correct instant of piston travel and hold it open just long enough to obtain the most efficient fuel admission and exhaust of the cylinder. See Fig. 2.

If the camshaft is chain driven, it rotates in the same direction as the crankshaft, but if driven by a gear meshed with the gear on the crankshaft the camshaft rotation is opposite from that of the crankshaft.

The rapidity with which the valves open and close is determined by the slope of the acting surfaces between the nose and hub, that is, the slope of the opening and closing faces of the cam. See Fig. 3.

HOW TO DESIGN A CAM

The operations necessary in laying out a cam to operate a valve under specified conditions are not complicated. Cam action has its limits and the conditions to be met may be such as to result in a noisy and quick wearing gear. There are some basic or fundamental facts that should be considered in any cam design.

No matter how desirable is quick action in opening and closing, the opening and closing faces should meet the hub circle tangentially, that is, there should be no abrupt change of direction of the acting surface.

This is especially important on the opening face because here the tension of the spring, 40 lbs. or more, opposes any sudden starting of the valve when it begins to open. Another point is the steeper the opening face the more the lateral thrust on the follower which increases the friction and wear.

Example: *Design a varying continuous motion cam to operate an admission valve under the following conditions: Lift $^{11}/_{32}$, nose arc $^{5}/_{16}$, admission period 240° cam to have quick acceleration in opening and closing. Since the admission period is 240° crankshaft degrees, on the camshaft which runs at half crankshaft speed the camshaft degrees will be $240 \div 2 = 120°$.*

Solution

In Fig. 4B, first draw the horizontal and vertical axes. Through the center O, draw Aa, and Bb, so that the angle AOB, is 120°, laid out so

VALVES AND VALVE GEARS

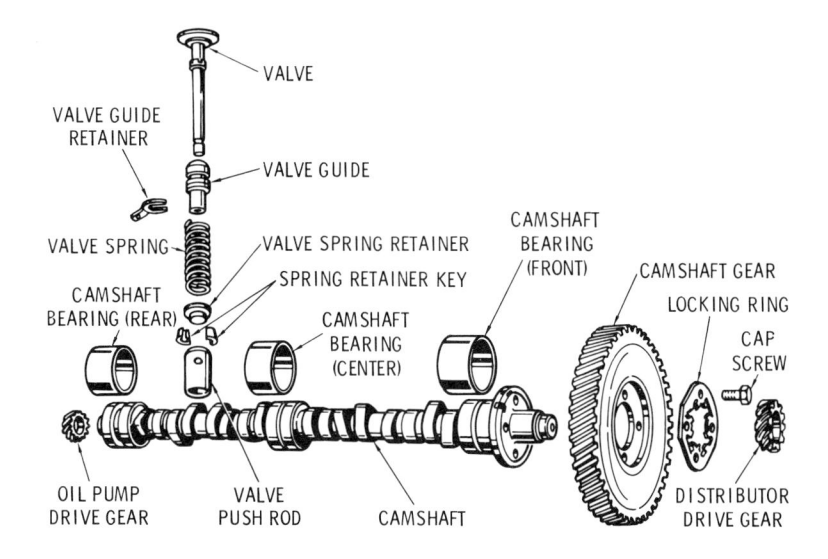

Fig. 2. Illustrating camshaft and associated parts.

that the vertical axis OM, bisects the angle AOB, that is, $<\text{AOM} = 60°$ and $<\text{BOM} = 60°$. Take diameter of cam hub circle, say 1 ½ in. From m lay off lift = $11/32$ in. With center on axis OM, describe through point n, nose arc with radius = $11/32$ in. On Aa, find by trial center of a circle that will be tangent to the hub circle and also tangent to the nose arc; describe the arc connecting them which gives the contour of the opening face. The shape of the closing face is identical and is obtained in the same way.

VALVE OPERATING MECHANISM

The function of the valve operating mechanism is to control the opening and closing of the engine valves at the correct instant of the cycle. The camshaft is rotated by the engine crankshaft by means of gears or chains, the two shafts are therefore kept in a direct speed relationship with one another.

The mechanism used to operate the valves is shown schematically in Fig. 5. Here the cam **A,** is the driving member and receives its motion from the main shaft through two gears although sometimes an idler is interposed between the crankshaft and camshaft gears. The cam turns half as fast as the main shaft since it takes two revolutions to complete the four-stroke cycle.

81

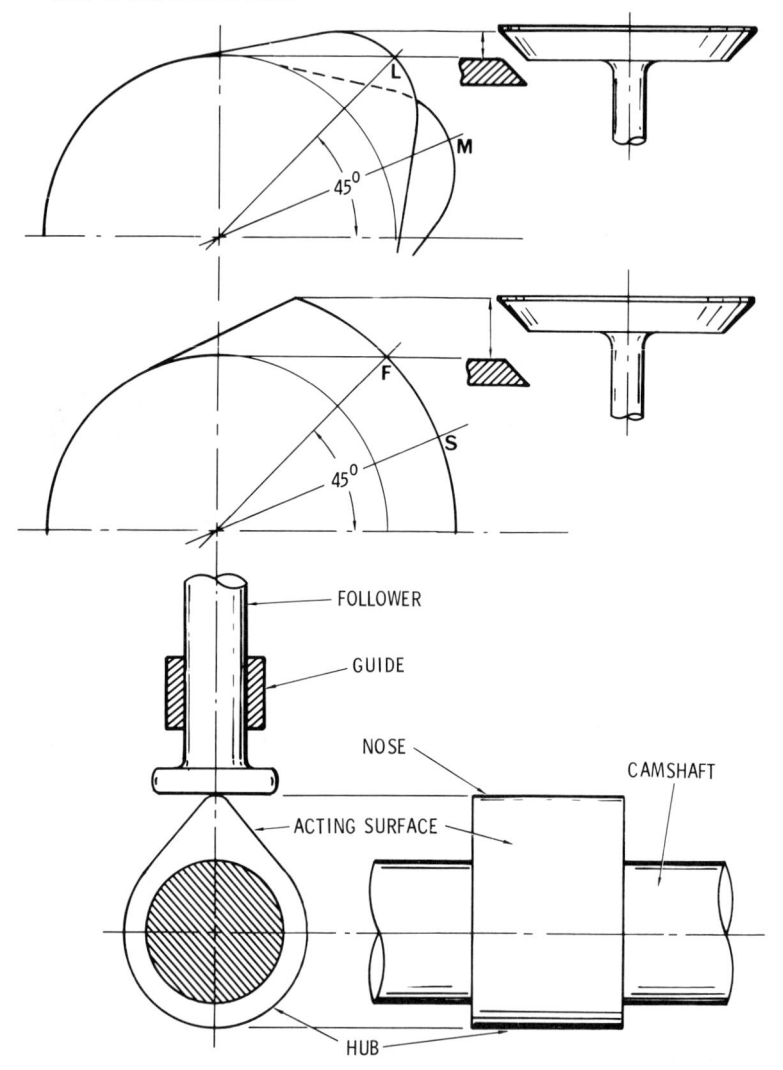

Fig. 3. Comparison of different cam lobs and valve openings.

The cam, as it revolves, transmits motion to plunger **B** which works in a closely fitting guide to prevent any lateral motion as the cam slides across the bottom of the plunger.

To prevent wear due to cam rubbing the same part of the plunger bottom, the cam is located off the center of the plunger so as to turn the

VALVES AND VALVE GEARS

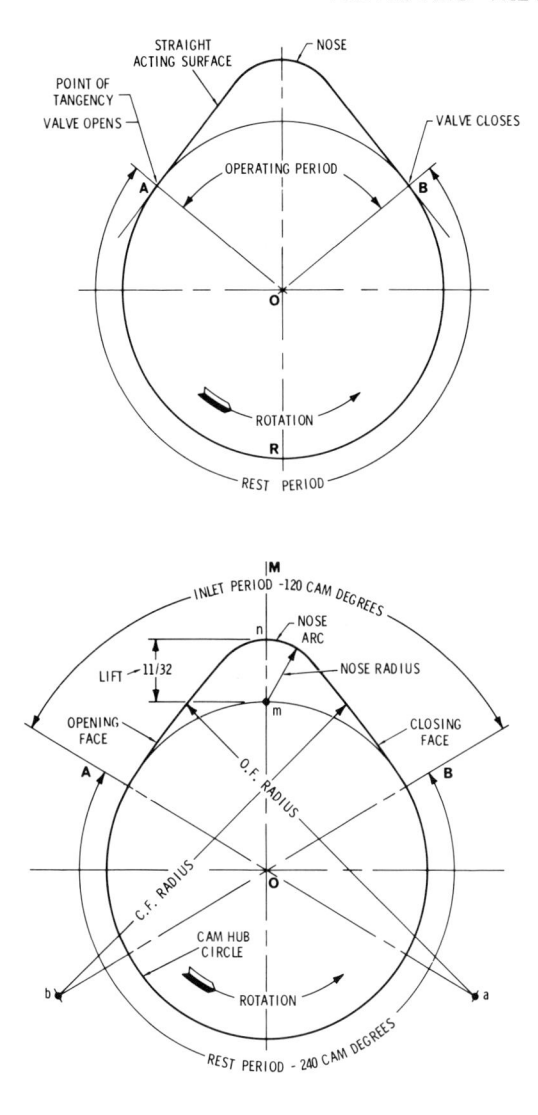

Fig. 4. Design of cam with curved opening and closing faces illustrated.

plunger to a new position each revolution of the cam. Thus the plunger not only reciprocates up and down, but turns through a small arc each time it engages with the cam. The plunger **B,** is variously called, tappet, lifter, etc.

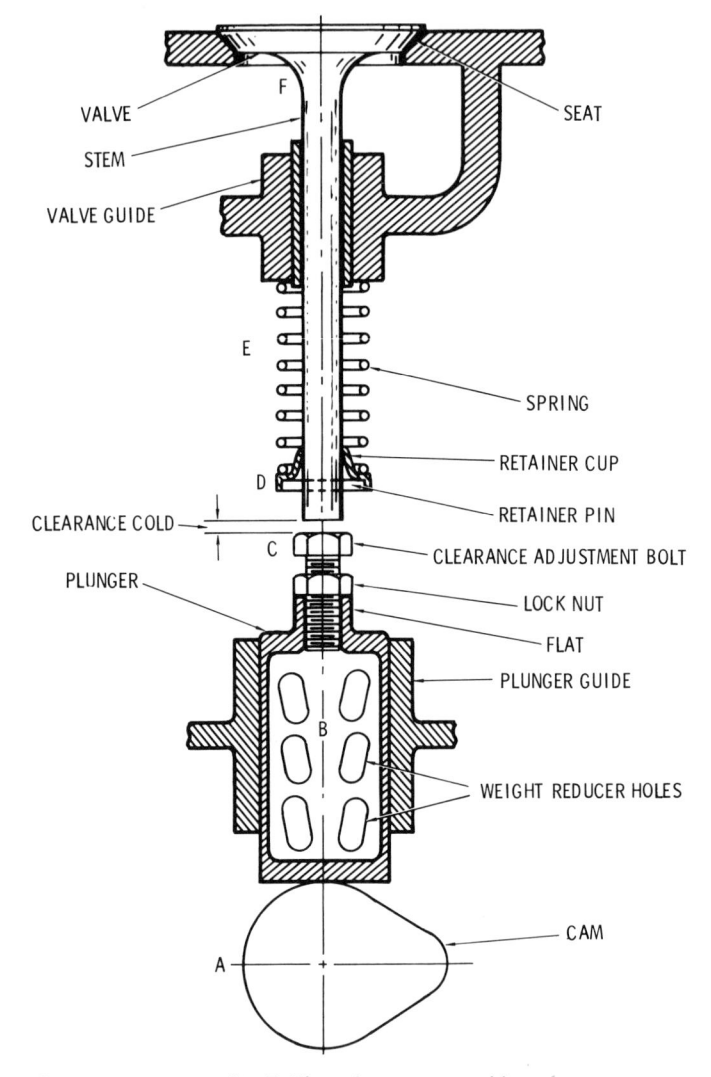

Fig. 5. The valve gear assembly.

Forming a part of the plunger is the clearance adjustment mechanism consisting of a case hardened bolt **C**, threaded into the top of the plunger and secured by a lock nut.

The requirements of a valve spring are that it will exert sufficient force to seat the valve firmly to insure tightness and to give sufficient accelera-

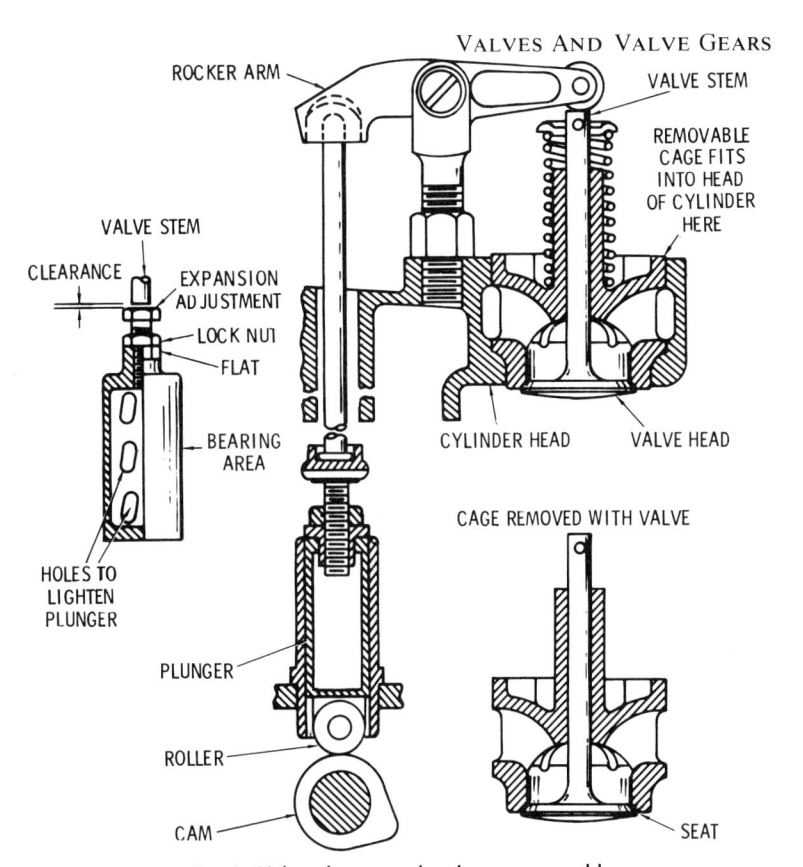

Fig. 6. Valve plunger and rocker arm assembly.

tion in closing so that the valve will follow the motions imparted by the gear at highest possible speed. The strength required brings a sizeable load on the plunger and cam, hence it should not be any stronger than necessary to avoid excess friction and wear.

Various locking devices are used to retain the spring in a partly compressed position on the valve stem. Fig. 5 shows a spring cup or seat held in position by a pin passing through a hole in the valve stem. **E,** is the spring and its function is to hold the valve closed except when pushed off its seat by the plunger. To properly operate the valve in closing, the spring must be quite stiff.

The valve is shown at **F.** The stem passes through a guide which serves to hold the valve in a central position with respect to its seat, thus assuring that the valve will properly seal in closing. Fig. 6 illustrates the valve plunger and rocker arm assembly.

CHAPTER 10

Valve Timing

The expression "timing the valves" of a gas engine applies both to the valves and the ignition system. As previously noted the valves must be timed to open and close at precisely the proper instant; otherwise the sequence of events of the working cycle would be disturbed and the engine would not operate satisfactorily or not at all, depending upon how much the timing was off.

Valve timing, therefore, refers to the exact times in the engine cycle at which the valves trap the mixture and then allows the burned gases to escape. The valves must open and close so that they are constantly in step with the piston movement of the cylinder which they control.

The position of the valves is determined by the camshaft and the position of the piston by the crankshaft. Correct valve timing is obtained by the proper relationship between the camshaft and the crankshaft. See Fig. 1.

T-HEAD

Fig. 1. Valve gear assembly for T-head design. This assembly requires two camshafts.

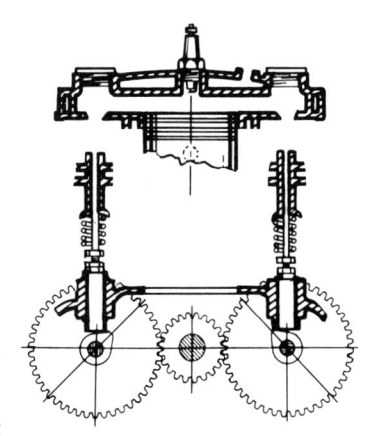

GAS ENGINE MANUAL

The extreme accuracy with which it is desired to open and close the valves may be comprehended when consideration is given to the speed at which the valves of a modern high speed engine operates. This is the primary reason for "valve overlap", which simply means that the admission and exhaust valves may both be open at the same time in any one cylinder. The valve overlap is necessary in order to compensate for the time required by the air or gas to flow through the manifolds.

A valve diagram for a typical high speed engine is shown in Fig. 2. It will be noted in the diagram that the admission or intake valve opens 15 degrees before upper dead center, while the exhaust valve does not close until 10 degrees after upper dead center. Both intake and exhaust valves are, therefore, open during 25 degrees of the crankshaft rotation.

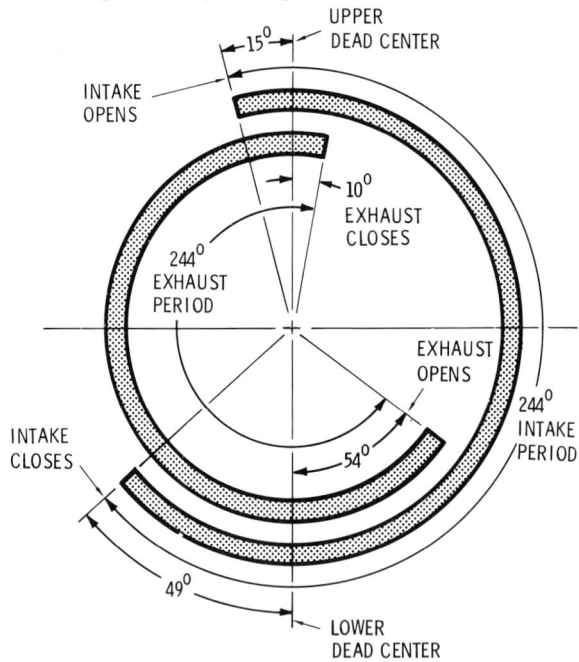

Fig. 2. Typical valve timing diagram.

The instant at which a valve begins to open or when it closes (that is, leaves or comes into contact with its seat respectively) may be expressed in terms of:

1. Distance moved by the piston from top or bottom dead center.
2. Degrees around the flywheel referred to dead centers.

VALVES TIMING

Timing is usually expressed in degrees and marks are placed on the flywheel and the camshaft gear, Fig. 3, by the engine manufacturer to facilitate the timing operation. On most engines, the timing marks on the flywheel and camshaft gear respectively are so located that a straight line drawn through their centers will intersect both.

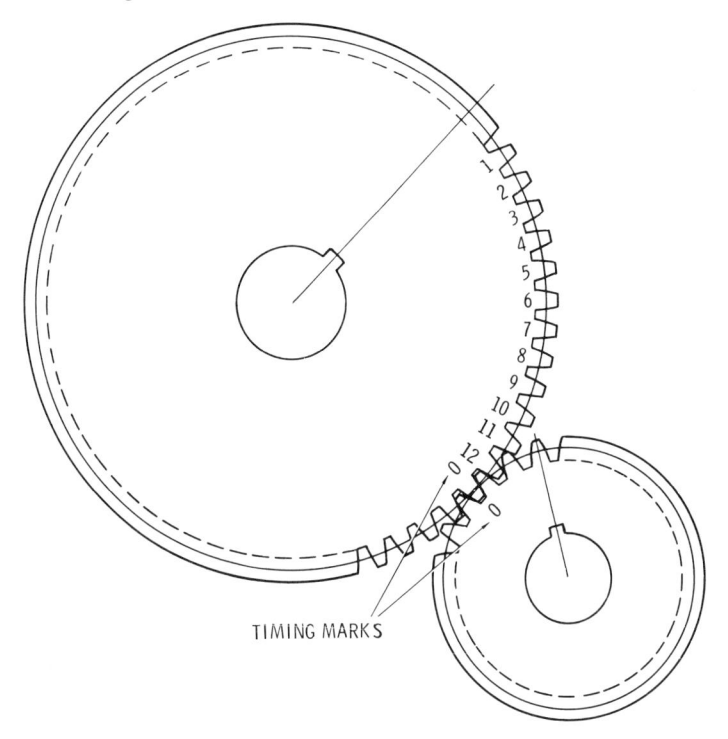

Fig. 3. Timing gear assembly showing timing marks.

HOW VALVES ARE TIMED

The instant of opening and closing of a valve is controlled by the angular position of the cam. Since the half speed camshaft is driven by a gear on the main shaft, evidently the timing may be changed by altering the position of the camshaft gear with respect to the main shaft gear.

The correct meshing positions of the two gears have already been determined by the manufacturer and the two gears marked so that when they are meshed with the marks opposite each other they are correctly set. See Fig. 3. The half speed camshaft gear and the crankshaft gear are known as the timing gears.

89

Where the admission and exhaust valves are operated by separate camshafts, there are two half speed gears and evidently each must be timed separately. With such arrangement as on T-head engines both sets of valves may be out of time with each other, and at the same time being out of time with the pistons, or one set may be in time with the pistons, while the other set is out of time.

The valves may all be considerably out of time and the engine will still run but at the expense of reduced power, increased fuel consumption and excessive vibration.

As just explained, timing of the valves is controlled by the relative positions of the timing gears. Allowance, however, must be made for the clearance between the tappet and valve stem, that is, the cam of the valve to be opened must turn through an arc sufficient to raise the tappet a distance equal to the clearance before the valve will begin to open.

With tappets adjusted to minimum clearance, this lost motion arc is very small and need not be considered unless it is desired to time valves to the last degree of precision. Such precision is not necessary in view of the considerable variation in practice as to values given to the abnormal events of the cycle. The valves of most engines are timed to preadmit the charge vary from 2½° to 13°.

The operations necessary to timing the valves on an L head engine, that is, one having admission and exhaust cams on the same shaft are:

1. Locating engine on dead center.
2. Determining correct position of the camshaft.

As a preliminary, adjust tappets to the specified clearance. After the valves are timed, this clearance will have to be readjusted with engine hot to allow for expansion—unless the clearance is given for cold engine.

In practically all cases there are markings on the flywheel and a pointer provided to locate engine on dead center. In addition there are marks on the timing gears to determine the correct position of the camshaft. In the absence of such markings the dead center and angular position of the camshaft must be determined.

HOW TO FIND THE DEAD CENTERS

Anchor at some convenient point on the engine an indicator or pointer terminating at the rim of the flywheel. Turn flywheel till crank is 10° to 20° off top dead center and measure from a convenient point of reference distance of piston to this point. Make a mark on the flywheel opposite pointer. Again turn the flywheel till crank comes on other side of top dead center and piston has traveled the same distance.

VALVES TIMING

Put a permanent mark on the flywheel rim half way between the pointer and the first mark. Identify the permanent mark by stenciling the letters T.D.C. and also the cylinder number, thus if it be cylinder No. 1, the marking will be T.D.C.1.

This mark is the dead center and when the engine is turned till the mark registers with the pointer, it is on the dead center.

In turning the flywheel each side of the dead center, it should be turned in a direction such that the crank pin will bear against the same half of the bearing to avoid any error due to lost motion.

DETERMINING CORRECT POSITION OF THE CAMSHAFT

Turn flywheel till piston is at beginning of the admission stroke. This is indicated by the movement of the tappet in opening the admission valve. Note from manufacturers instructions the number of degrees before or after T.D.C. for timing of the admission valve. These degrees must be converted into inches on the flywheel rim to determine how far the flywheel must be turned to correspond.

Example *If the valve setting is to be for 8° preadmission and diameter of flywheel is 12 ins. how many inches on rim must the wheel be turned for this setting?*

Solution

Inches to be turned = $\dfrac{\pi D \phi}{360}$, in which:

D = diameter of flywheel in inches
π = 3.1416
ϕ = admission degrees.

Substituting

Inches to be turned = $\dfrac{3.1416 \times 12 \times 8}{360}$ = .84

Accordingly, on flywheel lay off an inch on either side of the T.D.C. permanent mark and divide into tenths as in Fig. 4A. Since 8° crank rotation corresponds to .84 in. on the flywheel, the T.D.C. mark must be turned this distance from the pointer. Since, in operation, the top of the flywheel rotates from left to right, the T.D.C. or zero mark must be turned in to the left the required distance as in Fig. 4B because admission begins 8° or .84 in. on flywheel before top dead center.

Gas Engine Manual

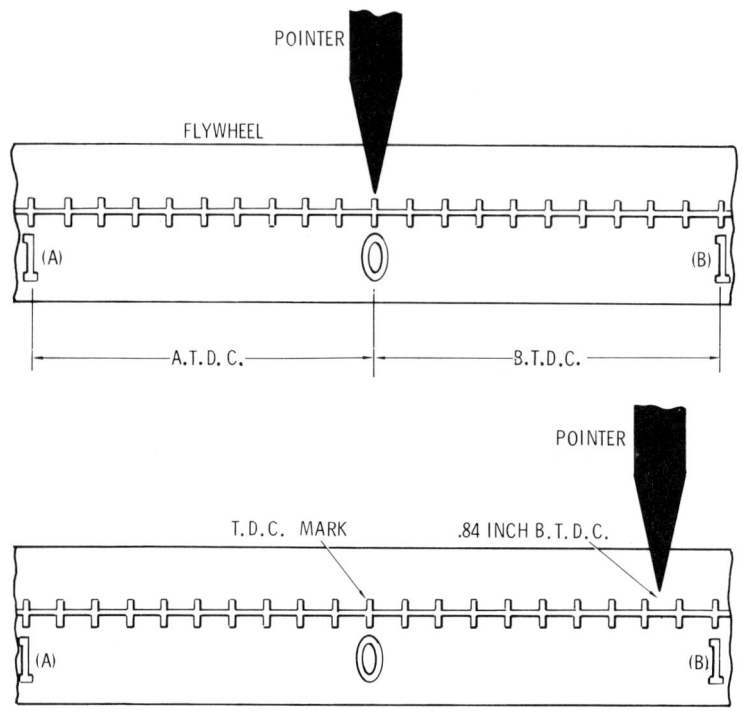

Fig. 4. Degrees converted to inches on flywheel.

CHAPTER 11

Lubricating Systems

The lubricating system forms an important part of any engine, because if not lubricated properly, the engine cannot run for any length of time without serious damage.

The primary function of engine lubrication is to reduce the friction between the moving parts. Lubrication also assists in carrying heat away from the engine, it cleans the engine parts as they lubricate, and forms a seal between piston rings and cylinder walls to prevent "blow by" of combustion gases.

The various types of lubrication systems employed in internal combustion engines, are:
   1. Forced lubrication,
   2. Splash lubrication,
   3. Oil feed with fuel.

Forced lubrication by pump pressure is the prevailing method in large high speed engines, although some engines use a combination of splash and force. In the splash system, the engine strikes the oil in the reservoir and splashes it around over the other parts in the crankcase. The lubrication thus supplied is part fluid and part oil mist.

Two stroke cycle engines have no oil reservoir and no specific oiling system. It consist of mixing oil with fuel in certain specified amounts, and as the fuel circulates in the form of vapor in the engine crankcase, the oil is carried to the working parts in the form of an oily mist. The main bearings are usually lubricated by grease cups.

FORCED LUBRICATION

In this system of lubrication the oil is forced into the crankcase of the engine through a suitable strainer. It is drawn from the reservoir in the sump of the engine by a circulating pump usually of the rotary gear type and forced under pressure through oil pipes or ducts to the camshaft bearings and to the main bearings.

From the main bearings the oil is forced under pressure through holes bored in the crankshaft to the crank pin bearings. From the latter bearings it is again fed through oil pipes attached to the connecting rod to the wrist pins. Oil escaping from the wrist and crank pins lubricate the cylinders, piston and piston rings.

Some engines employing this system of forced lubrication have provisions for additional oil, feed under pressure to each cylinder, the feeds being controled by the engine speed.

After having passed through the various bearings the oil is returned to the sump through a strainer where it enters the pump and circulates again as previously described. For the guidance of the engine operator an oil pipe usually connects the oil pressure line with a pressure gauge mounted in a suitable location.

COMBINATION LUBRICATING SYSTEMS

In the combination system, oil is under pressure directly to the main bearings through oil passages. Positive pressure also is provided for lubrication of the camshaft bearings. Connecting rod bearings are lubricated by dippers on the rod bearing caps, which dip into oil filled troughs in the oil pan.

The dippers also splash oil up into the cylinders and over the pistons and cylinder walls. Lubrication of the valve mechanism in an overhead valve engine is accomplished by oil pumped to the hollow rocker arm shafts.

OIL PUMPS

Oil pumps are mounted either inside or outside of the crankcase, depending upon the design of the engine. They are usually mounted so that they can be driven by a worm or spiral gear directly from the camshaft.

There are three general classes of oil pumps, namely: The *gear, vane* and *plunger*. In the gear-type pump, oil is forced into the pump cavity, around a gear set, and out the other side into the oil passages. The pressure is derived from the action of the meshed gear teeth, which

Lubricating Systems

prevents oil from passing between the gears, and instead forces it around the outside of each gear. See Fig. 1.

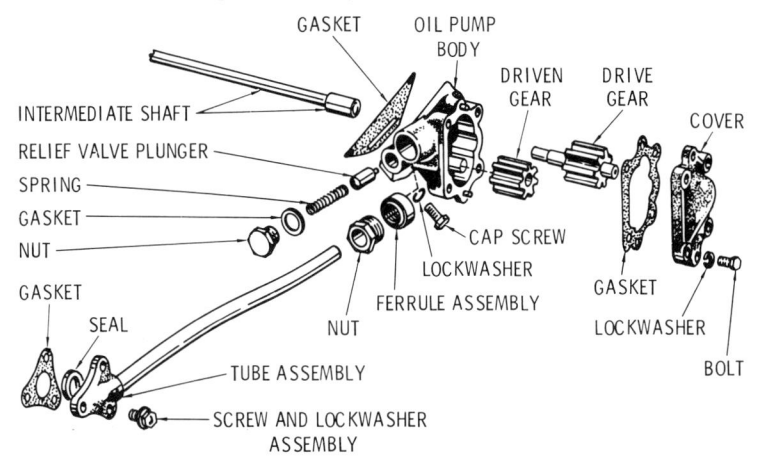

Fig. 1. View of a typical gear-type oil pump.

The general arrangement of an oiling system is shown in the elementary illustrations, Figs. 2 and 3. Fig. 2, shows oil pump and connections for lubricating the main and crank pin bearings including the oil filter. The method of lubricating the crank, wrist pins and cylinder walls is shown in Fig. 3.

In Fig.2, it will be noted that the pump is driven by the camshaft through spiral gears. This gear also serves as drive for the distributor which is coupled to the pump shaft.

Oil Strainers

The oil strainer is usually a fine mesh bronze screen located so that oil entering the pump from the oil pan must flow through it. The strainer will usually be hinged to the oil pump inlet so that it floats on top of the oil. Thus, all oil taken into the pump comes from the surface. In this way it prevents the pump from drawing oil from the bottom of the oil pan where dirt, water and sludge are likely to collect.

Most oil strainers are designed so that they will be collapsed by high oil pressure if they become clogged, thereby preventing the possibility of a complete stoppage of oil flow and consequent damage to the engine. Although some engines are equipped with two strainers, most manufac-

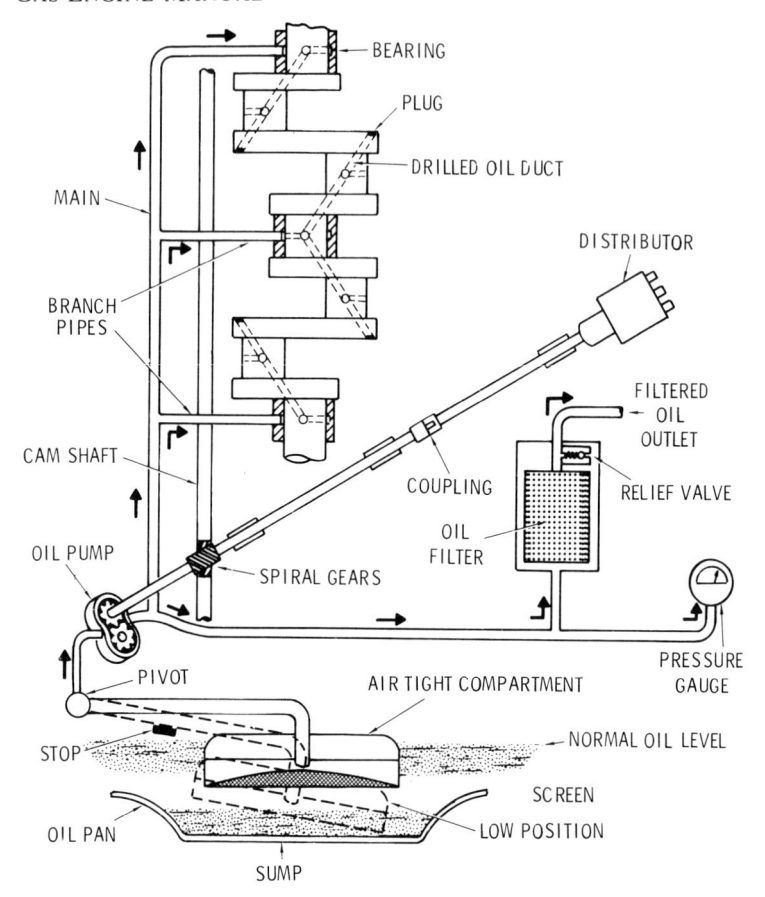

Fig. 2. Elementary diagram of a full pressure lubrication system.

turers of in-line and V-type engines place at least one oil strainer in the lubrication system.

OIL FILTERS

A second oil purification device is known as the oil filter or oil cleaner. The oil filter is placed in the oil line above the pump. It filters the oil and removes most of the impurities, such as sand, dirt and metal particles, that have been picked up by the oil during its circulation through the engine, and escaped the strainer.

LUBRICATING SYSTEMS

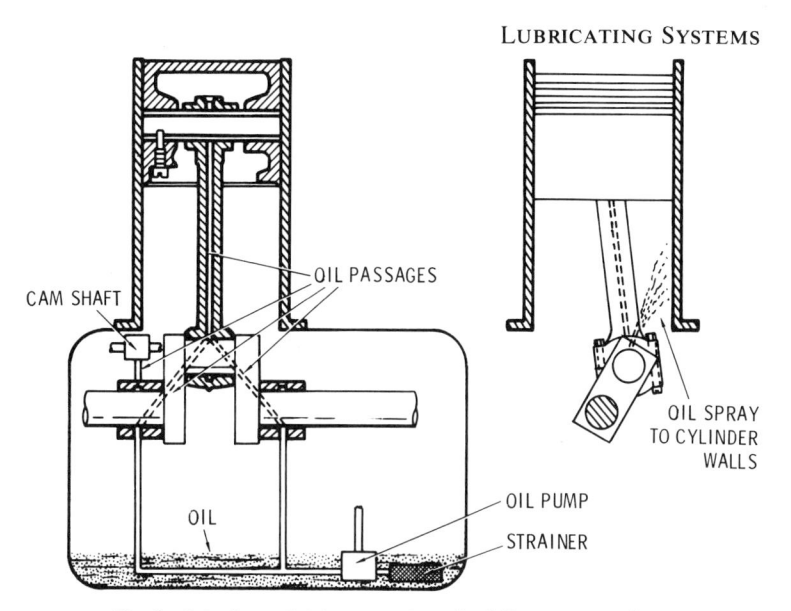

Fig. 3. Cylinder wall lubrication from the full pressure system.

The filter is usually mounted outside the engine and is connected so that part or all of the oil passes through the engine. Some filters, termed full-flow filters, are designed to handle the full output of the oil circulating pump, and all of the oil passes through them before being distributed to the engine parts. Other types divert only a small amount of the oil each time it is circulated, and after filtering, it returns directly to the oil pan.

A typical oil filter is shown schematically in Fig. 2. Here the filtering element consists of an arrangement of screens and a filtering material capable of retaining impurities as the oil is forced through. In time, filters will eventually become blocked with impurities so that oil cannot pass. For this reason, most filters are provided with relief or by pass valves which allow the oil to flow around the filter when the back pressure caused by clogging becomes greater than the tension of the relief-valve spring. Some filters must be replaced when clogged, in others the filter element can be removed and cleaned. See Fig. 4 for the complete lubricating system on a V-8 engine.

OIL GAUGES

Internal combustion engines are normally equipped with two oil gauges, one to indicate the oil pressure in the lubricating system, and the other to measure the oil level in the oil pan.

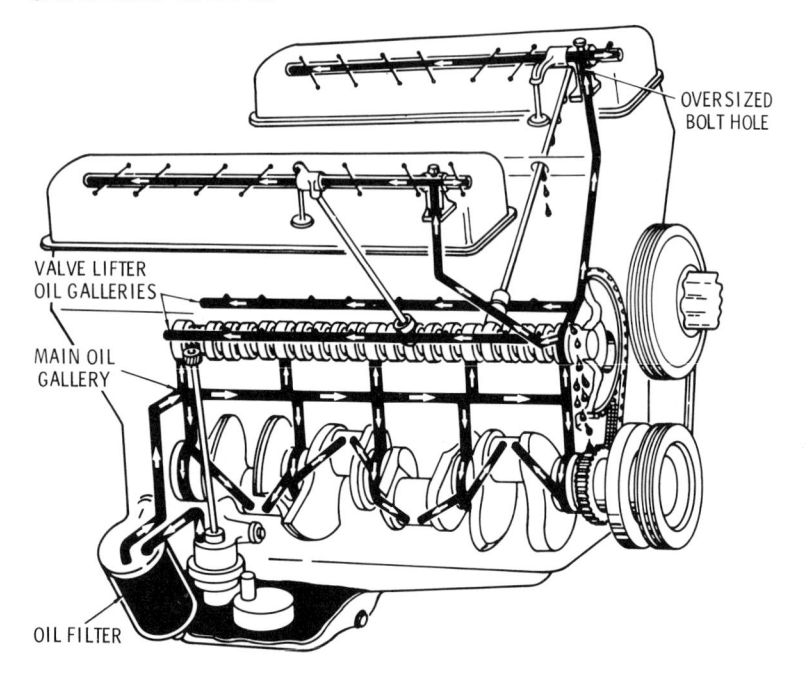

Fig. 4. Full pressure lubrication as applied to a typical V-8 engine.

The oil pressure gauge is normally mounted on the engine instrument panel, and is calibrated in pounds per square inch or some other comparative system to indicate the pressure in the lubrication system. It is connected to an oil line tapped into the main oil supply passage leading from the pump. The pressure of the oil in the system acts on a diaphragm within the gauge, causing a needle register on a suitable graduated dial.

Oil pressure gauges are of two distinct types and are termed according to their method of operation as:
 1. Pressure expansion type,
 2. Electrical type.

The pressure type gauge is similar in principle to the well known water pressure gauge. In the oil pressure expansion type of pressure gauge, however, oil under pressure passes from the engine unit up the connecting tube to the dash unit. As the pressure through the tube builds up, it has a tendency to straighten the C-shaped Bourdon tube in the dash unit and thus move the pointer attached to the free end of the tube.

The electrical type, Fig. 5, consists of a dash unit and an engine unit. The dash unit is connected to the ignition switch and in series with the

LUBRICATING SYSTEMS

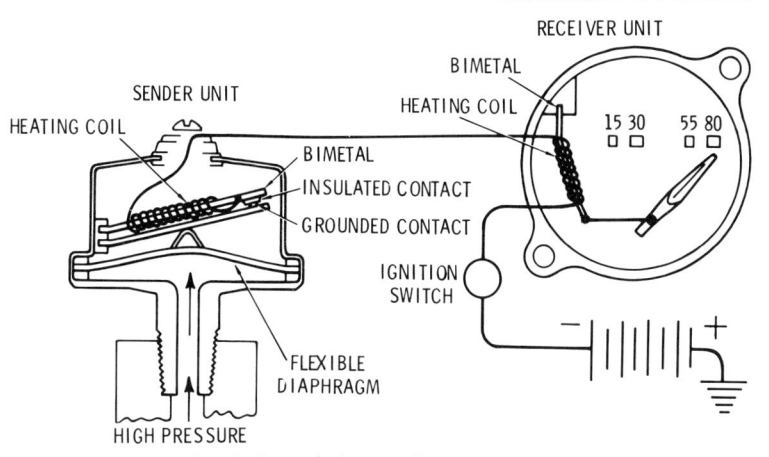

Fig. 5. Typical electric oil-pressure gauge.

engine unit. When the ignition switch is turned off, the pointer will rest at the extreme left position.

As noted in the illustration, the engine unit contains a diaphragm which is deflected in proportion to the pressure of the oil in the line. When the diaphragm is deflected, an electrical circuit is closed, allowing current to flow through a heating coil, wound around a bimetal strip. Heat, generated in this coil, deflects the bimetal to the point where the contact is opened. The bimetal then cools and returns to its original position, which again closes the electrical circuit. This cycle of opening and closing is repeated continuously.

The dash unit contains a similar heating coil formed around a bimetal, and connected in series with the coil in the engine unit. As heating takes places in the engine unit, heating also takes place in the dash unit, causing the bimetal strip in each unit to deflect simultaneously.

The pointer indicator is linked to the bimetal strip and oil pressure is indicated by the amount of deflection actuating the pointer. Increased oil pressure causes greater deflection of the diaphragm in the engine unit, therefore a greater amount of current is required to open the heating coil circuit. This increased current is transmitted to the dash unit, causing a corresponding increased bending of the dash unit bimetal and resultant indication of the oil pressure.

If some part of the engine lubricating system become clogged, the pressure indicated on the gauge will rise abnormally. Cold and heavy oil will also produce a high pressure reading, whereas very thin oil, under high temperature conditions, will produce a low pressure indication.

The oil level gauge is actually a measuring stick usually of the bayonet type and consists of a small rod of rectangular cross-section which extends into the oil pan through a small hole on the side of the crankcase near the oil filler opening. It is usually graduated to show the actual oil level, thus giving a reliable indication of the necessity for adding oil to the engine.

Readings are taken by pulling out the gauge from its normal place in the crankcase, wiping it clean, replacing it, and again removing and noting the height of the oil on the lower end of the gauge. This gauge will also indicate the condition of the oil, since worn oil will normally become darkish in color, with particles of sludge and dirt adhering to the gauge when checking the oil level.

ENGINE OILS

Oils used for engine lubrication are principally derived from petroleum. The petroleum oils are compounded with animal fats, vegetable oils and other ingredients to produce satisfactory oils and greases. Among the several important requirements for a lubricant are:
1. Body,
2. Fluidity or viscosity,
3. Freedom from gumming,
4. Absence of acidity,
5. Stability under temperature changes,
6. Freedom from foreign matter.

The body of a lubricant indicates a certain consistency of substance, that prevents it being entirely squeezed out from the rubbing surfaces. The particles of the lubricant should adhere to the rubbing surfaces, thus securing effective separation. The body of a lubricant should be such as to prevent a too rapid running off, depending on the rubbing pressure.

Fluidity of a lubricant refers to a certain lack of cohesion between its different particles, which reduces the fluid friction. Fluidity, so far as it does not oppose body, is a desirable quality. Excessive fluidity allows the lubricant to run off too quickly, thus causing waste.

A lubricant that gums loses its fluidity easily, collects dust and grit, and thus increases friction and wear.

A lubricant that holds free acid would attack the bearing surface, destroy its smoothness, increase friction and lead to frequent and costly repairs.

Stability under temperature changes is important; lubricants should retain their good qualities, even when used under high temperatures as in a steam cylinder, or when used under low temperature, as in ice machines, or on exposed bearings. They should not evaporate, not be decomposed by heat, nor congeal by cold and should retain their normal body and fluidity as much as possible.

Foreign matter will increase friction, and clog feed tubes, thus causing heating and possible seizing of the rubbing surfaces.

Viscosity

Viscosity represents the flowing quality of an oil and is determined by noting the number of seconds taken to pass a certain quantity of the oil at a specified temperature through a standard size orifice made for the purpose. *The Saybolt universal viscometer* is employed in the United States for this purpose, while the Redwood, Engler and other similar instruments are used extensively in other parts of the World.

It is a well known fact that heat will thin oil, giving it greater fluidity or lower viscosity. At high temperatures, some of the heavier fractions retain a sufficient degree of viscosity to render suitable for service as lubricants. At low temperatures, oils become more viscous. For these conditions, therefore, the lighter fractions are more suitable for lubrication.

SAE Viscosity Numbers

These numbers constitute a classification of lubricants in terms of viscosity or fluidity, but without reference to any other characteristics or properties. The viscosity numbers are assigned by the *Society of Automotive Engineers* (SAE) in such a way that the higher the SAE number, the more viscous or heavy is the oil. Thus, an SAE 10 engine oil may be recommended for low temperature use, whereas an SAE 30 oil will be suitable for use in warmer weather.

The added designation of **W**, such as **10W**, indicates that the oil has the added ability to remain fluid or flow at a lower temperature. The manufacturer of an engine usually recommends the viscosity of oil to be used under various conditions.

During cold weather, engine oil selection should be based primarily upon easy starting characteristics, which depend upon the viscosity of the oil at low temperatures. Fig. 6, indicates the temperature range within which each grade can be relied upon to provide easy starting and satisfactory lubrication.

When the crankcase is drained and refilled, the oil should be selected, not on the basis of the existing temperature at the time of change, but on

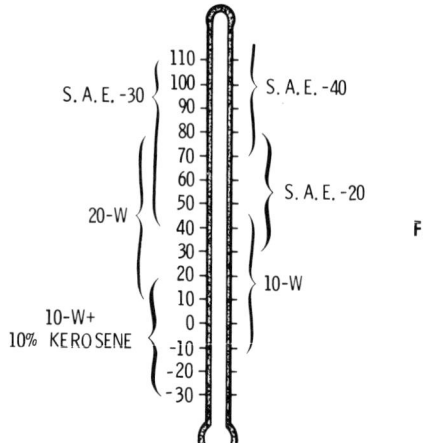

Fig. 6. Temperature range for various SAE oil numbers.

the anticipated minimum temperature for the period during which the oil is to be used. This is to prevent starting difficulty at each sudden drop in temperature.

Oil Dilution

When lubricating oil becomes diluted with gasoline, it loses its viscosity and some of its lubricating qualities. Excessive use of the choke causes an over rich mixture to be forced into the cylinders. This excess gasoline remains in a liquid state and drains by the piston rings into the crankcase where it mixes with the oil. When the engine operates at higher temperatures, this condition is corrected to some extent as the excess gasoline vaporizes in the crankcase and is carried off through the ventilation system.

Presence of gasoline will not lower the oil level but will maintain it or even raise it. The lubricating quality of oil however, is definitely reduced when diluted with gasoline.

Corrosion

Practically all present day engine fuel contains small amounts of sulfur, which in the state in which it is found, are harmless. This sulfur, however, on burning forms certain gases, a small portion of which is likely to leak past the piston and rings and when reacting with water form very corrosive acids. The more sulfur in the fuel, the greater the danger from this type of corrosion.

This condition cannot be entirely avoided but may be reduced to a minimum by proper care of the engine. As long as the gases and the internal walls of the crankcase are hot, no harm will result, but when an engine is run at low temperatures, moisture will collect and unite with the gases formed by combustion. It is in this manner acid will be formed and which is likely to cause serious etching or pitting.

High Temperature Operation

High temperature and hard service promotes oxidation of lubricating oil. This type of service may cause high temperature varnish and sludge deposits, stuck rings and scuffing of rings in all types of engines. It may also cause corrosion of some types of bearings.

This condition is aggravated by hard service particularly in hot weather. Under this condition, the crankcase oil is subjected to relatively high temperatures.

The nature of the fuel may have some influence on the severity of this condition, but its relative influence is less under these high engine temperatures than under start and stop conditions. In engine design especially adequate cooling of oil as well as of pistons, valve guide and seats, can minimize the effect on the oil.

Frequency of Oil Changes

The frequency of oil changes depends upon several factors. With reference to automotive type engines the frequency of oil changes are usually determined on a mileage basis. Visual inspection might result in an erroneous interpretation in cases where the filter is still in good condition, in which case the oil would appear fairly clean after a long mileage even though it might have accumulated acids which might prove harmful.

The recommendations of a 1000 mile change for winter, and a 1500 change for summer in some cases might be disposing of oil which is still satisfactory. There are, however, conditions where the roads are extremely dusty and where this mileage would be appropriate. It is better to change the oil too often than not often enough since keeping the engine oil in good condition is money well spent. In each instance, however, manufacturer's recommendation should be strictly adhered to.

CHAPTER 12

Cooling Systems

All internal combustion engines must be equipped with some type of cooling system, because of the great amount of heat generated by combustion of the fuel.

There is no dependable method of measuring the temperature in the combustion chamber during the burning of the fuel, but it is estimated to range from 2700°F. to 3200°F. for low and high compression engines respectively. Accordingly, it must be evident that the intense heat generated within a gas engine cylinder would very quickly overheat the metal within the cylinder to such an extent that it would become red hot, resulting in burned and warped valves, seized pistons in the cylinders, overheated bearings and a break down of the lubricating oil.

To avoid these conditions means must be provided to carry off some of the heat, that is, enough of it to prevent the temperature of the metal of the cylinder rising above a predetermined point and low enough to permit satisfactory lubrication and operation. The excess heat is carried off by some form of cooling system.

In this connection, it should be clearly understood that although heat is necessary as it causes expansion of the charge which acts on the piston head to produce power, a large part of it goes to waste through the exhaust ports and in the cooling system. Thus, for example, it is estimated that only about one third of the heat energy contained in the fuel is converted into useful power, whereas another third is dissipated through the exhaust, and the remaining third is absorbed in the cooling system.

There are four methods of cooling internal combustion engines; these are:
    1. By water circulation,
    2. By water and oil circulation,

GAS ENGINE MANUAL

3. By air cooling,
4. By air and oil circulation.

Water cooling is commonly obtained by means of a pump and associated piping, radiator, fan and a system of jackets and passageways through the engine within which the water circulates.

Water and oil cooling requires additional means for air cooling of the engine oil, generally used whenever the (water) radiator area is inadequate in size for the high-performance characteristics of the engine (as in racing cars).

Air cooled engines usually employ blades incorporated in the flywheel which acts as a fan to circulate air over the fins cast integrally with the cylinders. Engines of this type are used almost exclusively for small appliances such as lawnmowers, wood saws, or small foreign-automobile engines, etc.

Air cooling of the oil — to supplement the cylinder cooling — is also used with air-cooled engines in which the high-performance characteristics require additional cooling, or for which space and position limitations make ordinary air cooling (alone) inadequate.

WATER CIRCULATION SYSTEMS

Water is the most widely used coolant for liquid cooled engines. The main objection to the use of water is that because of its high freezing point it cannot be used without additives at temperatures below 32°F.

Cooling of the engine parts is accomplished by keeping the water circulating and in contact with the metal surfaces to be cooled. The pump draws the water from the bottom of the radiator, forces it through the jackets and passages, and ejects it into a tank on top of the radiator. See Fig. 1.

The water passes through a set of tubes to the bottom of the radiator and again is circulated through the engine by pump action. A fan, geared to the engine, draws air over the outside of the tubes in the radiator and cools the water as it flows downward.

It should be noted that the water is pumped through the radiator from the top down. The reason for this direction of flow is that the thermosyphon action aids the pump to circulate the water. This simply means that as the water is heated in the jackets of the engine, it expands slightly and as a result becomes lighter and flows upward to the top of the radiator. As cooling then takes place in the radiator tubes, the water contracts, becomes heavier and sinks to the bottom. This desirable

Cooling Systems

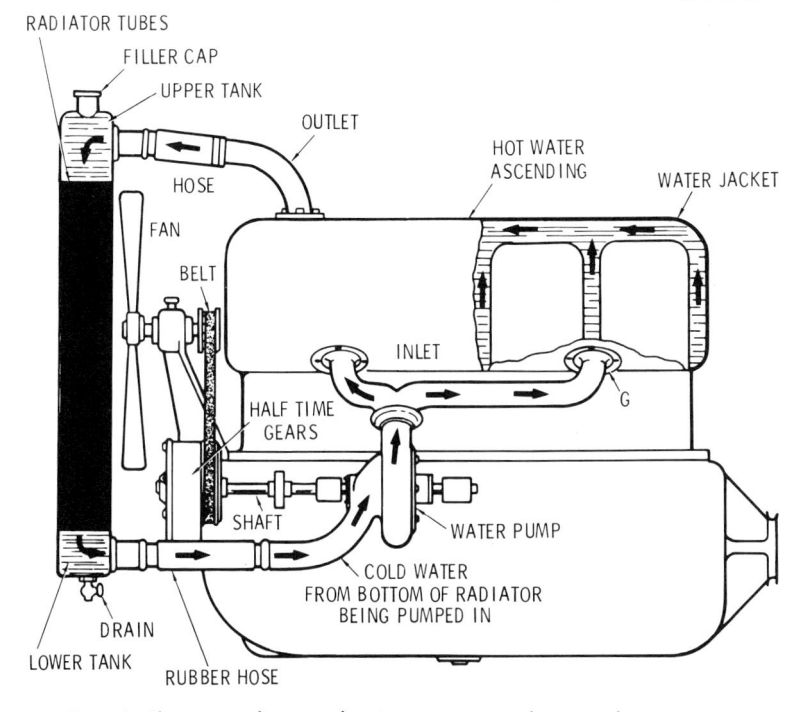

Fig. 1. Elementary diagram showing a water circulation cooling system.

action, however, cannot take place if the water level is allowed to become too low.

The fan that is used to cool the water in the radiator may be a constant-speed fan (one which produces a volume of air directly proportionally to the engine speed), or a fan that will vary the quantity of cooling air it produces in accordance with a formulated need. Actual need for fan-created air varies with the forward speed of the vehicle; a slow-moving (or stationary) vehicle has little or no motion-created airflow; but a faster-moving vehicle generates a proportionally increased airflow which, at some point, becomes greater that the airflow created by a fan. Furthermore, revolving a fan consumes motor power, and decrease of fan energy in relation to the decrease of its need obviously saves motive power.

There are two ways to decrease fan energy (motor-power consumption). One is to vary the rpm of the fan; the other, to vary the pitch (blowing angle) of the fan blades.

VARIABLE-SPEED FANS

Fan rpm is varied by means of a fluid coupling (fan clutch) in the fan shaft. Torque-carrying capacity of the coupling (and, consequently, the percentage of fan-pulley rpm transmitted through it to the fan) depends upon the amount of fluid (generally silicone oil) between the two coupling plates; the greater the amount of fluid, the greater the torque will be. Fluid is stored in a reservoir in the coupling body and is pumped between the plates or bled back into the reservoir by operation of a control piston at the coupling axis. This piston is moved by the expansion or contraction of a bimetal thermostat which may be either a flat-spring or a coiled-spring type, and which is located on the front of the coupling (between the fan and radiator) where it will be affected by the heat of the air passing through the radiator. See Fig. 2.

A VARIABLE PITCH FLEX FAN

This type fan has flexible blades which increase or decrease the blade "scoop" and thus vary the quantity of air the fan will blow at any given rpm. Since the blades are flexible they will tend to "flatten out" and lose their scooping action when operated at higher engine speeds.

ENGINE WATER JACKET

Water jackets in internal combustion engines consists of an outer casing surrounding the cylinders. The circulating water passes through the space between the casing and the cylinders. In modern construction, the jacket is so arranged that in the block, water completely surrounds the cylinder bores, the valve seats, and the valve stems.

Carefully sized water passages in the head aid in regulating the water flow and help to maintain uniform temperature throughout the block. Water completely surrounds the combustion chamber and spark plug bosses in the cylinder head. A baffle plate or water manifold is usually inserted in the block between the cylinders and valve stem guides to distribute the water.

RADIATORS

By definition, a radiator is an assemblage of numerous very small passages of circular or rectangular forms constructed of thin metal to form the cooling surface for the circulating water.

COOLING SYSTEMS

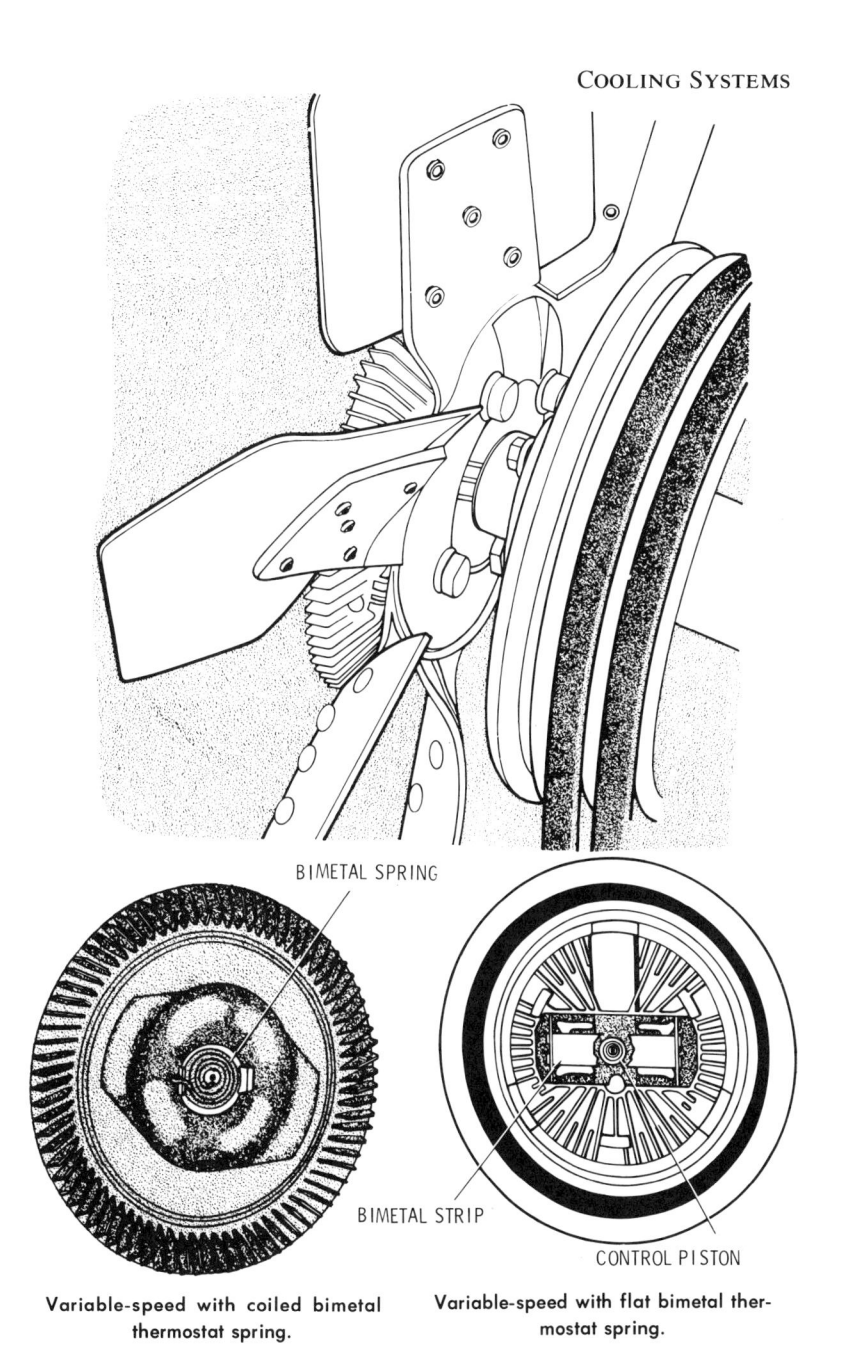

Variable-speed with coiled bimetal thermostat spring.

Variable-speed with flat bimetal thermostat spring.

Fig. 2. Typical variable-speed fan assembly.

The radiator must be strong enough to withstand vibration and the walls thin enough to reduce the weight to a minimum and offer the least possible resistance to the transfer of heat from the water circulating within its passages to the air current passing over the exterior or cooling surface. Radiators are built in two general classes, namely:
1. Cellular,
2. Tubular.

The tubular radiator cores or cooling surfaces, consists of a large number of vertical water tubes, and numberous horizontal air fins around the tubes. Water passages in the tubes are narrow, and the tubes are made of thin metal. The core divides the coolant into very thin columns or ribbons, thus exposing a large radiating surface to the volume of water to be cooled.

Cellular radiator construction consists in nesting individually formed hexagon tubes, cut to length according to the depth of the core from front to back and dipping both surfaces, front and back, in a bath of solder to make a complete unit. Such construction resembles a wax honey-comb made by bees.

If such cells are hexagon in appearance, the radiator is called a cellular true honeycomb. The water passes in channels between the tubes and the air passes through the cells in this type of core.

Another cellular construction may be accomplished by the juxtaposition of crimped or corrugated ribbons of brass, copper or bronze of the width corresponding to the depth of the core from face to rear.

Each ribbon forms the half of a cell and when joined such halves make up a complete series of cells of honeycomb shape. Such structure, in its final form, also being a multitude of cells through which the air passes and having a channel for the water passage, but not being created on the principle of a wax honeycomb, is called a cellular honeycomb type. Thus, it will be seen that only one specific construction is a true honeycomb but that all honeycomb radiators are cellular.

WATER CIRCULATING PUMP

All modern cooling systems have pumps to obtain forced water circulation. The water pump is usually of the centrifugal type and has an impeller with blades to force the water outward as the impeller rotates. The pump and fan are usually driven from a common V-belt which is driven by a pulley at the front end of the crankshaft. See Fig. 3.

Advantages of the centrifugal type pump are that it circulates a great quantity of water for its size, is not easily clogged by small particles of

COOLING SYSTEMS

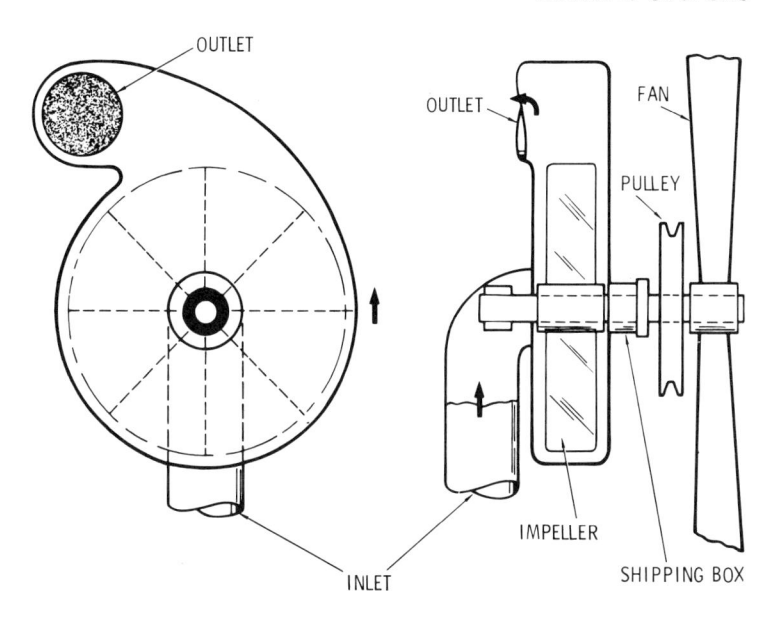

Fig. 3. Illustrating the elementary centrifugal water pump.

dirt and is simple in construction. Another advantage is that it permits limited circulation by thermo-syphon action even if the engine is not running. See Fig. 4.

TEMPERATURE CONTROL

Radiators are provided with sufficient cooling surface to adequately cool the engine even during hot weather at normal load. At starting and during cold weather, however, some method must be found to regulate the temperature since if the engine remains too cool unsatisfactory operation will result.

To keep the engine as closely as possible at a predetermined temperature, a thermostat is inserted between the water jacket and the radiator, usually at the housing at the cylinder head water outlet as shown in Fig. 5.

The function of the thermostat is to regulate the engine temperature by automatically controlling the amount of water from the engine block to the radiator core.

The thermostat is merely a heat operated unit which controls a valve between the water jacket and the radiator. Thus, when the engine is cold, the thermostat valve remains closed, and the water is recirculated through

111

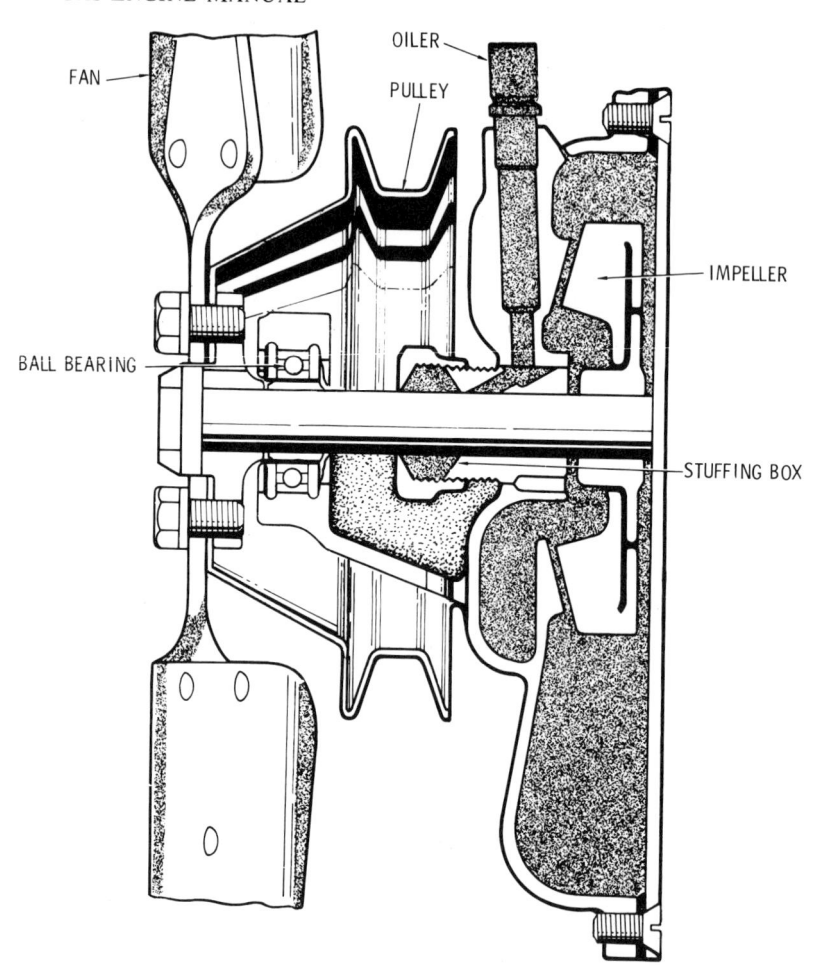

Fig. 4. Typical water pump and fan assembly.

the water jacket without entering the radiator. As the engine warms up, the valve slowly opens and some water begins to flow through the radiator, where it is cooled.

A typical thermostat consists of a flexible metal bellows attached to a valve. The sealed bellows, which is expandable, is filled with a highly volatile liquid. When the liquid is cold, the bellows chamber is contracted and the valve is closed. When heated, the liquid is vaporized and expands. As the chamber expands, the valve opens to permit water circulation between the engine jackets and the radiator.

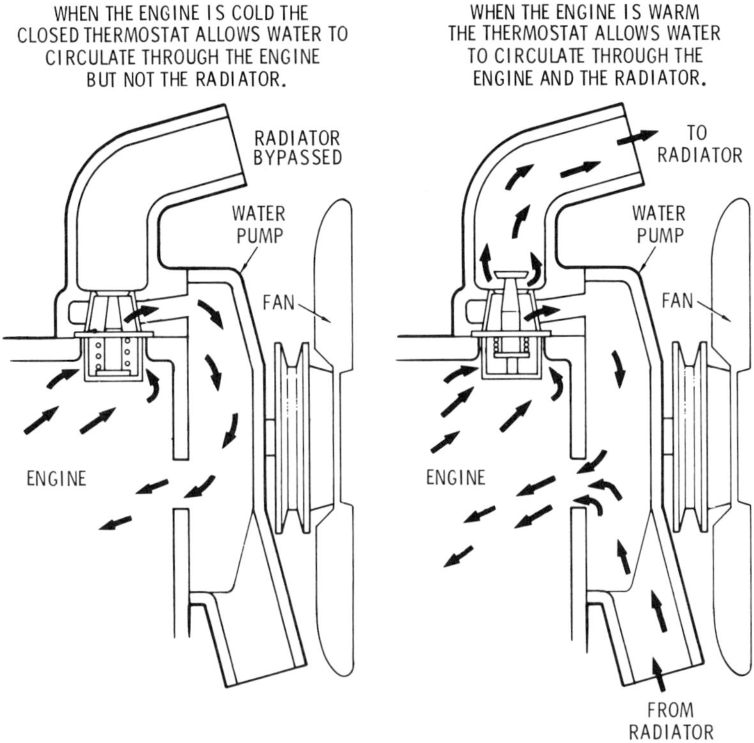

COOLING SYSTEMS

WHEN THE ENGINE IS COLD THE CLOSED THERMOSTAT ALLOWS WATER TO CIRCULATE THROUGH THE ENGINE BUT NOT THE RADIATOR.

WHEN THE ENGINE IS WARM THE THERMOSTAT ALLOWS WATER TO CIRCULATE THROUGH THE ENGINE AND THE RADIATOR.

Fig. 5. Showing general location and operating principle of a thermostat.

Other thermostat types include a sealed copper bellows containing only air. Another is bimetallic, and depends for its operation upon the difference in coefficient of expansion of the two metals.

RADIATOR PRESSURE CAP

Some engine cooling systems are sealed by a pressure type radiator filler cap which causes the system to operate at higher than atmospheric pressure. This higher pressure raises the boiling point of the coolant and increases the cooling efficiency of the radiator.

The pressure radiator cap contains two spring-loaded normally closed valves which seals the system. See Fig. 6. The larger of the two valves is a pressure valve, while the smaller valve is a vacuum valve. The pressure valve acts as a safety valve to relieve extra pressure within the system. The vacuum valve opens only when the pressure within the cooling

GAS ENGINE MANUAL

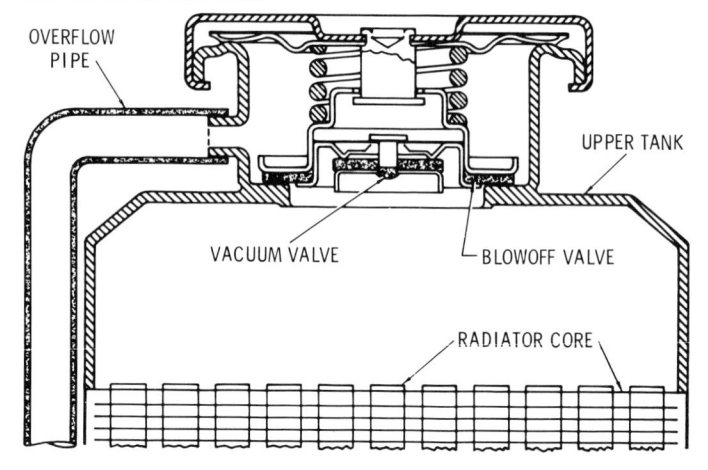

Fig. 6. Showing typical pressure-type radiator cap assembly.

systems drops below the outside air pressure as the engine cools off. Higher outside pressure then forces the vacuum valve to open, permitting air to enter the system by way of the overflow pipe.

ANTIFREEZE SOLUTIONS

When an engine is operated where the atmospheric temperature is below 32°F. an antifreeze solution must be added to the cooling water.

If a container is filled with water and allowed to freeze, it will burst. The same thing will happen to a cylinder block or to a radiator unless an antifreeze solution is added to prevent freezing.

There are several antifreeze solutions available that are satisfactory for engine cooling systems. Among these are, *methyl alcohol, ethyl alcohol, glycerin, and ethylene glycol.* The first two, prepared commercially as antifreezes, are the cheapest and provide adequate protection when used in sufficient quantities. The main objection to their use is that they boil and evaporate if normal operating temperatures are exceeded.

Glycerin offers the same degree of protection as alcohol and does not evaporate in use because of its high boiling point. Ethylene glycol (antifreeze compound) has an extremely high boiling point, does not evaporate in use, is noncorrosive, has no odor and furnishes complete protection when used in the proper amount.

When using ethylene glycol, it is necessary to clean the entire cooling system before putting in the antifreeze solution. It is also advisable to

Cooling Systems

tighten and replace all hose connections. It is important that the cylinder head gasket be kept tight to prevent leakage.

If there is leaks in the system, they should be located and stopped. If evaporation occur with the use of ethylene glycol, it is only necessary to add water to the solution. The cooling system should, however, be watched closely for leaks and should be tested when additional water is required.

The freezing point of an antifreeze solution may be determined by using a hydrometer. When testing the solution however, it should be tested at the temperature for which the hydrometer is calibrated, and the correct hydrometer for the solution should be employed.

Precautions

Solutions containing salt, calcium chloride, soda, sugar or mineral oils such as kerosene or engine oil should never be used in the cooling system as they either clog the water passages or damage the hose connections and other parts.

Rust Inhibitors

The use of inhibitors or rust preventors will reduce or prevent corrosion and the formation of scale in the cooling system. Inhibitors are not cleaners and do not remove rust or scale already formed; they merely added to the cooling liquid to prevent further rust or corrosion.

The logical time for flushing and introduction of the inhibitor is when antifreeze is installed in the winter and when it is removed in the spring.

Care must be taken, however, in the selection of an inhibitor as some of them contain strong acids or caustics that will react with the metal of the radiator core, eating holes through the metal and causing the radiator to leak.

The effectiveness of any inhibitor is limited to about six months, after which the cooling system should be flushed, refilled and new inhibitor added.

WATER AND OIL CIRCULATING SYSTEMS

High-performance (sports and racing) cars generally are so streamlined that the space provided for a radiator is limited and the resulting water-circulating system is inadequate for the cooling needed. To offset this design disadvantage, additional engine cooling is accomplished by cooling of the engine oil. This is done in different ways.

The simplest method is to provide an oversized oil pan having external fins exposed to the air so that heat carried by the oil circulating through the engine will be more readily dissipated to the atmosphere. Oil pan radiation area is carefully calculated to compensate for the reduction of water-cooling capacity.

Another method is to circulate the oil through an oil radiator in which it can be cooled prior to returning to its normal course through the engine. Generally, such a radiator (which is externally finned for maximum air cooling) is located within the airflow path of the water-circulation system fan. Some vehicle use both methods.

AIR COOLING SYSTEMS

An air cooling system, as the name implies, consists in forcing air under pressure, over special cooling flanges which are cast integrally with the cylinders. The circulating air fan is usually mounted on the flywheel.

In air cooled engines, the cooling flanges or metal fins, absorbs the heat of fuel combustion and diffuses it in the rush of air supplied by the fan. It should be noted, however, that although the cylinder fins increase the effective radiating surface of the cylinder to a considerable extent, larger horsepower engines are almost exclusively water cooled because air does not absorb heat as readily as water.

There are several physical characteristics peculiar to automotive engines of the air cooled type. The cylinder head and barrel are heavily finned for strength and adequate cooling. Air deflectors or baffles direct the air flow to the cylinders and increase its velocity. A streamlined ring-type cowling or shroud surrounds the engine as another means of controlling the air flow; finally a cooling fan is mounted on the engine flywheel to direct high speed air to the cylinders.

AIR AND OIL CIRCULATING SYSTEMS

As with a water-circulating system, preceding, some air-cooled vehicle designs (due to space limitations) do not permit adequate air alone for cooling, and additional cooling of the engine oil is required to maintain proper engine operating temperatures and oil lubricating qualities. An oil radiator, such as described for a water-and-oil circulating system, is employed. This is located within the engine shroud so that part of the air being blown past the engine cooling fins will be blown past the oil

Cooling Systems

radiator. A conveniently placed bimetal thermostat is generally used to adjust the air intake opening into the shroud so that total cooling effect can be controlled for all conditions of engine operation.

This combined air-and-oil cooling system is currently used for small car engines, and even for some bus and van engines. The system permits enclosure of the engine within the vehicle "shell" without visible protrusions of engine parts in order to obtain adequate air cooling.

CHAPTER 13

Fuel Systems

Two types of gas-engine fuel systems are in use:
1. A carburetor system,
2. A fuel-injection system.

Both systems require several basic parts (Fig. 1): A *storage tank* for the fuel; an *intake manifold* to distribute an explosive mixture of fuel and air to the cylinders; an *exhaust manifold* to dispose of the mixture residue after its use in the cylinders; and *tubing* to convey fuel from one system

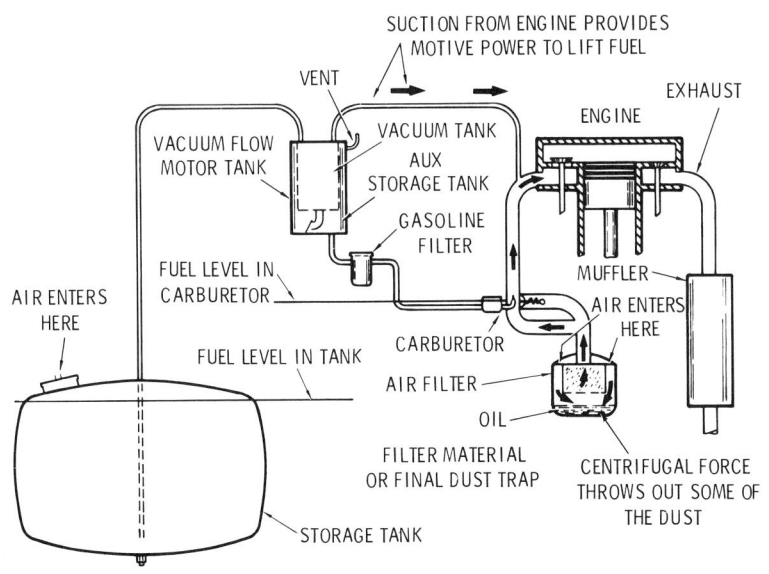

Fig. 1. Elementary diagram showing the essential parts of a fuel system.

part to another. Both systems also incorporate a *muffler* to silence the exhaust, a *fuel gauge* to register the amount of fuel remaining in the tank, some type of *fuel filter* to strain out impurities, and an *air cleaner* that accomplishes a like purpose. Other parts of the two systems differ, and require separate discussions. First, we will discuss all the components of a carburetor system.

CARBURETOR FUEL SYSTEM

This (most widely used) system employs a *carburetor* of some type to meter fuel as needed and spray it into an airstream for mixing with the air to become properly explosive. Mixing is begun in the carburetor (as the fuel spray enters the airstream) but is completed in the intake manifold through which the mixture is "tumbled" on its way to the cylinders. Some very simple carburetors do no more than to spray an amount of fuel determined by the throttle position into an airstream, the volume of which is principally controlled by engine speed. Other more sophisticated models are designed to meter the fuel as needed for a variety of conditions (such as at starting, when accelerating or under load, etc.) Carburetor functions and types are discussed in Chapter 14.

All of the components previously mentioned are needed with a carburetor system. In addition, all large industrial and automotive applications also require a *fuel pump* to deliver fuel from the tank to the carburetor. Some small engines, however, use either a *gravity-flow system* or a *pressure-feed system* in place of a fuel pump.

Gravity-Flow System

In this type of system there are *no* components in the fuel line between the tank and carburetor. The tank is simply exposed (by a vent hole in the tank cap or otherwise) to atmospheric pressure, and fuel flows by gravity to the carburetor, from whence it is "sucked" (as an air-fuel mix) into the cylinders. The tank must, at all times, be higher than the carburetor.

Pressure-Feed Systems

In the pressure-feed system, a special air-pressure tube, between the crankcase and the fuel tank, serves to feed the fuel to the carburetor. *In operation,* pressure built up in the crankcase, when the piston is on its downward stroke, is transferred to the gasoline tank by way of the engine *check valve* and tube. At periods of high crankcase pressure, the check valve is forced off its seat to permit transfer of pressure to the fuel tank.

FUEL SYSTEMS

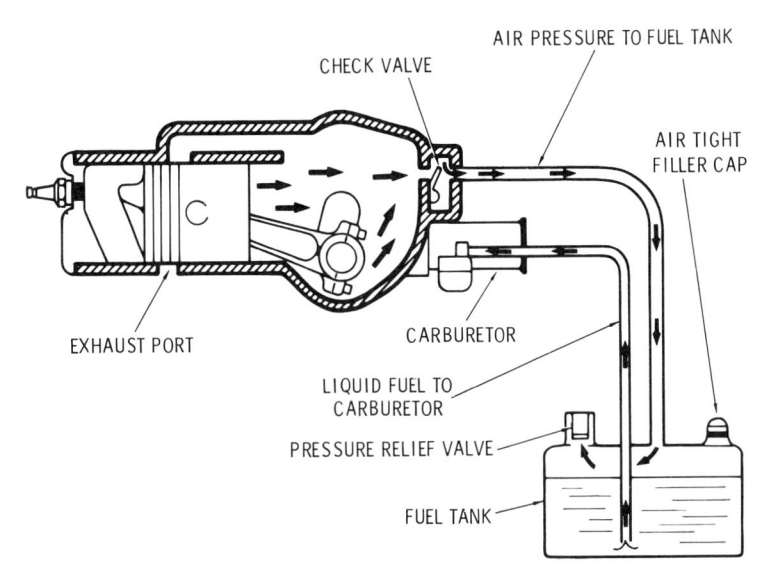

Fig. 2. Pressure-type fuel system as employed on a two-stroke cycle engine.

During low-pressure periods the valve closes, thus preventing pressure loss in the tank. An airtight fuel-tank cap is used so that the tank is *not* subject to atmospheric pressure — only to the pressures created within the crankcase. See Fig. 2.

PURPOSE AND TYPES OF FUEL PUMPS

Industrial and automotive engines having carburetors are always equipped with a fuel pump designed to supply fuel to the carburetor under all operating conditions. The pump must also maintain sufficient pressure in the fuel line between the pump and carburetor to keep the fuel from boiling, and to prevent vapor lock. Two types of pumps are currently in use: mechanically-actuated and electrically-operated.

Mechanically-Actuated Fuel Pumps

There are two types of mechanically-actuated pumps:
1. Single diaphragm,
2. Double diaphragm.

The single-diaphragm type is ordinarily used on gasoline engines. With reference to Fig. 3, illustrating a typical single-diaphragm pump, the rotation of an eccentric on the camshaft actuates the rocker arm,

Gas Engine Manual

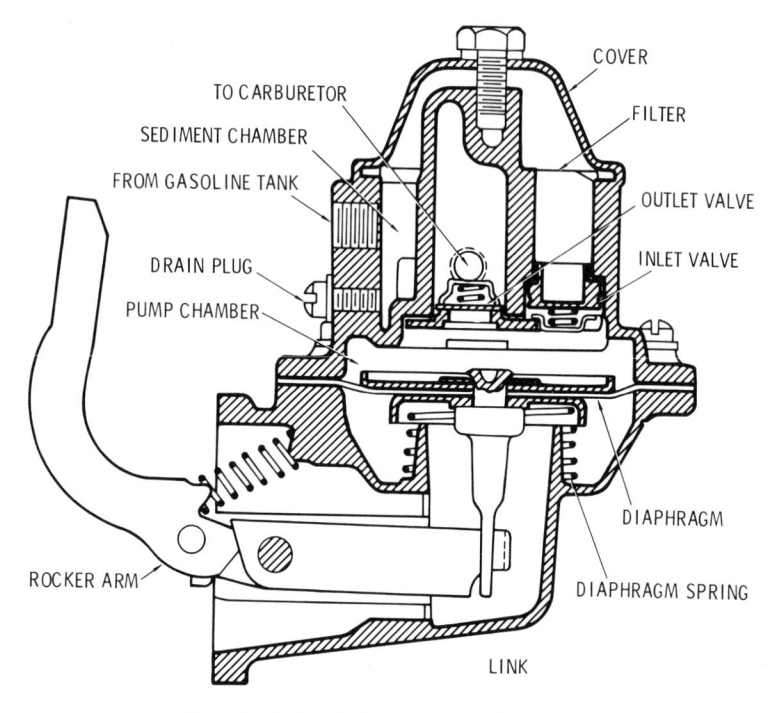

Fig. 3. Sectional view of a typical fuel pump.

which pulls the pump lever and diaphragm spring. This creates a pressure differential in the pump chamber which permits gasoline from the tank to be forced through the filter and inlet valve into the chamber.

The diaphragm is moved up on its return stroke by the pressure on the diaphragm spring, and gasoline is forced through the outlet valve to the carburetor. When the carburetor bowl is filled, a back pressure is created in the fuel pump chamber, holding the diaphragm down against the pressure of the spring. It maintains this position until the carburetor requires additional fuel and the needle valve opens.

The rocker arm and the pump lever are in two pieces which operate as a single part when the diaphragm is moving up and down. When fuel is not required, however, and the lower part of the rocker arm is held down at one end of the diaphragm pull rod, only the upper part operates in the normal way. This is possible because the rocker arm operates against the lower part only in the downward direction.

In the event that the fuel supply in the gasoline tank becomes exhausted, the fuel pump will require no priming when the supply is

Fuel Systems

replenished, as a few strokes of the pump at cranking speed will draw fuel from the tank to the carburetor.

The double diaphragm pump, employed on some automotive engines, has a vacuum booster section to provide windshield wiper operation. It is designed on the same principles as the single-diaphragm type. The booster section operates the windshield wiper only and has nothing to do with fuel system, except that it is operated by the fuel pump rocker arm.

Electrically-Operated Fuel Pumps

From the standpoint of automotive engine design, a mechanically-actuated fuel pump has the disadvantage that it must be positioned where the camshaft (by means of an eccentric on the shaft) can actuate the pump rocker arm. For an engine crowded with accessories, this limitation can pose severe design problems. On the other hand, an electrically operated pump can be located wherever it is convenient to run the fuel line. In some vehicle models the electric pump is positioned within the fuel tank.

One type of electric pump consists of a single diaphragm actuated by a spring-loaded battery-operated solenoid with breaker points to interrupt

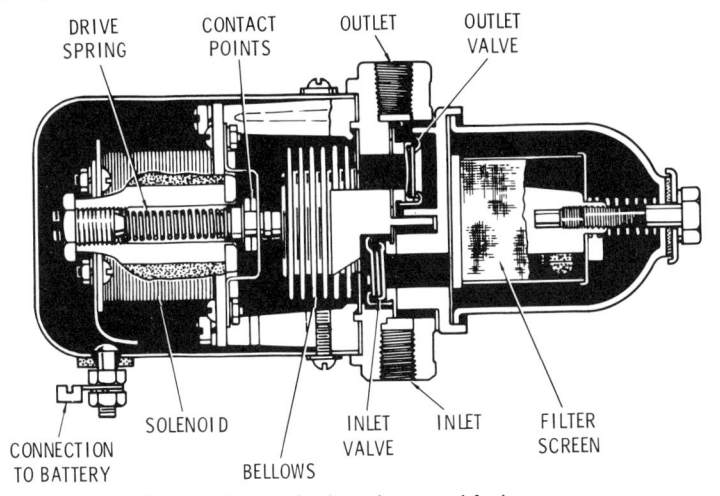

Fig. 4. A typical solenoid-operated fuel pump.

the current flow. Fig. 4 illustrates a typical pump of this type. Whenever the vehicle ignition switch is on, current passing through the closed breaker points energizes the solenoid magnet, which thereupon retracts the plunger and moves the diaphragm (connected to the plunger) outward to draw fuel into the pump body. At the end of this outward stroke the

GAS ENGINE MANUAL

plunger moves a rocker arm which, in turn, opens the points to break the circuit. This allows the spring to return the plunger to starting position, thus moving the diaphragm inward to force the fuel from the pump body into the outlet line and, at the same time, reclosing the points so that the cycle can be repeated. Spring-loaded, one-way valves at the inlet and outlet sides of the pump body keep the fuel moving forward.

Another type, designed especially for in-fuel-tank use, consists of a motor-driven turbine-type pump with an attached filter and (optional) an attached fuel gauge. The small d-c motor and integral pump blade are fully enclosed in one housing, and must be replaced as a unit if defective.

Both types of electric pumps generally are powered, when engine is running, through a switch controlled by the engine oil pressure. For engine starting, current is supplied through the ignition switch (in cranking position) or through the starter solenoid, but this circuit is cut out after the ignition switch is turned to run position. Dependence of the fuel pump

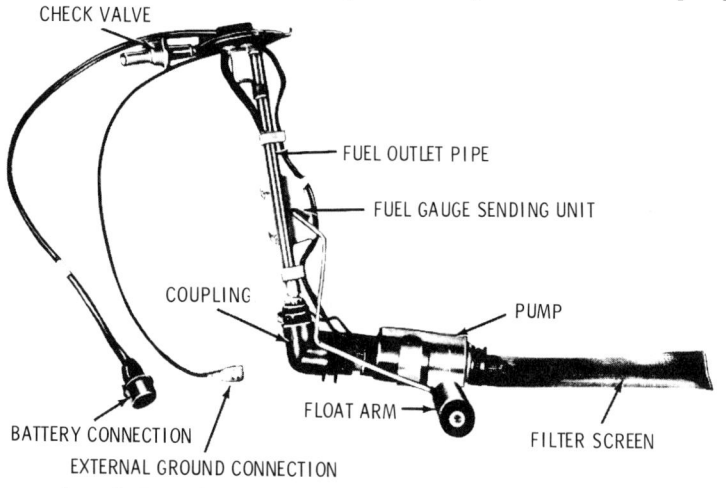

Fig. 5. Typical electric fuel pump and gauge tank-unit assembly.

for current on the oil pressure assures that the fuel supply to the carburetor will be cut off should engine oil pressure drop below a predetermined psi. See Fig. 5.

FUEL FILTER

The purpose of the fuel filter is to remove dirt and foreign matter from the gasoline. It may be located at any point between the fuel tank and the carburetor, but is usually placed between the fuel tank and the fuel pump.

Fuel Systems

The filter is often an integral part of the pump, especially if it is of the sediment bowl type. In a filter of this type, fuel enters the glass bowl and passes up through the filter screen before flowing through the outlet. Any water and solid matter caught by the screen falls to the bottom of the bowl, from where it can be removed.

Another type of fuel filter consists of a series of laminated disks placed within a bowl-type enclosure, the bowl acting as a settling chamber for the fuel and encloses the disks or strainers. (Fig. 6). In this type, fuel

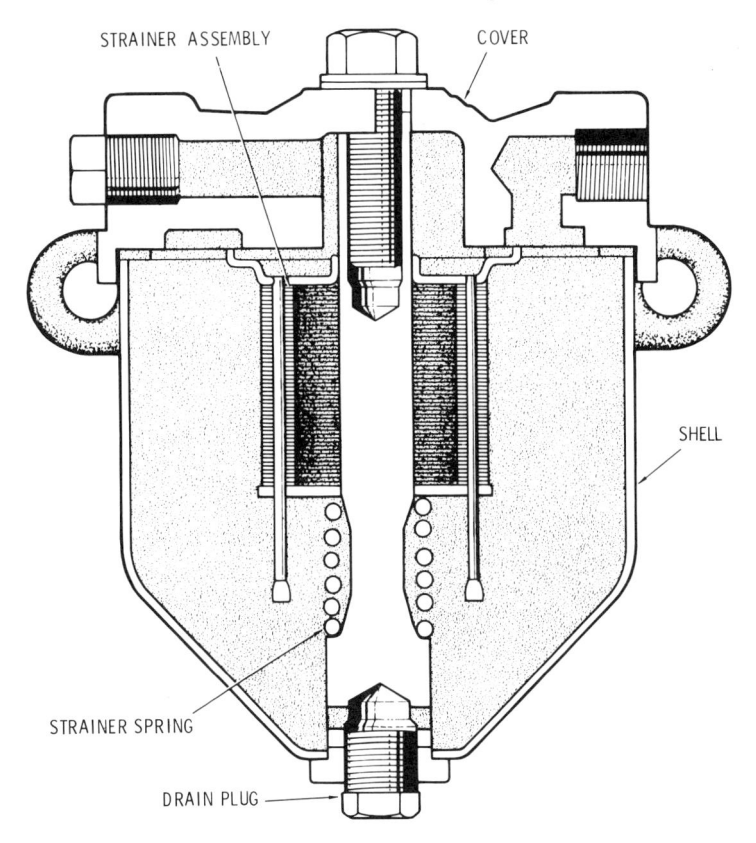

Fig. 6. Typical fuel filter showing laminated disk assembly.

enters the filter at the top inlet connection and flows down between the disks and then up through a central passage to the outlet connection at top of the bowl. Dirt and foreign particles cannot pass between the disks and are deposited at the outer rim.

Still another type consists simply of a material (screen), of which modern technology has created many. This is leakproof attached (like a sock) over the inlet end of the fuel line inside the tank.

Still another type of filter consists of a two-part casing inside of which

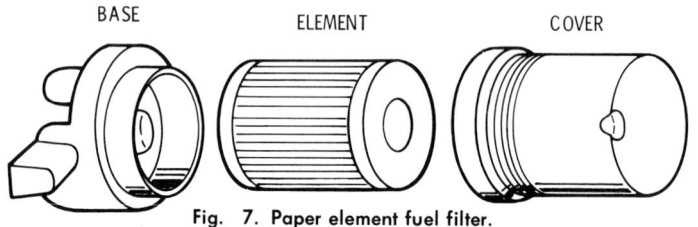

Fig. 7. Paper element fuel filter.

is a throw-away paper filter element (which must be replaced per manufacturer's specifications). See Fig. 7.

AIR CLEANER

The purpose of the air cleaner is to remove dust and other air impurities so that clean air is delivered to the engine cylinders. They are usually installed at the carburetor air intake. There are two types of air cleaners, classified as: *Wet type* and *dry type*.

The wet type, also termed *oil-bath cleaner,* consists of a main body, a unit filled with copper gauze and a cover plate with a felt pad, Fig. 8. Air

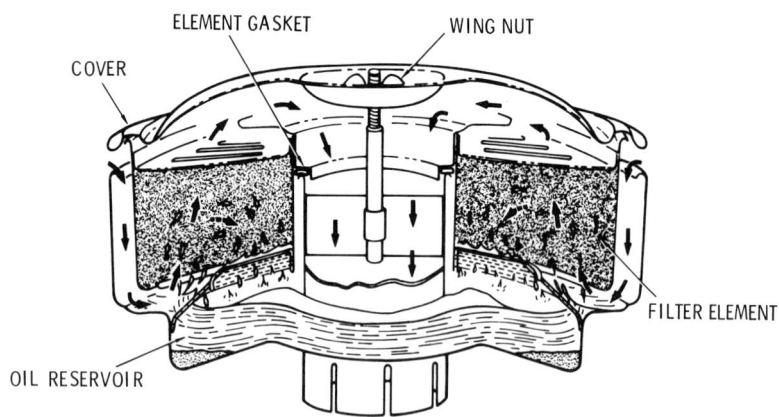

Fig. 8. Illustrating wet-type air cleaner.

entering the cleaner has to pass through small ports at the top of the main body and the copper gauze. After the air passes through the copper gauze, it hits the cover plate and is deflected down through a passage to the

FUEL SYSTEMS

carburetor. If a back-fire occurs, the flame is snuffed out by the felt pad before any damage results.

The dry type of air cleaner consists of a body, baffle plates to silence incoming air, loosely knitted copper gauze and a cover plate with a felt pad. The incoming air passes through a series of baffles, where dust and dirt are removed and then out into the carburetor.

INTAKE MANIFOLD

The intake manifold, being a part of the fuel system, delivers and distributes the fuel to the cylinders, prevents condensation, and assists in the further vaporization of the air-fuel mixture.

The intake manifold should be as short and as straight as possible to reduce the chances of condensation between the carburetor and cylinders. To assist in vaporization of fuel, some intake manifolds are designed so that part of their surface can be heated by hot exhaust gases. For this purpose a heat control valve is placed in the exhaust manifold, which deflects the gases toward the hot spot in the intake manifold when the temperature around the engine is low. As the temperature rises, gases are passed directly through the exhaust manifold without coming in contact with the hot spot.

All fuel in the intake manifold must pass over the hot spot before entering the cylinders. It does not contact the entire mixture at once, however, which helps keep the temperature down. Because of the hot spot, unvaporized particles that remain in the fuel mixture tend to vaporize instead of being carried into the cylinder as liquid particles.

EXHAUST MANIFOLD

The exhaust manifolds carries the exhaust gases of combustion from the engine cylinders. This must be accomplished with as little back pressure as possible. They may be single iron castings or cast in sections. They must have a smooth interior surface without abrupt change in size.

The shape of the exhaust manifold has an appreciable effect on the scavenging action of the engine, and for best results no more than three cylinders should be connected to one manifold.

MUFFLER

The muffler is designed to reduce the pressure of the exhaust gases and to discharge them to the atmosphere with a minimum of noise. It is connected to the middle or near the end of the exhaust pipe.

GAS ENGINE MANUAL

The pressure of gases in the cylinders of an internal combustion engine is still high enough when the exhaust valve opens to cause them to escape with a loud explosive sound.

An efficient muffler not only deadens the sound of the exhaust but also offers a minimum resistance to the gas escape. Any resistance to the gas escape causes a back pressure against the piston of the engine during the exhaust stroke and thus reduces the engine power and speed.

A muffler to be effective has several concentric chambers with openings between them. In operation, the gas enters the inner chamber and expands as it works its way through a series of holes into the other chambers and finally to the atmosphere.

FUEL TANKS

Fuel tanks serving internal combustion engines are made in a number of sizes and forms depending upon the capacity of the engine. The fuel tank may be located at any convenient point, except in the gravity flow system where the tank must be mounted above the engine.

Fuel tanks are usually made of thin-gauge metal, and are covered with terneplate to prevent corrosion. An inlet for refilling and an outlet in the side, or top of the tank leading to the fuel pump completes the fuel connections.

The outlet pipe, fitted for the fuel line connection, extends down into the tank to about an inch from the bottom. This prevents sediment accumulating at the bottom of the tank from being drawn into the fuel line.

Large engines such as are employed for industrial and automotive use are provided with baffle plates to provide additional strength and to prevent fuel from surging and splashing around the tank due to movement. As noted in Fig. 9, notches or perforations are provided in the baffle plates to permit fuel to flow freely through the compartments. A drain plug is usually provided for draining and cleaning the tank when necessary. The tanks are vented by a small hole in the filler cap.

FUEL GAUGES

Early types of fuel gauges were mechanically operated, and the indicator was incorporated as a single unit within the tank, with the fuel tank float and operating mechanism. The modern gauge type is electrically

Fuel Systems

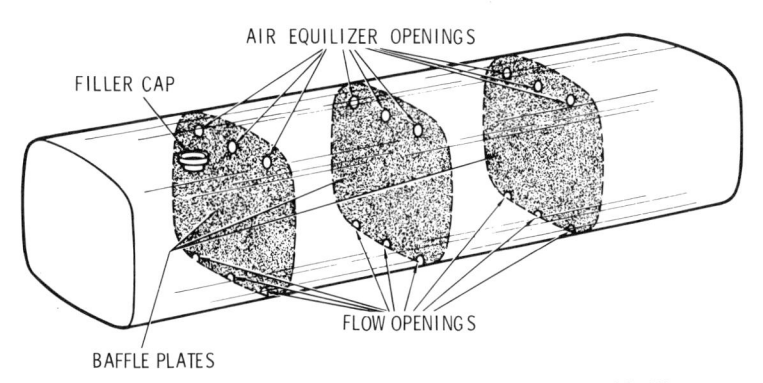

Fig. 9 Typical fuel storage tank showing filler opening and baffles.

operated and has the operating mechanism and float within the tank, but the indicator is located on the instrument panel.

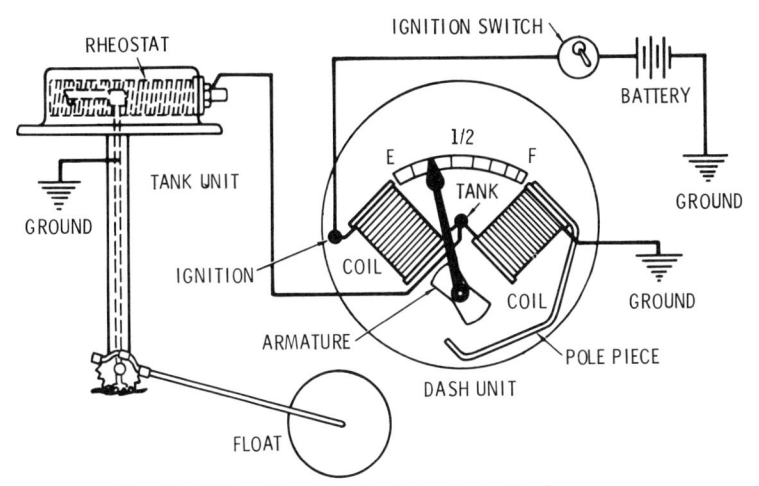

Fig. 10. Wiring diagram of typical actuated fuel gauge.

A typical electrically operated fuel gauge is shown in Fig. 10. In this arrangement the fuel indicator consists of two coils spaced 90 degrees apart with an armature and integral pointer at the intersection of the coil axis. The tank unit consists of a housing enclosing a rheostat or resistance unit with a brush which contacts the resistance unit. This contacting brush is actuated by the float arm movement, which in turn is controlled by the height of the fuel in the tank.

Variations in resistance (height of fuel) change the value of the indicating unit coils so that the pointer indicates fuel level. A calibrated friction brake is included in the tank unit to prevent wave motion of the fuel in the tank from oscillating the pointer on the indicating unit.

A FUEL-INJECTION SYSTEM

This type of system, because it affords a means of precisely controlling the air-fuel ratio for every designated engine operating condition (and, thereby, of controlling exhaust emissions to limit polutants), is rapidly coming into wider use for industrial and automotive applications, despite the fact that it is much costlier than a carburetor system and more critical to maintain. Originally developed for diesel engines, the first applications to gas engines were for sports and racing cars that benefited from the superior acceleration and top-speed characteristics afforded by the improved mixture-ratio qualities. Development of compact electronic controls made this system feasible for average over-the-road gas vehicles.

A typical American-design system is called *Electronic Fuel Injection* (EFI). All the components mentioned at the beginning of this chapter are also used with this system (i.e.: a storage tank, an intake manifold, an exhaust manifold, tubing, a muffler, a fuel gauge, a fuel filter and an air cleaner). In addition, this system (also) requires a fuel pump. Most of these components are the same as (or similar to) those used in a carburetor system, except that both fuel and air filtering must be as good as possible (better filters and, in some cases, two or more fuel filters are used)—and

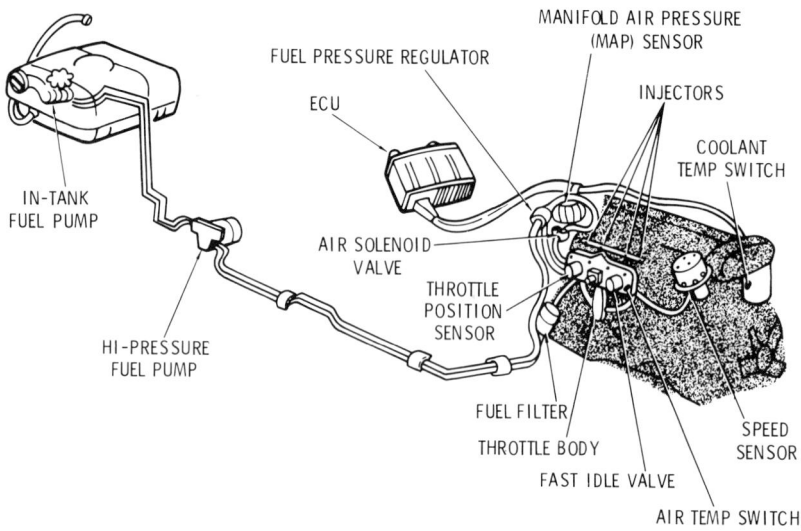

Fig. 11. A typical electronic fuel injection (EFI) system.

FUEL SYSTEMS

the intake manifold may also be modified by the addition of controls. See Fig. 11.

Other components — in addition to the preceding — are also needed. The basic component is an *electronic control unit (ECU)*, the "brain" of the system. Various operating conditions are monitored, the information is continuously fed to the control unit, and the control unit correspondingly determines the amount of fuel being fed into the air-fuel mix. As with a carburetor system, the throttle controls air intake, but, unlike a carburetor system, the amount of fuel injected into the airstream is metered entirely by the control unit (*not* by the suction of a venturi). In addition to metering fuel for all operating conditions, the control unit also increases the airflow (and fuel) for cold starting (provides a "fast idle"). On most vehicles having pollution controls it also operates the exhaust-gas-recirculation (EGR) solenoid.

Engine operating conditions are monitored by a variety of sensors and switches which transmit electrical data to the preprogrammed analog computer that serves as the control unit. Principal among these is a *manifold absolute pressure sensor (MAP)* that monitors changes in the intake manifold pressure and thereby signals the control unit regarding variations in engine speed and load, and barometric pressure and altitude (factors that affect the required richness or leanness of the air-fuel mix). This unit may be mounted in the control unit or elsewhere, but is always connected with the throttle body or the manifold through a pressure line.

Additional monitoring devices generally are: (1). *Temperature sensors* (for coolant and intake air — and, in some cases, for crankcase oil), each of which is mounted somewhere within the area to be monitored. (2). A *throttle-position switch* — always operated by the throttle linkage in some manner — which "reports" both throttle movement and its position in order to transmit the operator's demand for more or less fuel (vehicle speed). (3). A *speed sensor,* the duty of which is to synchronize fuel injection with cylinder-valve operations, and which is geared to the camshaft. (4). In some systems, a *fast-idle valve* that operates to by-pass additional air into the manifold for cold starting (only) and is usually located in the throttle body but may be located elsewhere — and which may or may not be supplemented by an additional *air solenoid valve* (which responds to engine coolant temperature).

The *fuel injectors* — one for each cylinder — are mounted in the intake manifold with the nozzles "aimed" at the respective intake valves (called "port injection"). Gasoline, under a constant high pressure, is fed to the injectors by a fuel line called a *fuel rail*. Fuel feed to the rail is accomplished first by an *electric solenoid pump* in the fuel tank, then by a (second) motor-driven, *constant-displacement, roller-vane pump* that is generally mounted at a convenient place in the line to the fuel rail. The first pump serves to prevent vapor-lock in the line by assuring a full-line supply for the second pump, which serves to assure the high pressure required. A *pressure regulator,* mounted in the fuel rail, maintains the exact constant pressure needed by returning excess fuel directly to the tank through a return line.

Each injector is solenoid operated to be fully open or fully closed. Therefore, and because fuel pressure in the rail is always the same, the time elapsed from start of injector opening until closing is the sole factor in determining the amount of fuel injected. Hence, while the engine is operating, the control unit converts the various information fed to it into electrical impulses that determine: (1). When an injection solenoid is to be opened. (2). How long it is to remain open.

The control unit is activated by turning the ignition switch to on or to start and, in turn, activates the fuel pumps and other electrically-operated devices in the system.

In some systems, the injectors are operated in groups rather than singly. Also, fuel injection may be directed into the intake ports or directly into the combustion chambers, instead of into the intake manifold. Then, too, instead of metering fuel in the manner described above (by timed injector openings and a constant-pressure fuel feed), some systems use an *injection* (a third) *pump* that varies the quantity of fuel being fed to the injectors in response to electrical and/or mechanical "signals" — and *no* other control unit is used.

In this latter type system, the injection pump is geared with the camshaft to operate in-time with the intake valve openings. Fuel is fed to the pump under pressure, and the quantity of fuel ejected from the pump is varied by varying the size of an outlet-valve opening. Opening size is mechanically determined by the position of a toothed rack *(control rod)* that rotates a *control sleeve,* and is primarily controlled (in response to engine speed) by a flywheel governor. There is an overriding control, effected by mechanical linkage with a solenoid (cold-start magnet) and the control throttle. The governor is fitted with adjustments for idle, partial and full loads, and is partially controlled by manifold vacuum through a suction line to the throttle body. The solenoid is controlled by electrical impulses from a *coolant thermostat* and an *inlet-air thermostat.* Excess fuel is fed from the pump back to the tank through a return line, and fuel ejected from the pump passes through a *fuel metering unit*

FUEL SYSTEMS

(have a precise size orifice) on its way to the injector (which is simply a nozzle). See Fig. 12.

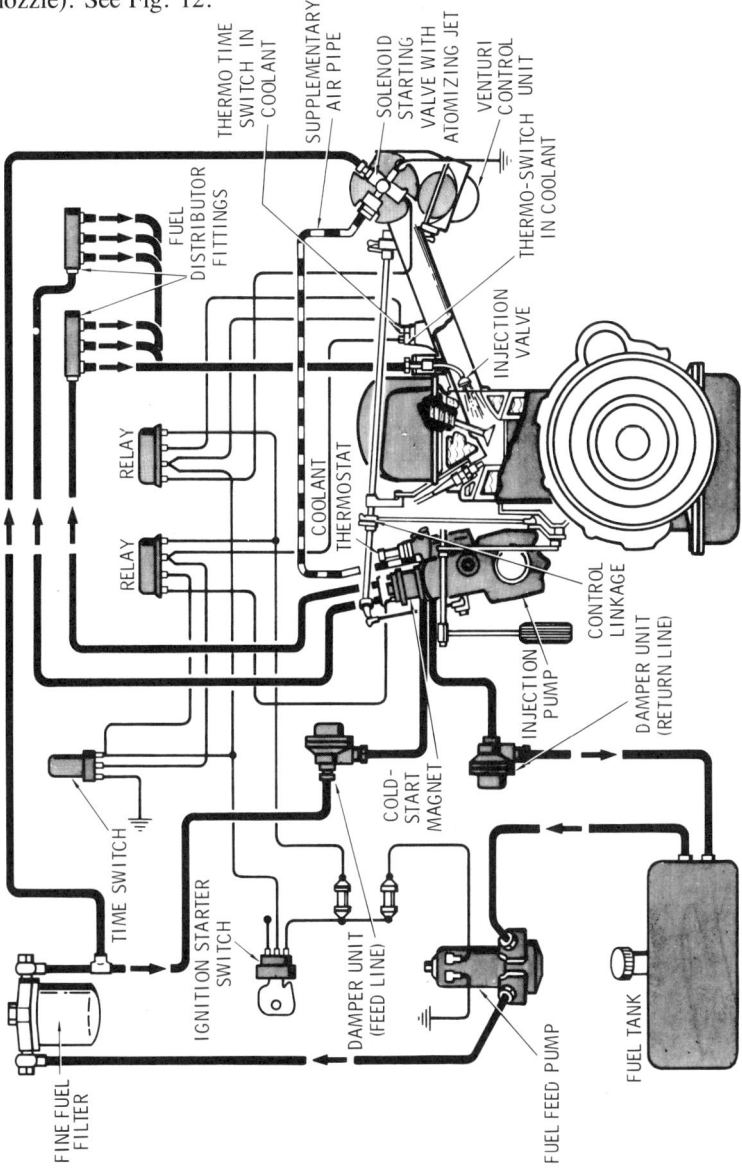

Fig. 12. Details of a fuel injection system used in a model 300 SE *Mercedes-Benz* automobile.

NOTE: In the particular system illustrated a small amount of fuel is, during cold starting, fed directly to a venturi jet in the manifold by operation of a *solenoid starting valve*.

GASOLINE

Petroleum or crude oil is the common source of fuel for internal combustion engines. It is a liquid bituminous substance composed essentially of carbon and hydrogen. After petroleum is pumped from the oil wells it is the task of the refinery to extract from it the clean liquid known as gasoline.

Petroleum contains many impurities which must be removed during the refining process, before gasoline suitable for use in engines is produced. In the gasoline making process, petroleum is first placed in a closed vessel and heated. During the heating process, the most volatile and lighter parts evaporate first after which it is cooled and condensed. It is in this manner that gasoline is obtained as the first distillation product.

As the temperature to which petroleum is subjected is increased, the heavier fractions such as kerosene, gas oil and lubricating oils which are composed of hydrocarbon molecules containing successively more carbon atoms, are passed off in the same way, leaving a final residue such as fuel oil, asphalt or coke, the molecules of which contain a relatively high precentage of carbon atoms.

The various products obtained in percent from the distillations are about as follows; gasoline 41, kerosene 5; gas and fuel oil 42; lubricating oil 3, etc.

Cracking Process

The large yield of gasoline is made possible by the process called "cracking." In this process some of the products obtained by previous distillation, such as gas oil are subjected to high pressure and high temperature. This breaks up the arrangement of atoms within the molecules of the substance so treated and results in the formation of hydrocarbon molecules of the type which constitute gasoline, thus making possible a further yield of this fluid.

Combustion of Gasoline

There are two ways in which gasoline burns:
 1. By combustion
 2. By detonation.

By definition, *combustion* is a more or less rapid chemical union of carbon with oxygen, whose combination is sufficiently energetic to evolve heat and light. In gas engines, combustion is the steady progressive burning of the charge at a uniform rate.

Detonation is very rapid oxidation, that is, an explosion which causes a very sudden rise in pressure. In gas engine operation this occurs after the flame has traveled part way across the combustion chamber. This results in loss of power and over heating because the rise in pressure is almost instantaneous and the resulting energy cannot be transmitted to the piston so efficiently as when the pressure rise is more gradual. When detonation takes place it produces a hammer-like blow against the piston head and the engine is said to "knock."

Detonation may harm an engine in several ways. In extreme cases, pistons have been shattered and rings and cylinders damaged. Other effects of detonation may be overheating, broken spark plugs, overloaded bearings, high fuel consumption, loss of power and frequent need for overhaul.

Octane Rating

In the development of anti-knock fuels by improved refining processes it became necessary to establish some standards of rating resulting in the octane method. In this method two hydrocarbons, octane and heptane, were used.

The ability of a fuel to resist detonation is measured by its octane rating. The octane rating of a fuel is determined by matching it against mixtures of normal heptane and octane in a test engine and under specified test conditions until a mixture of the pure hydrocarbons is found which gives the same degree of engine knocks in the engine as the gasoline being tested.

The tendency of a fuel to detonate, however, varies in different engines, and in the same engine under different operating conditions.

Since engines are designed to operate within a certain octane range, it follows that their performance is improved with the use of the higher octane fuel within that range, if timing of the spark is changed accordingly.

In this connection it should be noted, however, that if an engine operates satisfactorily at the upper limit of its fuel octane range, its performance will not be improved if fuel that exceeds the designed octane range is used.

Tetraethyl lead is the most popular of the compounds added to gasoline to suppress knocking. Improved refining methods also have produced fuels of greater anti-knock quality. Tetraethyl lead and other anti-knock compounds are effective because they reduce the rate of burning of the fuel, thus tending to prevent explosive burning or detonation.

GAS-OIL MIXTURE FOR TWO-STROKE CYCLE ENGINES

Two-stroke cycle engines require an oil-gasoline mixture in a certain volumetric ratio for their proper operation. The best oil to use is that recommended by the engine manufacturer, which is normally an SAE 30 or 40 grade. Either grade can be used provided it is mixed with gasoline in the proper proportions.

Most engines operate satisfactorily on regular automotive type gasoline which is available almost everywhere. Gasoline containing a large amount of tetraethyl lead *should not be used,* since it tends to foul up the spark plug gap, making frequent cleaning necessary.

Manufacturer's instructions as to the grade and amount of oil to be mixed with each gallon of gasoline should be carefully adhered to. In cases where such information is not readily available a fuel mixture chart such as shown in Fig. 13, will be helpful.

Desired Mixture: (pints of oils per gallons of gasoline)	For 2½ Gals. of gasoline add pints of oil as shown	For 5 gals. of gasoline add pints of oil as shown	For 10 gals. of gasoline add pints of oil as shown
⅓	⅞	1⅝	3¼
½	1¼	2½	5
⅔	1¾	3½	7
¾	1⅞	3¾	7½
1	2½	5	10
1¼	3⅛	6¼	12½
1½	3¾	7½	15

Fig. 13.— Table showing various fuel mixtures for two and one-half, five and ten gallons of gasoline.

With reference to our table, it will be noted that the volumetric ratio between the desirable amount of oil to be added to each gallon of gasoline differs a great deal, depending upon the size of engine and other factors dealing with its construction.

CHAPTER 14

Carburetors and Fuel-Injection Components

The proportions of air to gasoline (each by weight) for gas engine operation under different running conditions may be said to have an approximate range from 12 to 17. Hence, the average (about 15 parts of air to 1 part of gasoline) may be called a "medium mixture" and, on this basis, a *rich* mixture may be defined as one having a proportion of air less than 15 to 1 while a *lean* mixture is one having a proportion of air greater than 15 to 1.

AIR-FUEL RATIO

A richer mixture is needed for starting an engine (and the mixture must be enriched in relation to the coldness of the engine) — and is also needed to some extent (which depends upon engine temperature, atmospheric pressure and other factors) when a running engine is stressed by acceleration or an overload (such as when the vehicle is climbing a steep hill). On the other hand, a too-rich mixture not only wastes fuel, it also becomes a pollutant (because an insufficient amount of air cannot supply the oxygen needed for complete combustion). On the whole, it is desirable to operate an engine on as lean a mixture as possible for the immediate circumstances under which it is operating.

Unfortunately, there is no simple formula for the "proper mixture" for all vehicle or industrial gas engines under all operating conditions, nor is there a simple solution to the problem, expecially if fuel economy and pollution effects must be taken into account, together with engine performance. For maximum safety of locomotion an over-the-road vehicle must be capable of sufficient acceleration and deceleration to run-around or avoid dangerous situations. For maximum fuel economy it must not waste fuel by an excessive hp-to-weight ratio. To avoid the increasingly stringent pollution-control requirements it must either burn all its fuel, at all times, in a superlatively efficient manner, or must waste power (and fuel) converting the inefficiently-burned emissions. These objectives are, in many respects, also opposed, so that no single, simple method of accomplishment is apparent.

A carburetor depends primarily upon the suction of a venturi to meter fuel into an airstream and to thus maintain an approximate 15-to-1 air/fuel ratio throughout most engine operating conditions. The exact ratios for idling and normal operating conditions are adjustable, excess starting fuel is furnished by choking (disproportionately reducing the air supply), and the stresses of acceleration and overload are accommodated by pumping relatively more fuel whenever the accelerator is used to open the throttle. Air/fuel mixing (to assure optimum oxidation of all "particles" of the fuel) is accomplished by turbulence in the intake manifold.

An injector depends solely upon the preprogrammed commands of its system control unit (or injection pump) to meter the fuel in ratios required for all operating conditions monitored by the system, excepting only that there may be certain manual adjustments which can be made to the "commands." Air/fuel mixing is accomplished by high-pressure atomization of the fuel into the airstream, and by turbulence.

CARBURETOR OPERATING PRINCIPLES AND TYPES

By definition, a carburetor is a device for breaking up gasoline into a very finely divided state and mixing it with air in automatically varying proportions to meet the variable running conditions of a gas engine. Although there is a considerable difference in the appearance of carburetors used on various types of engines, they all operate on the same basic principle.

There is probably no part of a gas engine that has undergone a more useful improvement during the years than has a carburetor. The improvements which have taken place made it possible to obtain a great range of engine speeds, a reduction in fuel consumption, more nearly perfect combustion, and increased ease of starting.

The primary functions of a carburetor are: (1) To break up the gasoline into as small particles as possible, and (2) To mix these particles with air in the proper proportions.

Carburetors and Fuel-Injection Components

In this connection it should be noted that there must not be too much gasoline spray, as fuel would be wasted either by being decomposed into soot or unburned because of insufficient oxygen admitted in the air to burn it. Again too much air with ignition of the mixture, would result in unsatisfactory operation.

To progressively present the principles involved, two cases will be considered in which the air supply is:

1. Constant,
2. Variable.

The four essential elements of the first type are:

  a. Receiving chamber,
  b. Adjustable spray nozzle,
  c. Mixing chamber,
  d. Throttle.

These essential elements with exception of throttle are shown in Fig. 1.

The sectional view illustrates a receiving chamber and a mixing chamber or draft tube, the two being connected by a small pipe or duct

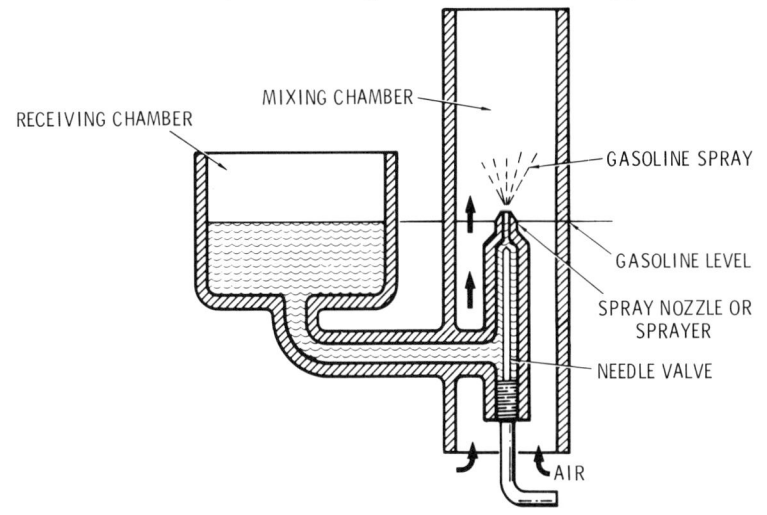

Fig. 1. Elementary carburetor principles.

which terminates at the spray nozzle or sprayer, arranged so that the supply of gasoline to the nozzle may be regulated by the needle valve.

The lower end of the draft tube is open to the atmosphere and the upper end connects with the inlet manifold of the engine.

In explaining the carburetor operation, it is assumed that the receiving chamber is filled with gasoline and the supply maintained at a level very

Gas Engine Manual

near the elevation of the top of the spray nozzle. Assuming the engine to be running with unvarying load and speed, the average vacuum produced in the manifold will remain constant and the excess atmosphere pressure outside the mixing tube will force air up through the mixing tube.

When the needle valve is opened, atmospheric pressure will also force gasoline through the fine bore of the nozzle, and will enter the mixing tube in a finely divided state, that is, in the form of a spray.

By adjusting the needle valve so that the gasoline and air admitted will be in proper proportions to suit the operating conditions of the engine, a correct mixture will be obtained and the engine will operate satisfactorily and economically.

In most engines, however, the speed and load are constantly subject to change. These conditions must be met with a mixture of different proportions, that is, the proportions of gasoline to air must also be changed.

To meet this requirement, the carburetor must be so constructed that the air supply may be automatically varied. In such arrangement there are two air supplies known as:

1. The primary air,
2. The secondary air.

Fig. 2 shows an elementary carburetor having means of automatically varying the amount of air which mixes with the gasoline.

Comparing this with Fig. 1, it will be noted that both carburetors are

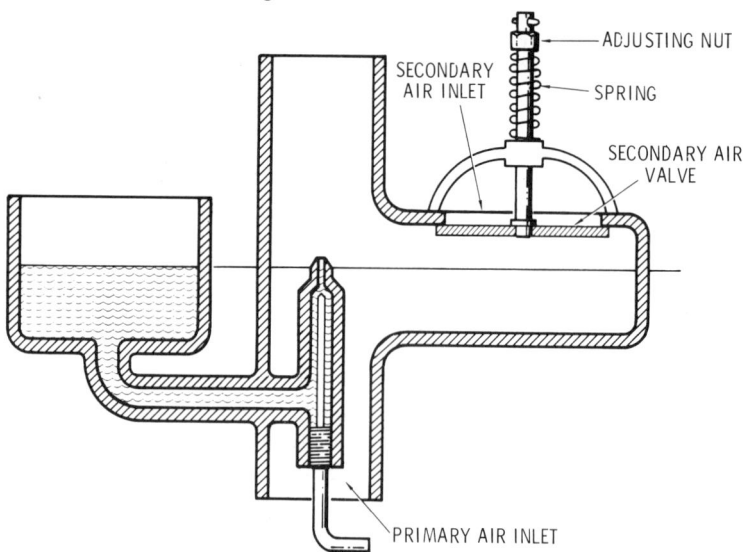

Fig. 2. Elementary carburetor with primary and secondary air supply.

Carburetors and Fuel-Injection Components

identical with the exception of the air valve shown at the right. This air valve is a form of check valve of light construction which is normally held closed by a spring.

In operation, when the reduction of pressure in the mixing chamber becomes great enough, the excess atmospheric pressure on the outside overcomes the force due to the spring and the valve opens admitting additional air supply which is called the *secondary air supply.*

The degree of opening of the valve and consequently the amount of secondary air admitted will evidently depend upon the difference of pressure inside and outside the mixing chamber. This secondary air supply can be adjusted by means of the nut on the valve stem.

In operation, if the engine is running at slow speed there will not be enough drop in pressure in the mixing chamber to open the air valve and sufficient air for a proper mixture will come in through the primary air inlet.

Now, if part of the load on the engine is removed so that it will run say twice as fast and if the air valve is held closed, the mixture will become too rich.

Although the higher speed of the engine will cause a further reduction of pressure in the air chamber which will cause more gasoline and air to be delivered to the cylinder, the excess air will not bear the same proportion to the excess gasoline as in the original mixture. That is, if the original mixture were say 15 to 1, the ratio would drop, say 12 to 1 and would be too rich for satisfactory operation.

The excess of gasoline is due to the fact that in order to get say twice the amount of air through the primary inlet, the degree of vacuum in the mixing chamber has to be more than doubled to compensate for the increased frictional resistance set up by the higher velocity of the incoming air.

Since the gasoline flowing through the nozzle is not subjected to such excess frictional resistance, more than double the amount of gasoline will flow into the mixing chamber causing the mixture to become too rich.

If now, the air valve is released so that it is held only by the spring, the higher vacuum in the mixing chamber due to the increased speed of the engine will cause the valve to open and admit additional or secondary air to make up for the inadequate supply. If the spring has been properly adjusted, the mixture will be in proper proportion for the higher engine speed, that is, neither too rich nor too lean. Fig. 3 shows the carburetor action at the higher speed with air valve partly open admitting secondary air.

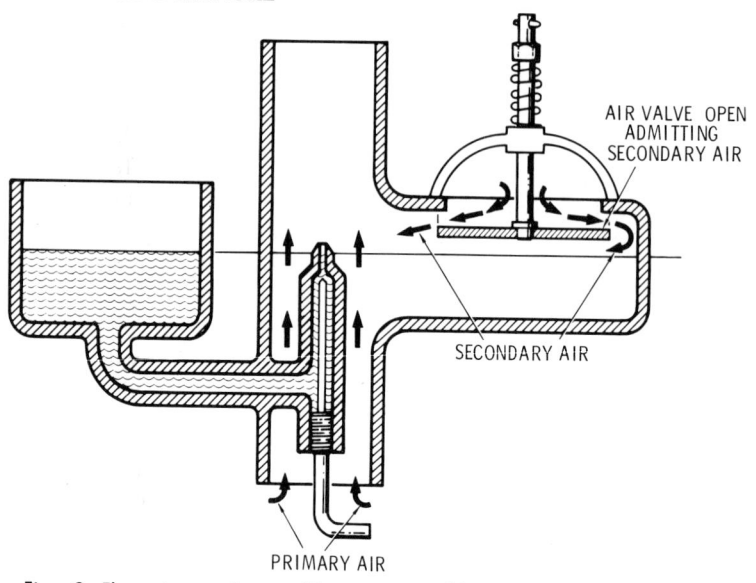

Fig. 3. Elementary carburetor illustrating variable secondary air supply controlled by an automatic air valve.

Choke Valve

In practically all carburetors, a valve is placed in the primary air passage (as shown in Fig. 4) so that the amount of air and vacuum can be

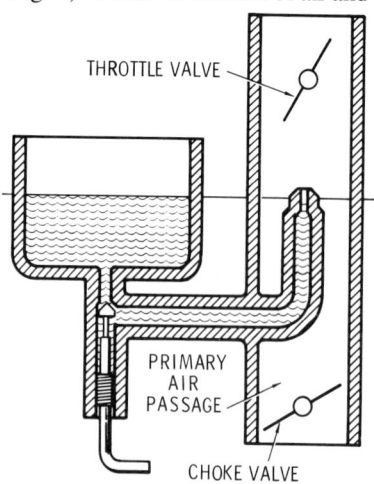

Fig. 4. Carburetor illustrating choke valve for regulating the primary air and manifold vacuum in starting.

Carburetors and Fuel-Injection Components

regulated. This is called the choke valve. This valve may be manually controlled, but is normally actuated by an automatically operated valve.

Carburetor Floats

The function of the carburetor float is to automatically maintain a supply of gasoline in the receiving chamber at very near the same level as the top of the spray nozzle. Since a float is generally used to accomplish this, the receiving chamber is popularly called the *float chamber*. In the float feed method of maintaining a constant level of gasoline in the receiving chamber, a cork or hollow metal float is placed in the float chamber.

It is connected so as to operate the gasoline inlet valve, usually by means of levers. These are arranged in such a manner that as gasoline enters the float chamber through the inlet valve, the float rises, and in so doing, closes the valve, thus shutting off the supply when the gasoline reaches the desired level.

In some instances the overflow method is used whereby gasoline is

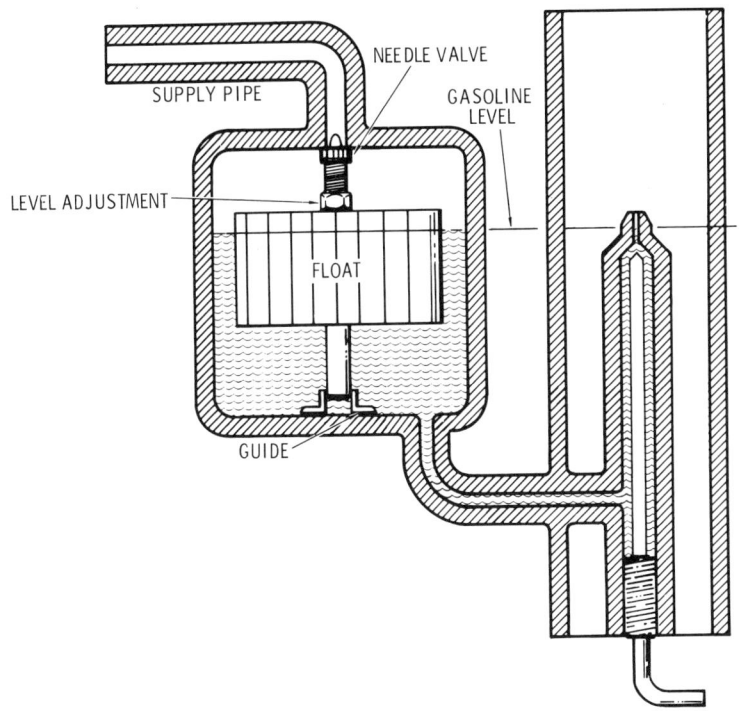

Fig. 5. Carburetor showing method of operation for direct connected flood feed.

GAS ENGINE MANUAL

maintained at the necessary level by a surplus volume being pumped or otherwise forced into a chamber whence the overflow returns to the main supply, the height and capacity for the return of the overflow maintaining the necessary level with reference to the spray nozzle. Figs. 5 and 6 show two elementary float feed systems of the offset type and illustrate down-flow and up-flow, respectively.

In Fig. 5, the float is attached to a central stem having at its upper end a needle valve connected by a threaded joint which provides means of adjusting the gasoline level. The lower end of the stem works in a guide which keeps it in a central position.

In operation, as gasoline in the float chamber is supplied to the carburetor, the float moves downward with the receding level of the liquid and the needle valve opens and admits additional supply which causes the float to rise and finally close the inlet when the liquid reaches the predetermined level controlled by the level adjustment.

The operation in Fig. 6 is similar except that it is arranged to admit the gasoline at the bottom of the float chamber. Fig. 7 illustrates the concen-

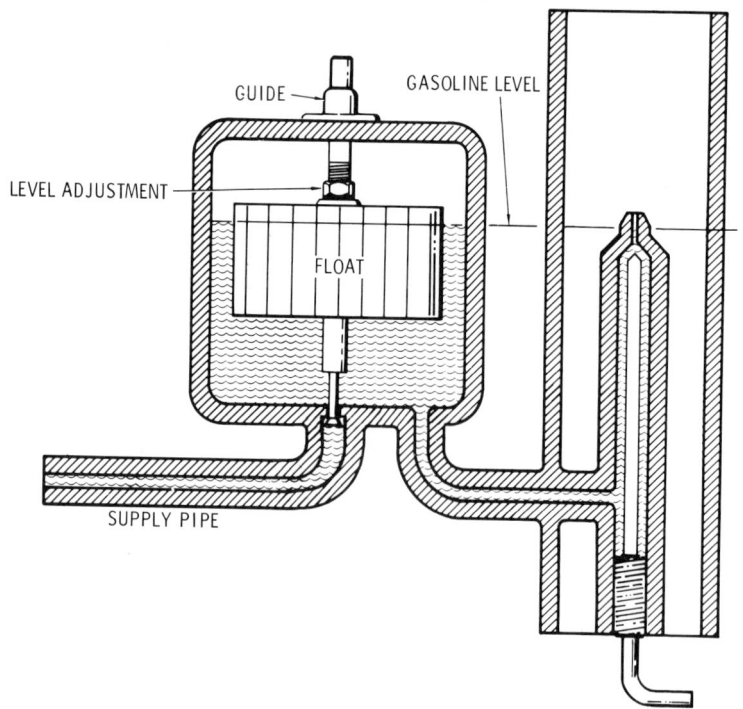

Fig. 6. Up-flow carburetor showing direct connected float feed.

144

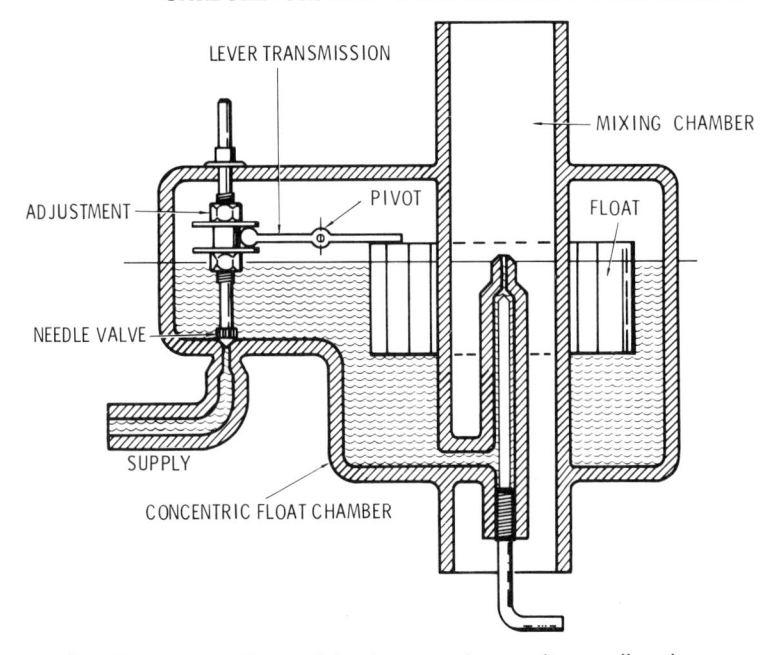

Fig. 7. Concentric float with level connected to gasoline needle valve.

tric location of the float whose movement is transmitted to the inlet needle valve by a lever. The arrangement shown in Fig. 8 employs double levers pivoted as shown, with their outer ends weighted and resting on top of the float.

In operation, on a receding level the float moves downward and the weight follows by gravity, thus opening the valve. As the level rises the buoyance of the float pushes up the weighted ends, thus closing the valve.

The Venturi Effect

In a mixing chamber of variable cross section the quantity of air or mixture which passes any section in a given time is the same, but its velocity is inversely proportional to the areas of the sections. The pressure is greatest at the largest section and least at the smallest. *This is known as the Venturi effect, or principle.* The Venturi principle has been applied to the carburetor design by shaping the mixing chamber like the familiar hour glass.

By locating the spray nozzle at the point of least cross section, the conditions are favorable for securing that marked economy of fuel which results from the use of high air velocities under low pressures. The greater

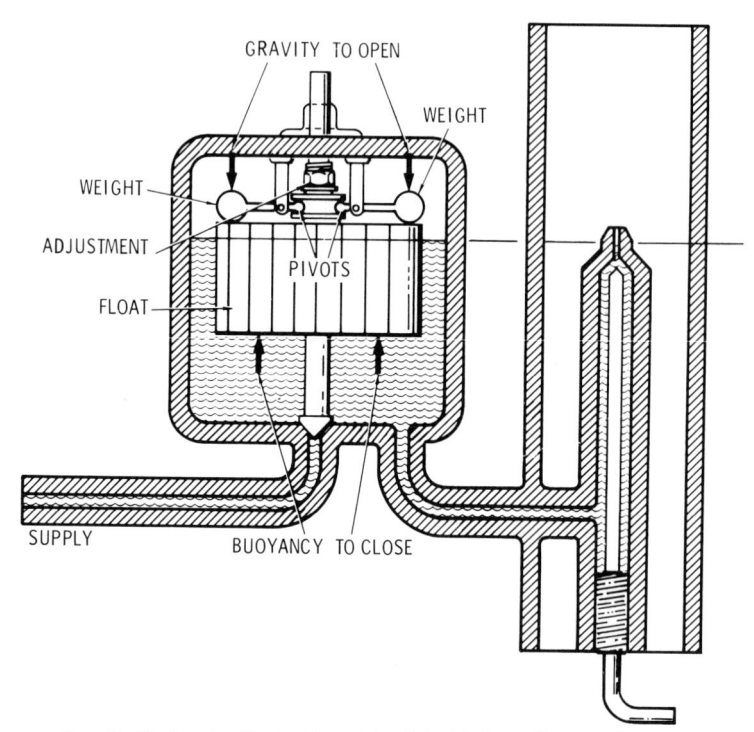

Fig. 8. Carburetor float with weighted double lever float mechanism.

the pressure drop at the nozzle, accompanied by a proportional increase in the air velocity, the finer will the gasoline be broken up or divided and the greater the percentage of the liquid that will be vaporized.

The very rapid agitation and internal motion of the mixture column, due to the restricted section of the Venturi tube, tends to produce a homogeneous fuel charge. A lowering of the pressure lowers the temperature of the liquid through vaporization, hence, in Venturi carburetors where any marked Venturi effect is sought, jacketing is advisable.

The advantages of the Venturi tube as applied to carburetors may be summed up as follows: Homogeneity of mixture; ease with which the mixing chamber may be jacketed, either by air or water; the mixing chamber may be placed in any plane, thus adapting it to varied engine designs.

The Venturi principle is illustrated in Fig. 9. Here the mixing chamber is considerably reduced at **A,** which in operation gives low velocity of air flow through the full sized portions of the mixing chamber and very little

CARBURETORS AND FUEL-INJECTION COMPONENTS

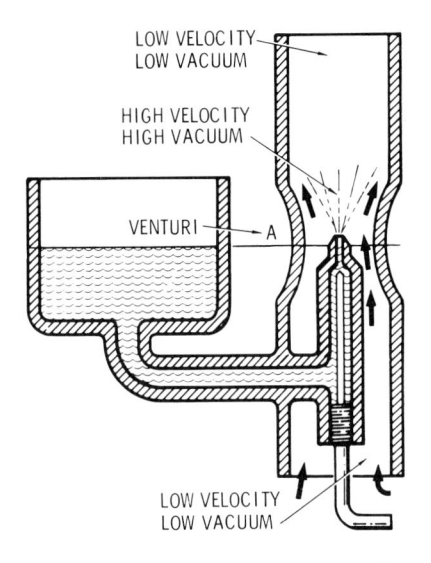

Fig. 9. Venturi principle illustrated by single venturi.

pressure reduction; at the restricted section **A,** the velocity is greatly increased (depending on the reduction in cross sectional area) which is accompanied by a considerable pressure reduction. This excess vacuum causes an increase in the percentage of the gasoline vaporized.

The Venturi principle has been extended to the use of two or more Venturi tubes arranged "in series" as shown in Fig. 10. This is called a triple Venturi.

In operation, spray from the nozzle is induced by the surrounding column of incoming air. The gasoline atomized in the first Venturi is kept centrally located in the air stream by the surrounding blanket of air passing into the second Venturi and again into the third or main Venturi, offering a triple protection against the liquid fuel coming into contact with the walls of the mixing chamber.

Spray Nozzles

The amount of liquid passing through the nozzle may be varied by an adjustable needle valve. Some carburetors are fitted with two or more simple nozzles, the idea being the several nozzles forming the unit, by coming into action progressively as the power demand increases, will provide the same effect as though several carburetors were used, each in turn being brought into action. See Fig. 11.

Gas Engine Manual

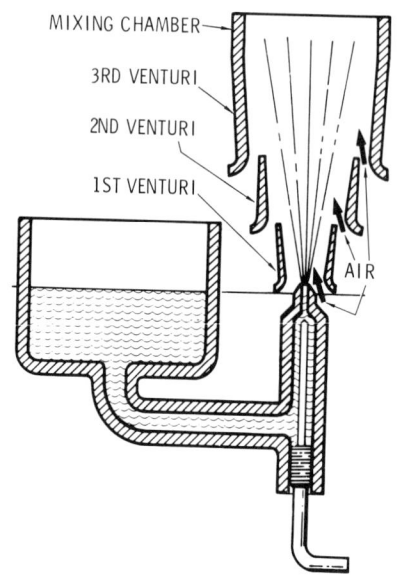

Fig. 10. Carburetor details illustrating the triple venturi.

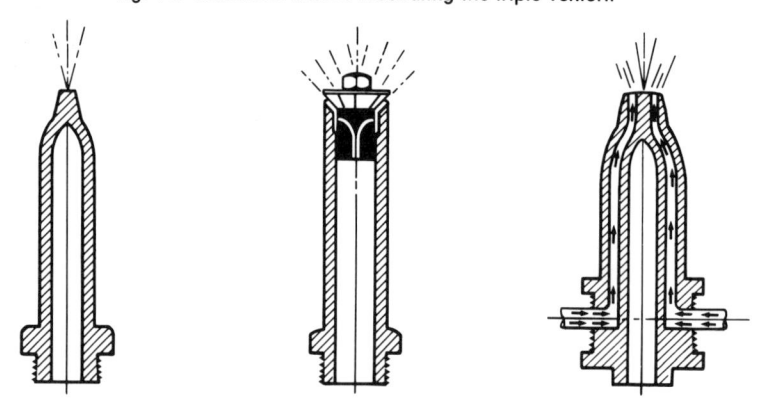

Fig. 11. Various carburetor nozzles.

Whatever form is given to the nozzle, the effectiveness with which it can break up the fuel varies as the difference between the pressures at its two ends, and this pressure difference varies throughout the speed range of the engines.

Since the nozzle has a very small opening, even for the largest engines, it is easily stopped up, and the construction should be such that it may easily be removed for cleaning.

CARBURETORS AND FUEL-INJECTION COMPONENTS

Air Bleed Principle

Air bleed is the admixture of a small amount of air with the fuel before it leaves the nozzle. The purpose of the air-bleed is to avoid too lean a mixture on low vacuum, and too rich a mixture on high vacuum.

Accordingly, the degree of vacuum in the mixing chamber may become so low, as when operating the engine at slow speed and light loads, that the mixture will become so lean as to cause the engine to miss or stop. to remedy this possibility the air bleed is provided.

The application of the air bleed principle to a typical *Stromberg* carburetor is shown in Fig.12. *In operation,* at closed throttle or slow engine speeds, the fuel is delivered through the idle system as indicated. The fuel is taken from the base of the main discharge jet, flowing into the bottom of the idle tube where it is metered. From the tube it flows through the connecting channel where air from the idle air bleed is mixed with it so that a mixture of air and fuel passes down the channel and is discharged from the idle discharge holes. The idle needle valve controls the quantity of fuel discharged from the primary hole, thereby effecting the mixture ratio.

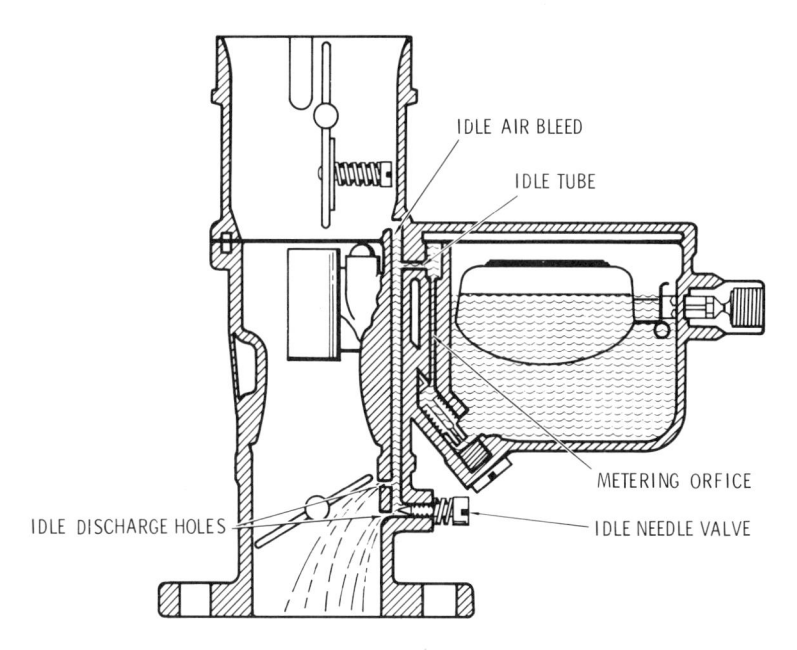

Fig. 12. Illustrating the air bleed principle as used on *Stromberg* carburetors.

Heating Methods

The object of heating is to promote vaporization. The methods of applying heat which is obtained from the exhaust gases or circulating water are by the use of jackets placed around the air supply pipe, the inlet manifold or around the float chamber bowl.

The latter method of heating is the best theoretically because heating the air supply or mixture results in raising the temperature of the incoming charge with the result that it is in an expanded state so that less charge is taken in than would at a lower temperature. The prevailing method, however, is to heat the mixture by jacketing part of the manifold and by passing some of the exhaust gases through the jacket.

Economizer

This name is usually given to an air bleed fitted with a needle valve so that the amount of air entering the nozzle can be varied to suit the engine running conditions.

The amount of economizer action is controlled primarily by the difference of the air bleed opening when the needle is open and closed, and this in turn is capable of adjustment by what is called an *economizer* reducer which consists of a needle valve placed in the air bleed tube between the air bleed valve and the nozzle. With these two adjustments the range of mixture that can be obtained is greatly increased.

Metering Rod

This device consists of a long metallic pin of graduated diameter fitted to the main nozzle or passage leading thereto in such a way that it measures or "meters" the amount of gasoline permitted to flow by it at various engine speeds.

It may be operated by suitable connection with the throttle so that it moves in fixed relation to the opening or closing of the throttle, or by air valve dash pots. It may be used as either a primary or secondary adjustment of the gasoline supply to the nozzle.

A typical metering rod will provide two economy steps and one power step of gasoline regulation as shown in Fig. 13.

In the construction shown in Fig. 14 the metering rod is controlled by the position of the throttle and is positive in its action. **In operation,** with throttle wide open the power mixture is delivered and in intermediate positions, the gasoline supply is cut down to give the economy mixture. The economy range covers all engine speeds, except wide open throttle.

CARBURETORS AND FUEL-INJECTION COMPONENTS

Fig. 13. Shape of a typical metering rod.

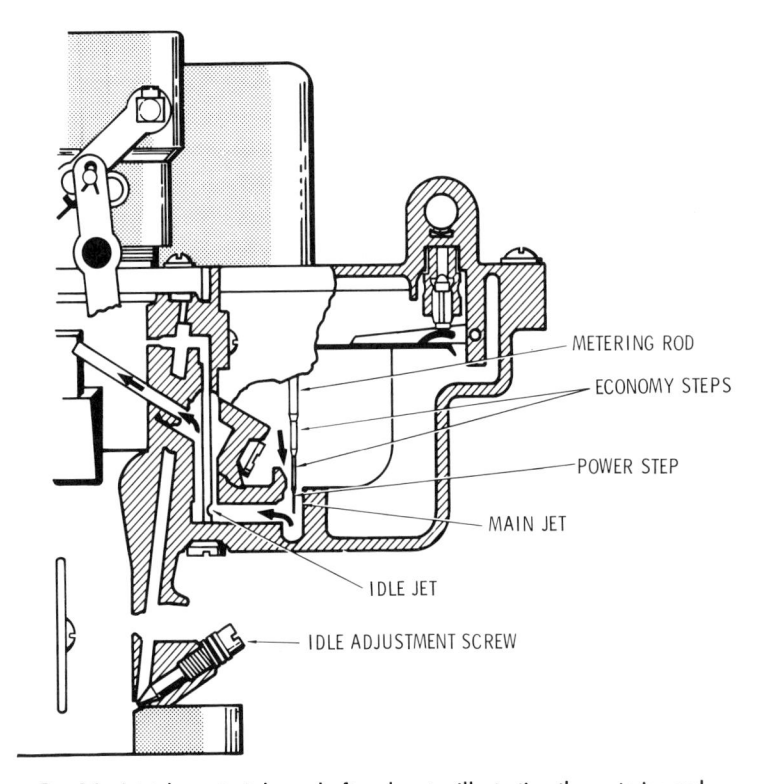

Fig. 14. A triple venturi down-draft carburetor illustrating the metering rod.

Acclerating Pump

To secure rapid engine acceleration, some carburetors are fitted with a specially designed pneumatic type accelerating pump.

A typical pump of this kind consists of a cylinder with a plunger containing an air ball and inlet and outlet valves. It is opened automatically by suitable connection with the throttle. See Fig. 15.

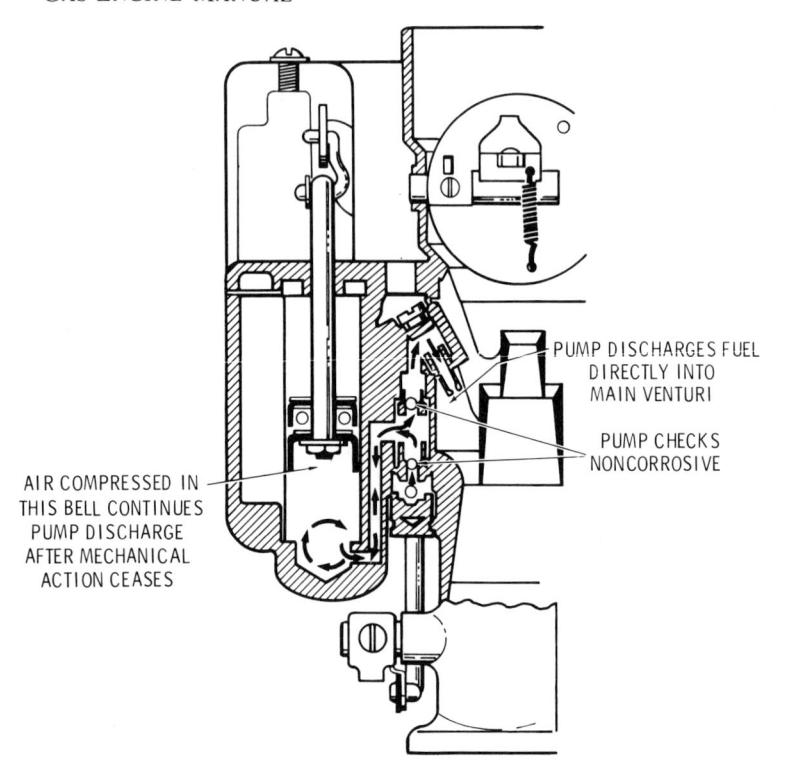

Fig. 15. Details of a carburetor accelerating pump.

In operation, the upward movement of the plunger when the throttle is closed draws a predetermined amount of fuel into the bottom of the cylinder, an air column always remaining between the gasoline and the plunger.

The slightest opening of the throttle compresses the air and causes a discharge of gasoline through the accelerating nozzle which points downward into the main Venturi. When the throttle is fully opened, the discharge is continued for a few seconds by the air compressed between plunger and the gasoline.

Updraft and Downdraft Carburetors

Carburetors may also be classified as *updraft* and *downdraft* according to their position with respect to the intake manifold. Thus, if a carburetor be mounted *below* the manifold it is classed as an *updraft* and if mounted *above* it is a *downdraft* type.

CARBURETORS AND FUEL-INJECTION COMPONENTS

In early fuel systems it was necessary that the fuel flow down by gravity to the float chamber, making the updraft carburetor essential because of its low position. The introduction of mechanical fuel pumps, however, has largely eliminated this problem, so that late model carburetors are usually of the downdraft type. In this design, the carburetor is placed above the engine. Air entering at the top passes downward, mixes with the fuel and then on to the manifold and engine.

Another type is known as the *horizontal* or *side outlet* carburetor. Here the air fuel mixture passes out to the engine manifold horizontally as noted in Fig. 16. It is used on air cooled lawn mower engines, on small auto engines, and on various stationary power applications. An advantage of this type is that it simplifies manifold construction, since it eliminates one right angle turn.

This carburetor is adequate for the engine installation in which it is used; it does not, however, have facility for providing an instantly richer

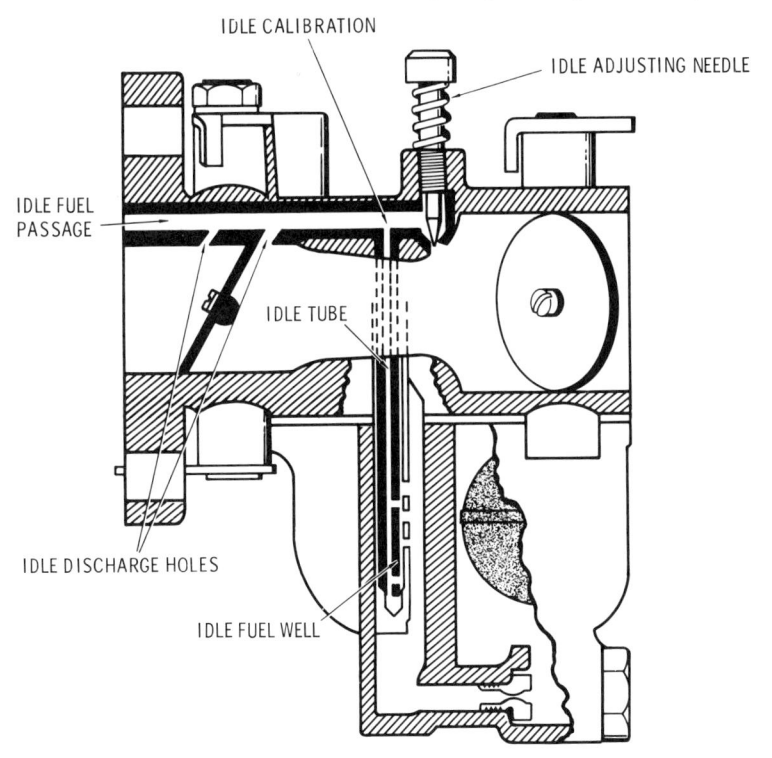

Fig. 16. Typical horizontal carburetor used on small industrial engines.

mixture for sudden acceleration. Also the calibration of the main tube or high speed circuit is of necessity a compromise between maximum power and best fuel economy.

FUEL FLOW CIRCUITS

In the foregoing, the various elementary principles of carburetors have been fully dealt with. In order to meet the varying air fuel ratios required by high speed multicylinder engine carburetors, however, numerous paths must be provided through which the mixture must flow before reaching the valve ports.

These various paths whose duty it is to properly prepare the mixture according to engine requirements are generally termed *carburetor circuits* or *fuel flow circuits*.

There are five circuits through which gasoline may flow through the ordinary multicylinder engine carburetor. They are:

1. Float circuit,
2. Low speed or idle circuit,
3. Main metering or high speed circuit,
4. Pump circuit,
5. Choke circuit.

Other circuits in a carburetor will be either a variation or a combination of these. Because the high speed circuit is unable to satisfactorily handle all of the varied conditions that a carburetor must meet, the other four circuits are necessary. The function of the various circuits together with a detail description will be provided on the following pages:

The float circuit maintains the correct level of fuel in the carburetor fuel bowl at all times. Proper float level together with proper venting of the bowl to the atmosphere assures availability of the correct amount of fuel to the other circuits.

The low speed, or idle circuit delivers the proper mixture of air and fuel when the throttle is practically closed. In some carburetors, it continues to function throughout the entire speed range, whereas in others, it merely overlaps the high speed circuit through a short range. The circuit delivers fuel from the bowl to a point below the throttle valve.

The main metering or high speed circuit meters and delivers the proper air and fuel mixture in the range above the low speed circuit. This circuit delivers fuel from the bowl to the Venturi. *The pump circuit* quickly provides a measured supply of fuel necessary for sudden acceleration. *The choke circuit* provides a method of enriching the fuel mixture when starting and warming up a cold engine.

CARBURETORS AND FUEL-INJECTION COMPONENTS

Float Circuit

The float circuit on all carburetors are practically the same. The float mechanism consists essentially of a float, a vent to maintain fuel at a predetermined height in the fuel chamber.

In operation, fuel enters the carburetor at the gasoline connection and flows through the needle valve seat into the float chamber. When the fuel reaches the prescribed level in the float chamber, the float presses the needle valve against its seat to shut off the flow of fuel. Thereafter, the fuel is maintained at the prescribed level by opening and closing the needle valve as required.

The float is usually hinged on a fulcrum pin which is retained in the float bowl by a pin as noted in Fig. 17. The float chamber is vented externally through a port in the air horn to allow fuel to be smoothly withdrawn through the various systems.

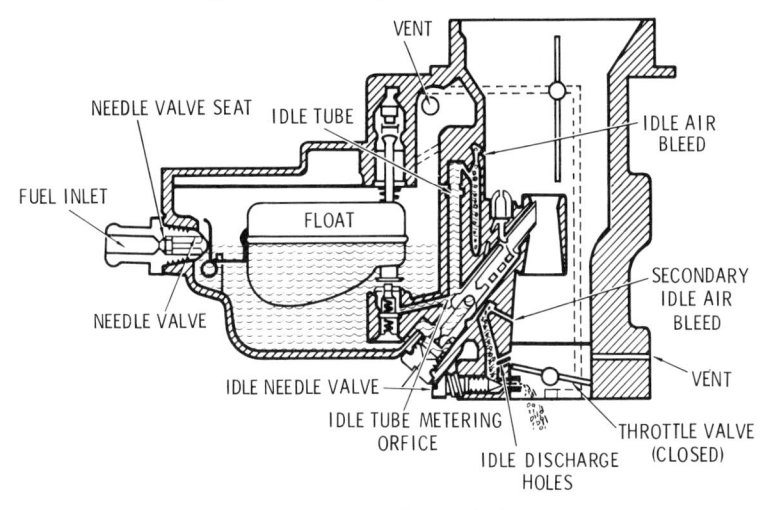

Fig. 17. Illustrating float and idle system.

Low Speed Circuit

The low speed, or idle circuit does not vary in any considerable extent in construction; some carburetors may, however, have economizer tubes incorporated in its general makeup. The low speed circuit is necessary because when the throttle valve is almost closed there will be very little air passing through the Venturi, that is, the difference in pressure between the bowl and the Venturi will not be great enough to cause fuel to enter the throat from the main nozzle.

In operation, fuel flows from the float chamber through the main metering jet and upward through the idle tube which meters the fuel. From the idle tube it flows through a connecting channel where air from the idle air bleeder is mixed with it so that a mixture of air and fuel passes down the idle channel to the idle discharge holes. Additional air is drawn into the air fuel mixture in the idle channel through the secondary air bleeder. See Fig. 17.

On idle or closed throttle operation, the fuel air mixture is drawn only from the lower or primary idle discharge hole due to high suction at this point. As the throttle valve is opened, suction is also placed on the upper or secondary idle discharge holes to feed additional fuel.

Fuel supplied through the idle discharge holes begin to diminish when throttle valve is opened to the point where the main metering system begins to supply fuel, until a throttle position is reached where the idle system ceases to function.

The idle needle valve controls the quantity of fuel that is supplied through the primary idle discharge hole, thereby affecting the final fuel air ratio supplied to the engine while the idle system is in operation.

Main Metering Circuit

The main metering system controls the flow of fuel during the intermediate or part throttle position. *In operation,* the fuel flows from the float chamber into the main metering jet and then into the base of the main discharge jet. Air is bled through the high speed bleeder into the main discharge jet so that the mixture of air and fuel is discharged from the main discharge jet into the carburetor barrel. See Fig. 18.

The main discharge nozzle is designed so that if any vapor bubbles are formed in the hot gasoline, the vapors will follow the outside channel around the main discharge nozzle instead of passing through the jet tube. These vapor bubbles escape through the dome-shaped high speed bleeder and thereby reduce percolating troubles.

Pump Circuit

The pump circuit is usually mechanically operated, and varies in construction, depending upon the type of carburetor, from a spring loaded piston with a mechanical release to a direct acting assembly. This circuit is basically designed to fill in the gap between the idle circuit and the high speed circuit.

The most widely used construction consists of a pump cylinder, a plunger mechanically actuated by a lever mounted on the throttle shaft, an

Carburetors and Fuel-Injection Components

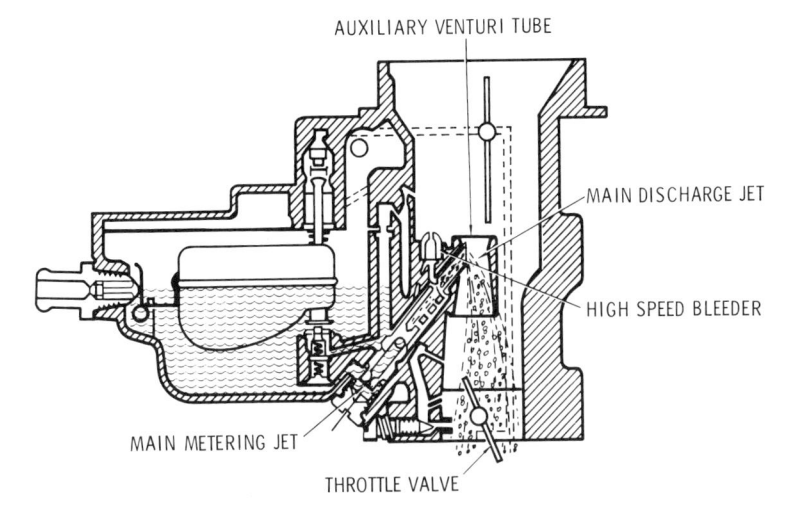

Fig. 18. Showing the carburetor main metering circuit.

intake check valve located in the bottom of the pump cylinder, a discharge check valve, and an accelerating jet to meter the amount of fuel used.

As previously noted, the pump circuit provides for smooth and rapid acceleration by means of an extra quantity of fuel when the throttle valve is suddenly opened. This is accomplished by operation of the accelerating pump piston, Fig. 19, which is directly connected to the throttle valve by means of a rod and pump lever.

When the throttle is closed, the pump piston moves up and draws a supply of fuel from the float chamber through the inlet check valve into the pump cylinder. When the throttle valve is opened, the piston on its downward stroke exerts pressure on the fuel which closes the inlet check valve, opens the outlet check valve, and discharges a metered quantity of fuel through the pump discharge nozzle. This, however, occurs only momentarily during the accelerating period since the pump duration spring provides a follow up action so that the fuel discharge carries out over a brief period of time.

When the throttle is held in a fixed position, the pressure on the fuel in the pump cylinder decreases sufficiently so that the outlet check valve closes and fuel ceases to discharge from the pump nozzle. With the throttle held in a fixed position, the fuel flows only through the idle or main metering system as previously described.

GAS ENGINE MANUAL

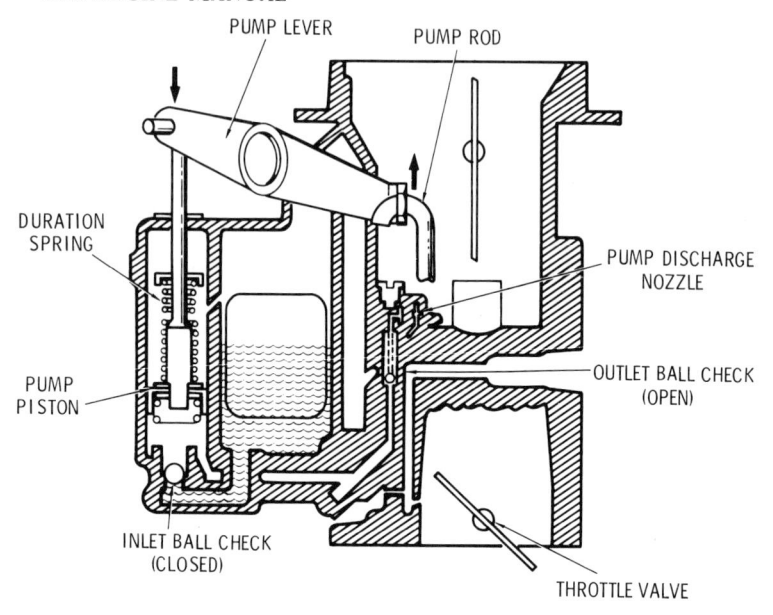

Fig. 19. Carburetor pump circuit.

Choke Circuit

When a cold engine is being started, much of the fuel discharged by the carburetor is unable to vaporize during its travel to the combustion chamber until sufficient heat is developed in the intake manifold to maintain a homogeneous mixture for efficient combustion. Therefore a much larger quantity of fuel must be supplied to compensate for this lack of vaporization when starting and operating a cold engine.

To compensate for this deficiency, an automatic choke has been provided. The automatic choke not only controls the air fuel ratio for quick starting at any temperature, but also provides the proper amount of choking to enrich the fuel mixture for all conditions of engine operation during the engine warm up period.

The automatic choke is an integral part of the carburetor and consists of a bimetal thermostatic spring and a vacuum piston which opposes the action of the spring. The spring is connected to the choke valve in such a manner as to close the valve when the spring is cold. The vacuum piston tends to open the choke valve when the engine manifold vacuum is high. See Fig. 20.

Therefore, under varying load conditions during the warm-up period, the position of the choke valve will be changed by operation of the

Carburetors and Fuel-Injection Components

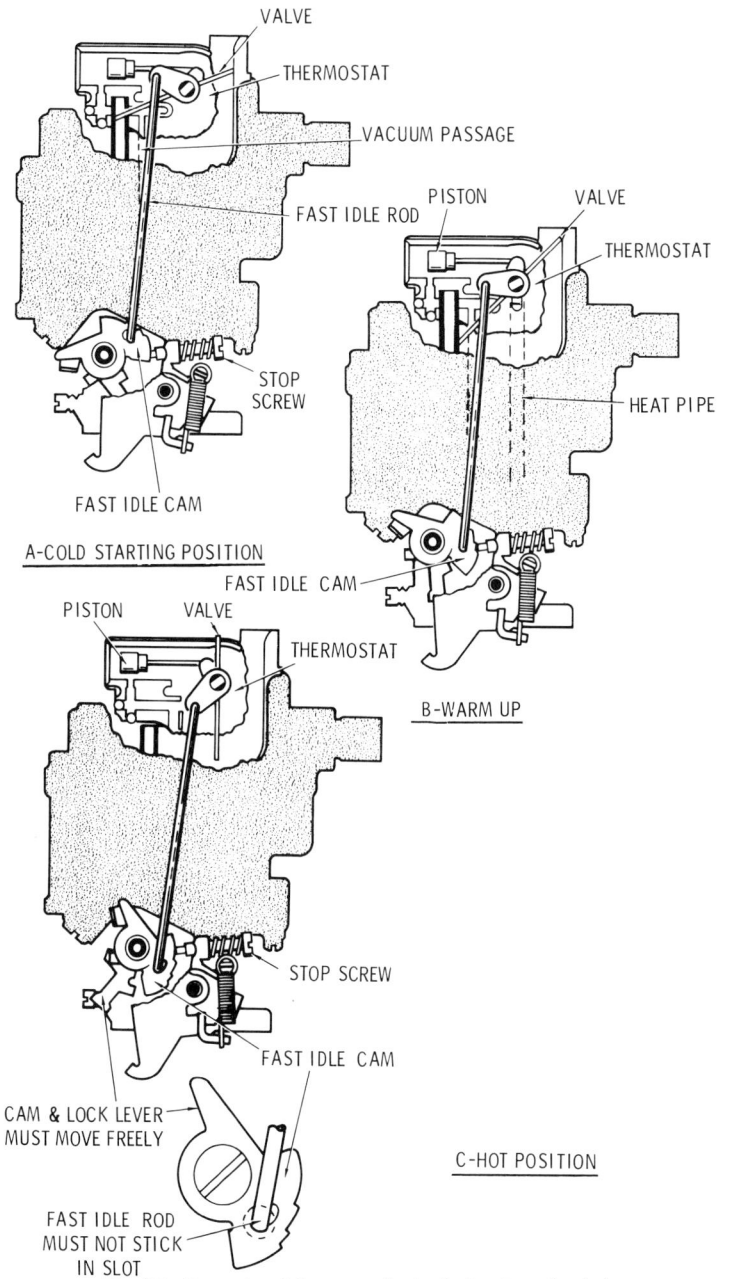

Fig. 20. Operational features of a typical automatic choke.

vacuum piston working against the thermostatic spring, and the air velocity in the air horn. Hot air from the exhaust manifold is directed to the thermostatic spring so that the spring loses its tension as the engine is heated. This permits the choke to open gradually, and after it reaches full-open position is held open by the action of the intake manifold on the piston.

FUEL INJECTORS AND ASSOCIATED COMPONENTS

Fuel injector valves are used with systems having a control unit (Chapter 13), these are solenoid-operated pintle valves with integral fine-mist nozzles that project into the intake manifold above the respective intake ports. Each valve is operated by a pulsed signal from the control unit (ECU) which opens the valve for the proper time interval (pulse width) to deliver the amount of fuel determined by the ECU. Fuel for the valves of an engine (one valve for each cylinder) is delivered by the *fuel rail* which is attached to the tops of the valve bodies, and in which fuel is kept at a constant predetermined pressure (so that the opening time of each valve is the only determinate of the amount of fuel injected at each valve opening.). Generally, the fuel rail is divided into two parts each of which serves half the total number of valves (i.e.: for cylinders 1, 2, 7, 8 and 3, 4, 5, 6). See Fig. 21. The two parts are so assembled that one pressure regulator, installed in the complete assembly, serves both halves. The valve assemblies are *not* serviceable; must be replaced as units.

NOTE: In systems not having a computerized control/unit the injectors do not have valves (are simply nozzles).

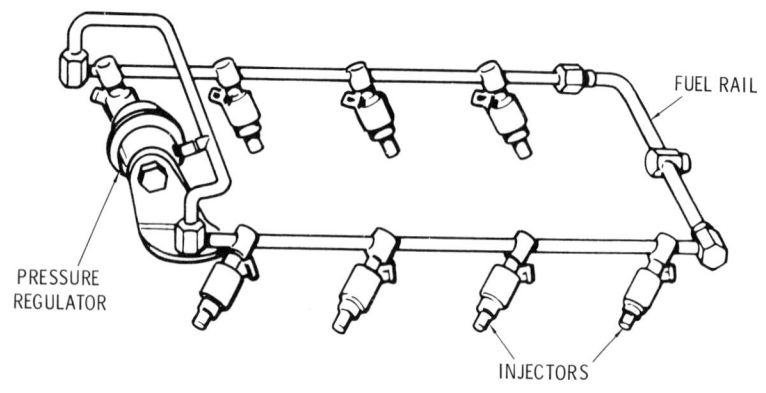

Fig. 21. Fuel rails and injector system.

CARBURETORS AND FUEL-INJECTION COMPONENTS

Fuel Pumps

Two electric fuel pumps are usually used. The first, located in the fuel tank, is generally a diaphragm—type booster pump (Chapter 13) that is integral with the fuel gauge and a simple, replaceable-element filter. The second, located somewhere in the fuel line ahead of the pressure regulator, is usually a roller-vane pump driven by a 12-volt motor, and is designed to produce a constant displacement. This (second) pump has a check (one-way pressure-opened) valve at the output side that prevents backflow to maintain pressure in the line when the pump is not operating. In general, pumps have built-in design factors that cannot be altered by service procedures.

Fuel Pressure Regulator

Because the (second, above) fuel pump operates continuously at maximum output, regardless of engine speed, load, etc., the engine seldom requires all the fuel the pump makes available. It is the function of the pressure regulator to return excess fuel through a by-pass line back to the tank, and to thus maintain a constant pressure in the fuel rail. The regulator contains two chambers separated by a spring-loaded diaphragm. One (air) chamber is connected by hose with the throttle body and is thereby subjected to an air pressure that is determined by the intake-manifold pressure and which, in turn, controls the position of the diaphragm. The diaphragm is mechanically connected with a valve in the other (fuel) chamber so that its position determines the amount of fuel by-passed to the tank. Thus, it is the differential between the diaphragm spring and the intake-manifold pressure that controls the percentage of by-passed fuel, and very accurately maintains a constant, predetermined pressure in the fuel rail. Generally, this differential is factory set and there is no manual adjustment.

A typical EFI system control contains the following units which are interconnected by an electrical harness:

1. A *manifold absolute pressure sensor (MAP)* that controls the *basic* quantity of fuel at each engine speed.
2. A *speed sensor* that determines exactly when each injector valve is to be opened (or, as is generally the case, two or four valves may be opened simultaneously).
3. A *throttle-position switch* that controls the amount of fuel needed for the engine speed called for by the throttle position.
4. *Temperature sensors* for engine coolant and intake air which varies the basic fuel quantity in accordance with immediate

engine operating temperatures (as for cold starting and warm up as opposed to a fully warmed-up condition).
5. A *fast-idle valve* that increases the normal (warm-engine) idle speed during periods of cold starting and warm-up. In some systems, an *air solenoid valve* is also used to supplement the fast-idle valve.
6. An *oil-pressure sensor* that closes down the system if oil pressure becomes dangerously low.
7. Finally, a *control unit (ECU)* that converts all the foregoing signals into the pulses that operate the injector valves.

In other (than EFI) systems the functions of a MAP may be accomplished by two units: a *pressure sensor* and a *pressure switch*. There may be more than two temperature sensors (at different engine locations). Also, a *fuel-injection pump*, to which the various system signals are transmitted, may replace the control unit and injector valves, and meter fuel directly to injection nozzles.

Electronic Control Unit (ECU)

This is an analog computer (with "printed" electronic circuits and transistors) that is preprogrammed by the manufacturer to "accept" certain signals from the various sensors of the system and to "translate" these into a pulsed signal for operation of the fuel-injection valves. See Fig. 22. Generally contained within a compact housing designed for mounting as desired and for connection, by a designed electrical harness, to the other components of a system. This is a non-serviceable unit which must be replaced if faulty. It may or may not contain the MAP.

Manifold Absolute Pressure Sensor (MAP)

This essential component of any fuel-injection system may be incorporated within the control-unit housing or may be a separate unit (mounted where convenient). Its function is to monitor the intake-manifold pressure, through a suction-tube connection to the throttle body, and to transmit an electrical impulse to the control unit whereby the duration of injector-valve opening is affected. Because intake manifold pressure is increased or decreased in relation to engine speed, engine load, barometric pressure and altitude, all these factors determine the "related" signal sent to the control unit. The unit contains an air-pressure diaphragm that mechanically operates an electronic control through which the strength of the transmitted impulse is varied.

Carburetors and Fuel-Injection Components

ELECTRONIC CONTROL UNIT
- ELECTRONIC CIRCUITS
- PRESSURE SENSOR

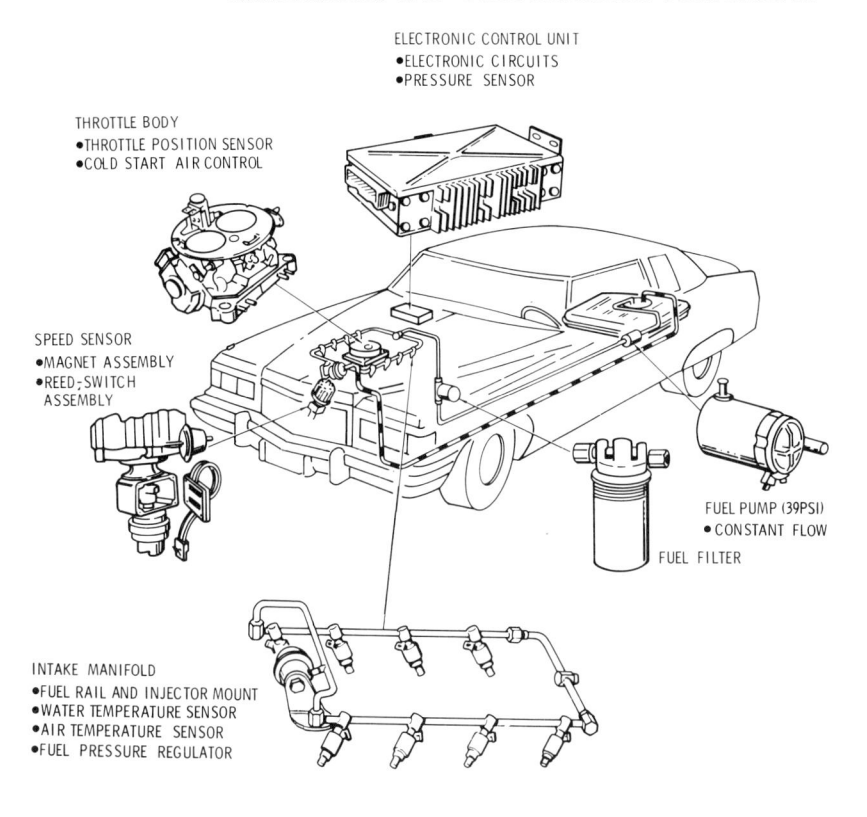

THROTTLE BODY
- THROTTLE POSITION SENSOR
- COLD START AIR CONTROL

SPEED SENSOR
- MAGNET ASSEMBLY
- REED-SWITCH ASSEMBLY

FUEL PUMP (39PSI)
- CONSTANT FLOW

FUEL FILTER

INTAKE MANIFOLD
- FUEL RAIL AND INJECTOR MOUNT
- WATER TEMPERATURE SENSOR
- AIR TEMPERATURE SENSOR
- FUEL PRESSURE REGULATOR

Fig. 22. A typical electronic fuel injector system.

Pressure Sensor and Switch

Together these function in the same manner as a MAP (preceding). The pressure sensor, which may be mounted in the throttle body, "senses" the manifold pressure and "signals" the switch (mounted elsewhere). The switch electronically measures the differential between the manifold pressure (signals from the sensor) and the ambient air pressure, and transmits signals to the control unit.

Speed sensor is used with EFI systems and the function of this component is to time the fuel injections both in relation to the firing orders of the cylinders and the speed of the engine. Therefore, the speed sensor operates with regards to the fuel-injection system in exactly the same manner that the distributor operates with regards to the ignition

Gas Engine Manual

system—and, to correlate the two, it is actually housed within the distributor.

There is a rotor, mounted on (and rotating with) the distributor shaft with two permanent magnets. There are also two electrically-magnetic reed switches that are mounted on a plastic "sensor" housing within the distributor body. Rotation of the rotor magnet within the field of the reed switches creates timed electrical currents in the switch circuits, which conduct the current (through the ECU harness) to the control unit. The position of the rotor on the distributor shaft and the speed of shaft rotation determine, respectively, the instant and frequency of the electrical impulses sent to the control unit—and these are subject to the identical factors (distributor advance and speed) that affect the ignition timing. See Fig. 23.

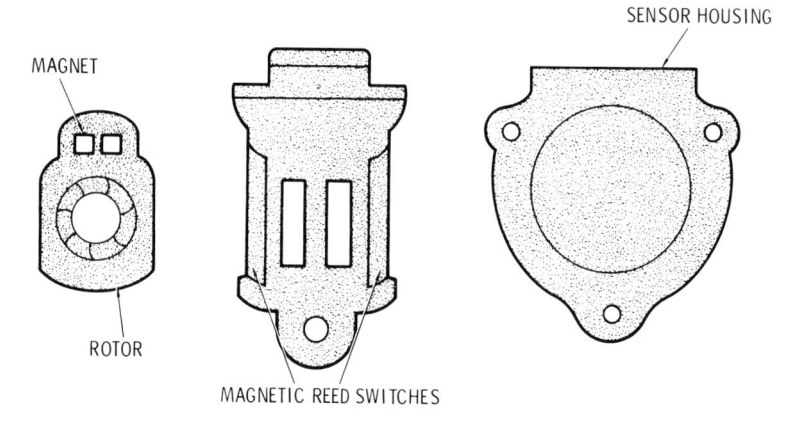

Fig. 23. A reed-type speed sensor.

NOTE: On engines not equipped with electronic ignition, instead of the magnetic rotor and reed switches, a duplicate rotor and breaker points (also within the ignition distributor) are used. In still other systems, a separate "fuel-injection" distributor (also operated by gearing to the engine camshaft) is used to accomplish the same purpose.

Throttle-Position Switch

Like the MAP, preceding, this is a mechanically - operated electronic device. The unit is mounted on the throttle body and operated by the accelerator linkage. It monitors both the movement and the position of the throttle shaft, and transmits appropriate signals to the control unit.

Carburetors and Fuel-Injection Components

Temperature Sensors

These, again, are like the MAP—except that mechanical operation is furnished by a bimetal spring or strip instead of a diaphragm. Each operates independently to transmit signals to the control unit.

Fast-Idle Valve

There is an adjustable setscrew on the throttle body for normal (warmed-engine) idle speed, but a faster idle adjustment (equivalent to choking) is needed for starting and warm-up. The adjustment provided is a valve that admits additional air (and, therefore, also additional fuel) to temporarily increase the flow of air/fuel mix to the cylinders. This valve is mechanically operated by a thermal (bimetal) element that normally holds the valve open. An integral electric heating element serves to warm the bimetal element and close the valve. Turned on by the ignition switch, this heater requires more or less time to reach its "valve-closed" temperature, depending upon the ambient air temperature and the heat from the running engine. Once this temperature is reached the valve remains closed and no longer affects system operation. See Fig. 24.

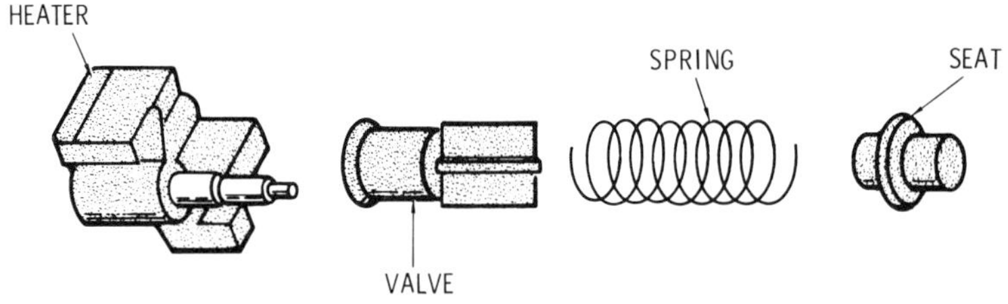

Fig. 24. Various parts of a fast idle valve.

The valve may be mounted in the throttle body or in the air intake from the air cleaner. In the latter case, an air solenoid valve may be used to supplement its operation. The pintle-type solenoid valve controls an auxiliary air-intake passage, and is operated by the control unit in response to signals from the coolant-temperature sensor. It therefore serves to introduce engine cooling-water temperature as a second factor in determining the length of the warm-up period.

Oil-Pressure Sensor

An oil-pressure sensitive, diaphragm operated unit, is otherwise similar to the preceding sensors. Its function is to shut off the fuel supply (kill the engine) if oil pressure drops below a safe level.

Fuel-Injection Pump

Used only on some European designed engines, this unit embodies all the functions of a control unit, injector valves, and all other preceding controls excepting for temperature sensors and a special fuel-injection starting valve (mounted in the throttle body). The pump is driven by the engine camshaft and utilizes both manifold pressure (through a section line) and a governor to determine the basic fuel requirements, which are metered by the pump in required amounts to the spray-type injector nozzles. See Fig. 25. Basic requirements are varied by sensors similar to those already discussed. A number of extremely sensitive manual adjustments are required.

ENGINE SPEED GOVERNORS

Governors are used on engines to regulate maximum speed and to prevent excessive wear. By experiments it has been found that the rate of wear in an engine increases as the square of its rpm. In the higher speed ranges, therefore, an increase of a few hundred revolutions per minute will result in a greatly disproportionate amount of wear.

Governors are classified according to their method of operation, as:
1. Centrifugal,
2. Velocity or vacuum.

Centrifugal Governor

This type of governor operates on mechanical principles, and in its basic form is made up of two weighted arms pivoted on a spindle which is connected by suitable linkage to the throttle valve.

In operation, the centrifugal governor is connected to the engine by means of a flexible drive shaft, driven from the camshaft or an accessory drive of the engine. As the engine and spindle rotates, the spindle weights will tend to fly outward actuated by centrifugal force but is retarded by the spring tension. A screw at the end of the spring controls the tension with which the spring holds the weight against the spindle. See Fig. 26.

It will be noted from the foregoing, that as the speed of the engine increases, the weights will tend to fly outward with sufficient force to actuate the throttle valve linkage and close the throttle. As the engine speed decreases, the weights are pulled inward by its spring, and the throttle opening is increased. It is in this manner that the speed of the engine is controlled automatically by a regulation of the throttle opening.

CARBURETORS AND FUEL-INJECTION COMPONENTS

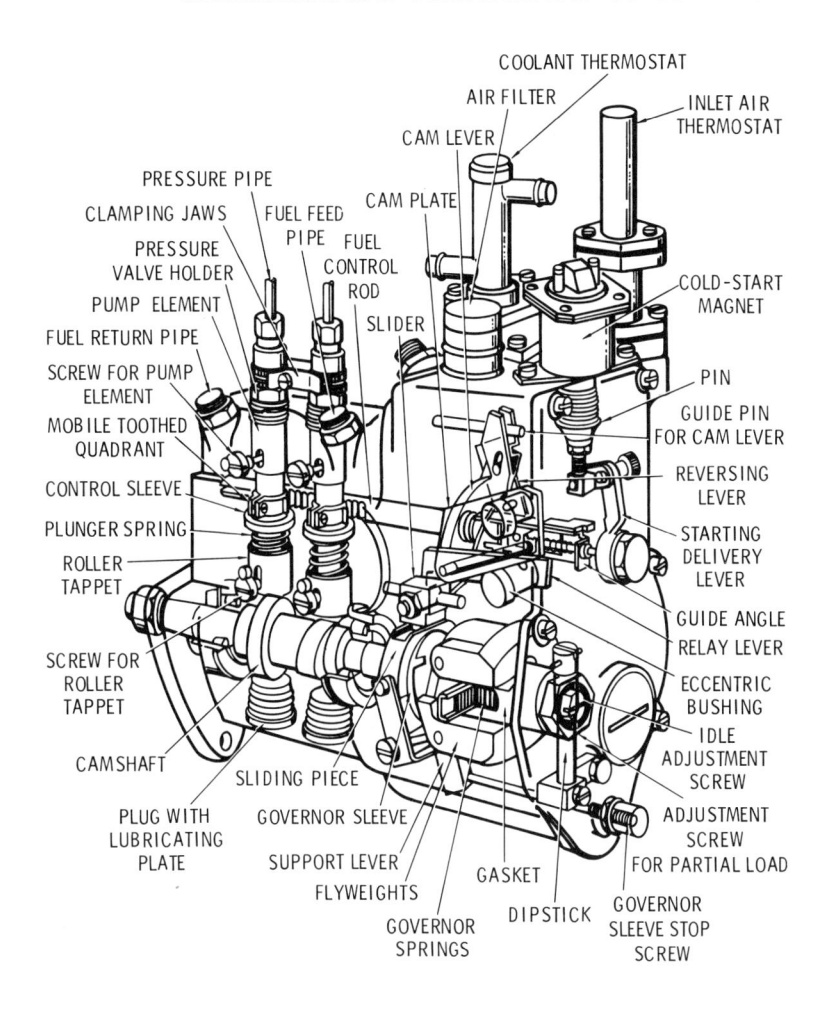

Fig. 25. A two-element fuel injection pump.

Vacuum Governor

The vacuum type of governor is not mechanically driven, but the intake manifold vacuum is used to regulate the engine speed. This is possible because the degree of vacuum decreases as the engine load is increased, and increases as the engine load decreases.

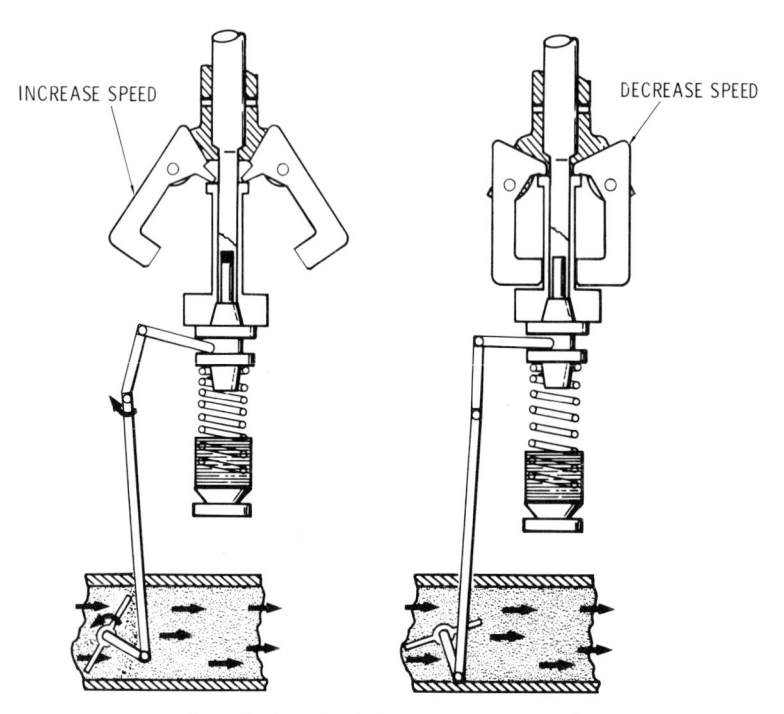

Fig. 26. Centrifugal force governor control.

CHAPTER 15

Emission Control Systems

The operation of a vehicle powered by a gasoline-burning, internal-combustion engine releases certain gases into the atmosphere which, in sufficient quantity and/or under certain atmospheric conditions, are harmful, and are referred to as air pollutants. Raw (unburned) gasoline releases pollutants by evaporation; an operating engine releases them in its exhaust.

The principal pollutants, for which the government has established acceptable levels of emission, are:

Carbon-Monoxide (CO). A compound consisting of one part carbon gas to one part oxygen. The combustion process, if complete, produces beneficial carbon dioxide (CO_2) instead; poisonous CO is produced when the combustion process is incomplete. It is also formed when raw fuel vapors mix with the atmosphere.

Hydrocarbon (HC) A compound of hydrogen and carbon gas that is also formed under the preceding conditions.

Oxides of Nitrogen (NO_x). Compounds of nitrogen and oxygen formed principally by too-hot burning of the air/fuel mixture.

CLASSIFICATION OF CONTROLS

In theory, there would be no problem with air pollutants if a vehicle were designed to prevent fuel evaporation and to completely burn the fuel it uses under controlled conditions. It follows that the control of emission levels presents two basic problems: elimination of raw fuel vapors and elimination (or reduction to acceptable amounts) of the final-exhaust pollutants. As discussed later, there is one "group" of **fuel-evaporation**

controls to deal with the first problem; the second problem is dealt with in a number of different ways, as follows:

One way to deal with the final-exhaust problem is to provide more efficient combustion during all the different conditions of engine operation. Ordinarily, CO and HC pollutant levels become excessive during engine starting and warm-up (especially, at cold ambient temperatures), during idling and whenever the engine is decelerated from a high speed.

Cold starting and warm-up increase the levels because choking is required to enrich the mixture and the cold air does not provide thorough vaporization and mixing-in of the fuel. Idling, even with a warmed mixture, can increase the levels because the compression ratio is decreased (less content to be compressed in a cylinder) and a normally advanced spark cannot fire all the reduced-density content. Also, during deceleration, the content density is decreased, but conventional operation produces a too retarded spark for the engine rpm due to the drop in carburetor vacuum. Deceleration may be the result of accelerator operation or the change in engine load due to hill climbing.

The control of combustion therefore necessitates mixture controls that will effectively maintain a more combustible mixture during the critical engine operating periods. It also necessitates controls that will retard the spark as needed, and throttle controls to prevent too rapid diminishment of cylinder content.

Car manufacturers have devised a variety of "systems" (also called "packages") under various trade names to deal with combustion-control problems. Some of these systems involve little more than modifications of the engine and components already familiar; others require the addition of several distinctly new control devices. Later, under the heading "Combustion Control Systems" we will discuss typical systems—then, we will discuss typical new devices used in the systems.

Since any unburned or partially-burned mixture exhausted from the cylinders must travel through the exhaust manifold, pipe and muffler before entering the atmosphere, another method of reducing CO and HC emissions is to more nearly complete the combustion process for gases being exhausted. Not wholly effective, this cannot be relied upon as a complete solution, but is used in some cases to supplement the foregoing. New kinds of devices are needed to implement this type of control, as will be discussed later under the heading "Combustion Control Systems."

There are two basic ways of controlling nitrogen oxides emissions: (1). by limiting the heat rise within the cylinders; (2). by limiting the spark advance to periods of engine cruising-condition operations (when per-

Emission Control Systems

formance is normally at its peak). Method 1 requires new and different kinds of controls; method 2 is based upon new controls similar to the preceding spark controls. Both are discussed under the heading "NO_x Controls".

Final-Exhaust Controls

Insofar-as all other methods have practical limitations, the final control for CO and HC emissions is to chemically convert these emissions, at the point of final exhaust just ahead of the muffler, into harmless compounds. This type of control will be discussed under the heading "Catalytic Converters" at the end of this chapter.

Fuel Evaporation Controls

There are several different methods used for controlling gasoline evaporation from the fuel tank, lines and crankcase. Generally, the controls are combined into one system whereby fuel-tank vapors are vented into the crankcase, and the combined fuel-tank and crankcase vapors are then "fed" into the intake manifold to mix with the air/fuel mixture. Such a system typically requires several vehicle modifications and new devices.

Fuel Tank and Lines Modifications

The vented fuel-tank cap generally is replaced by a *nonvented cap*. Prior to this change, a vented (to atmosphere) cap has been used to prevent internal pressure, created by heat expansion of the fuel vapor, from bursting the tank and to prevent the vacuum created, when the fuel pump sucks fuel from the tank, from causing the tank to be collapsed by atmospheric pressure. To obviate these problems in an unvented tank, the tank cap may be fitted with a two-way check valve that will open at predetermined pressures, or, the cap may have a one-way check valve (to route vapors into the control system) while a vent line from the carburetor or air intake supplies air to replace used fuel. The tank may also be designed to have an *expansion chamber or area,* (which cannot be filled with fuel) that is vented (for air replacement) as above.

Because ordinary fuel lines are subject to decomposition, special material fuel lines are used in the venting systems. Substitution of an "ordinary" vented tank cap will make the system inoperative; substitu-

Gas Engine Manual

tion of "ordinary" fuel line hoses may clog the system and result in tank bursting or collapse. This can also happen if a check valve incorporated in the tank cap (or system) malfunctions. See Fig. 1.

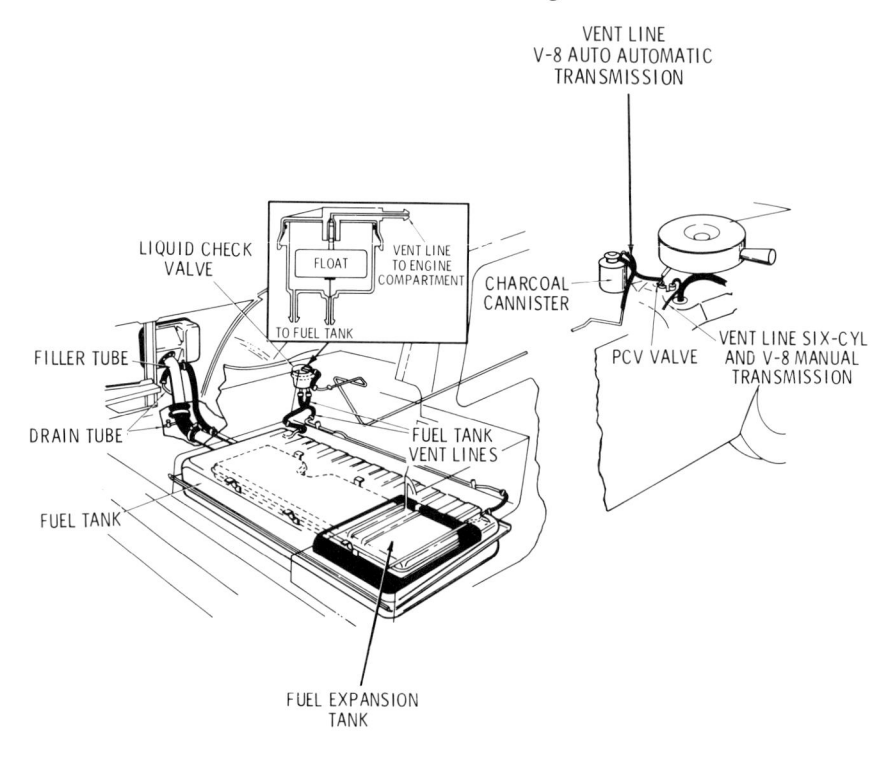

Fig. 1. Principles of a tank vapor control system.

Tank Vapor Venting

Tank vapor may be vented to the crankcase (to be vented from there as explained later) or to the air intake (to slightly enrich the carburetor mixture). Most systems include a *liquid check valve* (also called a *liquid/vapor separator)* in the vapor line, which functions to prevent raw fuel from entering the system (it has by-pass tubes through which any raw fuel splashed or sucked into the system is returned to the tank). All systems also have a *fuel-vapor-storage canister* in which the vapors, arising from the tank, are stored until they can be dissipated. See Fig. 2.

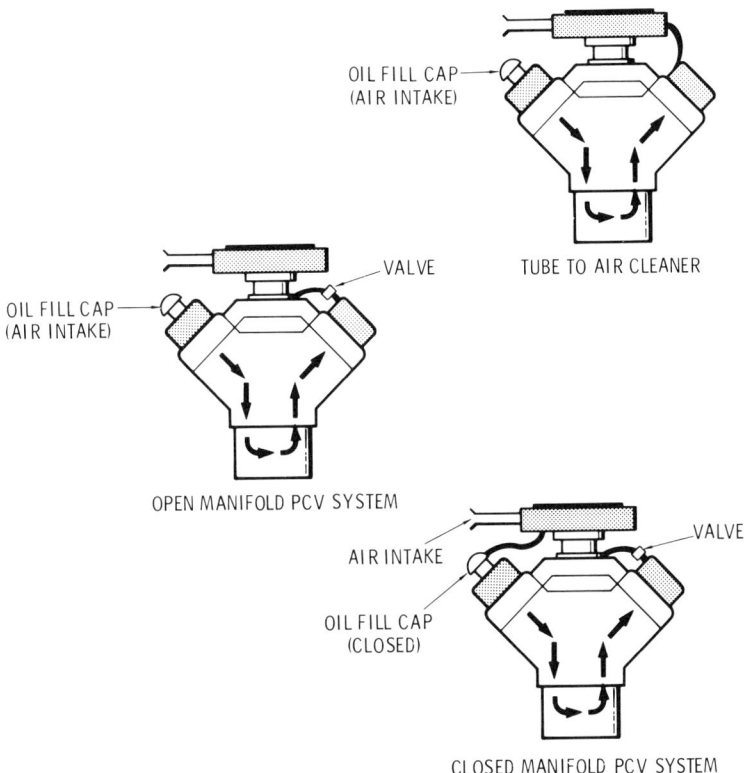

Fig. 2. Typical PCV installation.

This canister may be filled with charcoal or carbon granules (which absorb and hold the vapors until a "draft" carries them off) and may or may not have check-valve controlled vents to other parts of the system. On the whole, the purpose is to collect tank vapors, then disperse them as intended.

From the canister, vapors are dispersed either to the crankcase (as already noted) or directly into the air-intake of the carburetor.

Crankcase-Vapor Venting

Unburned fuel that leaks past the piston rings into the crankcase also presents a vapor evaporation problem. Ordinarily, a crankcase is vented (through the oil-filler pipe) to atmosphere, to dissipate these vapors. In a pollution control system, vapors are not permitted to escape to the atmosphere. Instead, air is circulated through the crankcase, entering either by way of a tube from the air intake or through an opening in the oil-filler cap, and is directed by tube to the manifold side of the air intake. Most systems have a spring-loaded valve that remains closed when manifold vacuum is high, but opens whenever manifold vacuum drops sufficiently.

This is called a *PCV (positive crankcase ventilation) valve*. By its operation the crankcase vapors (mixed with air and, as previously told, with fuel-tank vapors) are introduced into the air/fuel mixture during most periods of engine operation. The carburetor must be adjusted for this mixture enrichment, and any malfunction in the PCV system will adversely affect the mixture and engine performance.

Combustion-Control Systems

As previously noted, various systems (or "packages") have been created to minimize CO and HC emissions resulting from the combustion process. Although different types and/or arrangements of individual control devices are used, all the systems are designed to eliminate, insofar-as-possible, mixture enrichment beyond the 15 (air) to 1 (fuel) ratio that provides a satisfactorily-complete combustion under normal running temperature conditions. All systems also include some type of spark-advance control together with, in many cases, throttle-position controls and other accessory and/or engine modifications.

Air/fuel mixture controls are used for controlling the extent of mixture enrichment can be classified as follows:

(1). *Cold engine starting controls* that operate to assure a rapid choke opening after a cold engine has been started. Two basic types are in use: An *electric assist choke* and a *choke hot-air modulator*.

(2). *Intake air temperature controls* that operate to provide rapid warming of the mixture after a cold start so as to assure soon-as-possible maximum vaporization of the fuel. Practically all systems have some form of *thermostatic air-intake control* whereby exhaust gases are used as necessary to preheat air entering the air cleaner.

(3). A *control that circulates hot exhaust gas* through the intake manifold to hasten mixture warm-up as above.

(4). A valve that admits *additional mixture into the intake manifold* during high-speed deceleration (which, in effect, lowers manifold pressure), to reduce the enrichment of the mixture caused by high manifold vacuum at the carburetor idle-speed fuel nozzle.

(5). A *faster idle-speed throttle stop* that serves two purposes: (a) reduction of fuel enrichment at idle, by admitting proportionally more air past the throttle; (b) smoother engine idling on the leaner mixture.

(6). *Recalibration of the carburetor* to provide a leaner mixture, either for all operating conditions or for certain operating conditions such as at idle and slow speed.

(7). Substitution of a *fuel injection system* (instead of a carburetor). An electronically-controlled fuel injection system provides better (than carburetor) mixture control for various operating conditions.

Spark timing controls are used to improve combustion under different operating conditions can be classified as follows:

(1). Over-riding *distributor vacuum-advance controls* that retard the spark from conventional setting for idle operation, or at least, for idle operation except during cold-engine operating periods (when an advanced spark is needed for combustion of the enriched mixture). The conventional idle spark advance increases engine rpm and manifold vacuum at the fuel idle-speed nozzle, resulting in a richer mixture.

(2). *Vacuum-advance controls* that either and/or: (a) delay conventional spark advance during acceleration, (b) provide an advanced spark or a more slowly retarded spark during deceleration. A conventional distributor advances the spark in relation to engine rpm and throttle opening. During a fast acceleration, the resulting too-rapid spark advance tends to increase the rpm disproportionately with the throttle opening, resulting in over-enrichment of the mixture. During deceleration, the closed (or closing) throttle cuts off carburetor vacuum so that the spark is instantly (or too rapidly) retarded, and combustion occurs too late to be completed in the cylinders.

(3). *Redesign of the distributor centrifugal-advance mechanism* to delay and/or retard its operation—to accomplish one or more of the foregoing purposes.

(4). *Ambient- or coolant-temperature sensing controls* that are used in conjunction with the spark-retarding controls to advance the spark when engine (or air) temperature is below a predetermined degree—and also, in some cases, whenever the temperature exceeds a (different) predetermined degree. Advancing the spark for a cold engine increases the rpm and (as previously mentioned) enriches the mixture to improve vaporization during warm-up; advancing it when an idling engine becomes overheated reduces the temperature by increasing the idle speed.

(5). Substitution of *computerized or electronic ignition* for the conventional distributor. Such a multicontrol system can accomplish some or all (depending upon design) of the foregoing purposes.

Throttle control generally consists of an increased *fast* idle setting (obtained by resetting the idle-stop screw), together with a solenoid control that retards return of the throttle to idle position during high-speed deceleration. The faster "curb" idle setting allows the engine to run cooler on a leaner idle mixture and/or with a retarded spark, and also avoids rough idling and carburetor icing. Retardation of deceleration helps curb the mixture enrichment previously explained. This is also accomplished (in some cases) by controls that by-pass a measured amount of mixture around the throttle during deceleration.

Another throttle control frequently used is one which will positively close the throttle when the ignition switch is turned off. A thinned (hot-burning) mixture and a faster-idle throttle position can result in the engine continuing to run a short time after the switch is off (called *dieseling*). By providing an effectually closed throttle setting and a device *(idle-stop solenoid)* to return the throttle to this position with the turning off of the switch, dieseling is eliminated.

Because prolonged idling (even "fast" idling) with a thinned mixture and/or retarded spark can cause engine overheating, in some cases the coolant system is modified, either by complete redesign or by recalibration of the coolant thermostat. Use of a no-lead fuel together with a reduced compression ratio also improves combustion to reduce the pollutant levels.

Combustion control systems generally incorporate several of the controls mentioned above. For instance, most systems include either an electric assist choke or a choke hot-air modulator, some type of thermostatic air-intake control, together with one or two of the spark advance controls plus, in some cases, throttle and coolant-circulation controls. On the other hand, several systems rely more upon recalibration of the carburetor (in one case, to provide an 18 to 1 air/fuel ratio), recalibration of the distributor, cooling-system modifications, and just one control device (a vacuum-advance control valve) to control — and retard — distributor vacuum advance during idle, slow speed and decelerations.

Electric assist chokes are designed to prevent prolonged enrichment of the mixture at engine starting and warm-up, this is an integral, non-serviceable unit attached to the carburetor that has *no* effect on conventional carburetor calibrations or adjustments. It serves only to hasten opening of the choke. See Fig. 3.

A typical unit contains a *thermostatic spring,* a bimetal *switch* and a ceramic *positive-temperature-coefficient heater (PTC).* It is connected into the vehicle electrical system so that current is supplied to the switch whenever the ignition switch is at start or on. At temperatures below 60° F the switch remains open, but at about 60-65° F it closes to pass current

EMISSION CONTROL SYSTEMS

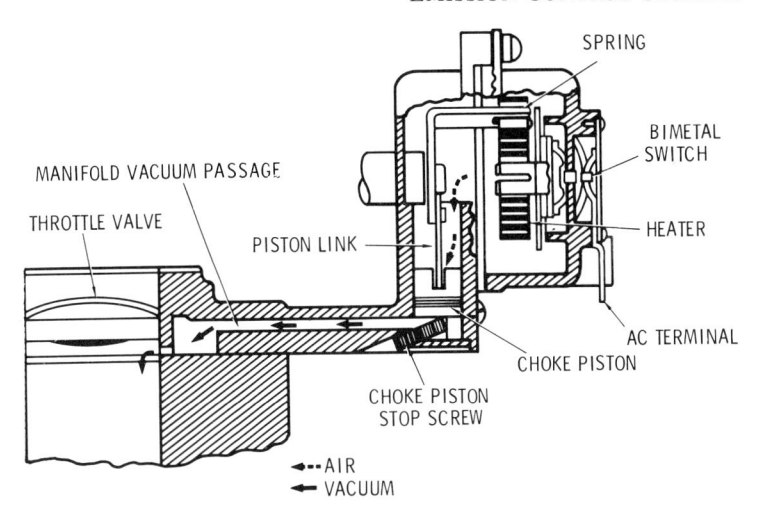

Fig. 3. A typical electric assist choke.

through the heater. Warming of the heater moves the spring to pull the choke open.

From cold start at 60°F to choke open takes only about one to one-and-a-half minutes, thus assuring a leaned mixture much sooner than with a carburetor not so equipped. At ambient temperatures below 60°F the time is extended until engine heat closes the switch.

Some units have a second bimetal switch that opens when engine temperature reaches approximately 110°F, to open the circuit of the first switch and turn the heater off. Others are designed to provide two or more stages of choke opening.

NOTE: Cars with fuel injection are fitted, instead, with a *fast-idle* valve which functions in a similar manner.

Choke Hot-Air Modulator

This system controls the choke (for starting and warm-up) by sensing the temperature of the air being drawn through the air cleaner. Air, drawn through a *modulator (a bimetallic valve) located at the bottom of the air cleaner, is heated in a heater coil* located in the exhaust manifold, and is passed through a bimetallic *thermostatic coil* which is also subject to manifold vacuum and which controls the choke. See Fig. 4.

At temperatures below approximately 68°F (ambient), the modulator opening is negligible to permit very little air to be heated by the heater coil and affect the bimetallic coil, which consequently holds the choke open.

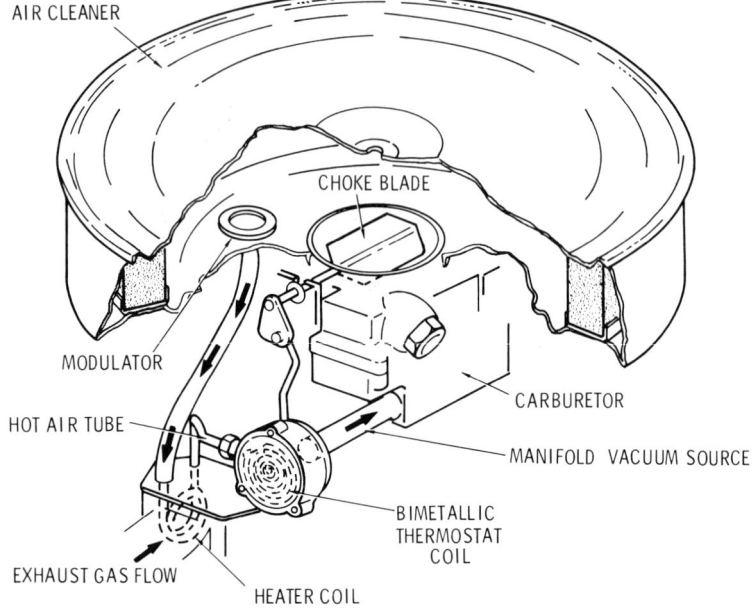

Fig. 4. A hot-air modulator choke system.

At higher ambient temperatures the modulator opens to pass more air; also, as exhaust temperature rises, the heater coil increases the temperature of the air to the bimetallic coil. Thus, choke opening is hastened both by a higher ambient temperature and by engine warm-up.

Thermostatic Air-Intake Control

The idle adjustment of a carburetor must normally be somewhat richer than the high-speed adjustment, to compensate for the lower temperature of the mixture at starting and warm-up. By hastening the mixture temperature rise, thermostatic control of the intake air (into the air cleaner) permits a leaner carburetor idle-speed adjustment.

The pre-heated air needed to rapidly increase the mixture temperature is drawn through a *shroud* (an encircling duct) around the exhaust manifold, and is conveyed by an air-duct to a *snorkle* (a projecting attachment) on the side of the air cleaner. This snorkle, or the side of the air cleaner, also has another opening through which unheated air (referred to as *under-hood air)* can enter. Unit controls are designed to allow only heated air to enter when the under-hood-air temperature is below a predetermined degree (generally, 100°F or more), and to switch from heated to under-hood-air at higher temperatures. See Fig. 5.

EMISSION CONTROL SYSTEMS

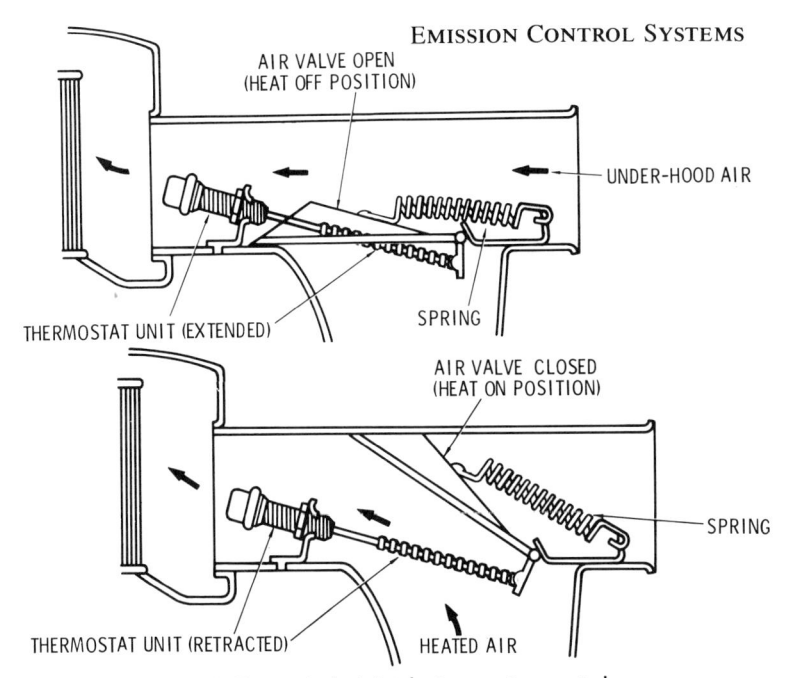

Fig. 5. Two typical air intake temperature control.

The simplest control is a thermostaticly operated air-duct valve arranged to close off the under-hood air passage into the snorkle, then to open this passage while closing off the heated-air passage. A *spring* holds the valve in its first position; a *thermostat* moves it to its second position as the air temperature rises, and holds it fully open whenever the temperature exceeds approximately 130°F.

In order to provide sufficient air for the mixture (should the engine be accelerated while still cold,) a *vacuum override motor* (or similar) may be used. The type illustrated (Fig. 6) operates the foregoing valve and is attached to the snorkle. Another type of control used is a separate unit which opens into the air cleaner through which additional fresh air can be allowed to enter. In either case, the "motor" is a spring-loaded diaphragm exposed (through a hose) to the intake manifold, and arranged to operate a lever connected to the air-duct valve. During acceleration, the decrease in manifold vacuum moves the diaphragm against its spring, and the lever overrides the thermostat unit to open the air-duct valve and admit more under-hood air into the air cleaner.

More complicated units have a vacuum motor (Fig. 7), instead of a thermostatic element, to operate the air-duct valve—and a *bimetal switch (or sensor)* controls the vacuum operation of the motor. This switch is

Gas Engine Manual

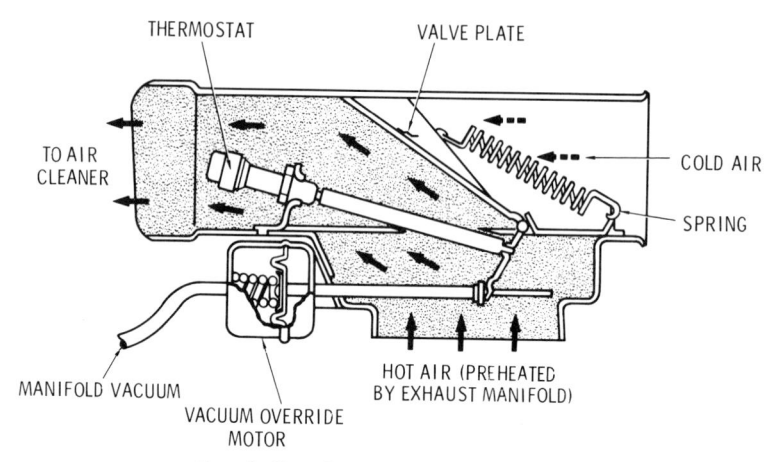

Fig. 6. Use of a vacuum override motor.

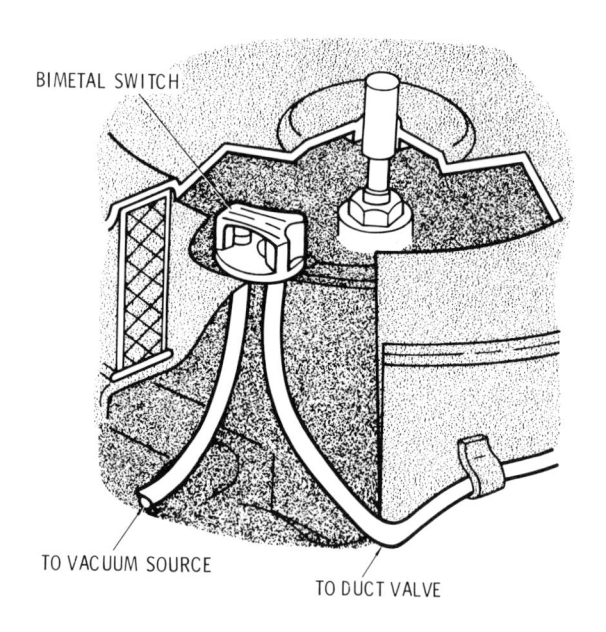

Fig. 7. Typical bimetal sensor switch.

installed inside the air cleaner and is connected into the vacuum hose between the manifold and motor. By sensing the air temperature it functions to open or close the vacuum hose, thus serving the same purpose as the thermostatic element.

Emission Control Systems

Some units have two (instead of one) air-duct valves (called "doors") to separately control the flow of hot (heated) air and cold (under-hood) air into the air-cleaner snorkle. These doors are connected to the motor to work in opposition so that one opens as the other closes.

A cold-weather modulator may be used (instead of a vacuum override motor) to provide under-hood air during cold-engine acceleration. The modulator contains a bimetal switch and is located in the vacuum line to the vacuum motor. At temperatures below about 55°F the modulator shuts off vacuum to the motor to hold the under-hood air valve open.

Heat Control Valve

A different method of accomplishing fast mixture warm-up is the passing of exhaust gases through a pipe in the intake manifold and/or to the base of the carburetor, at temperatures *below* a predetermined degree. This is accomplished by a *heat control valve (HCV)* or an *early fuel evaporation valve (EFE)* located between the exhaust manifold and pipe with a bypass through the intake manifold. The valve is operated by intake manifold vacuum through a coolant-sensing switch which closes the vacuum line (to close the valve) when engine coolant temperature reaches the design degree.

Deceleration Valve

Mounted on the intake manifold, this valve operates to meter an additional amount of air/fuel mixture into the manifold during high-speed deceleration. The unit contains a spring-loaded valve that is opened by movement of a diaphragm and, during deceleration, the high manifold vacuum serves to move the diaphragm. With the valve open, mixture is drawn from the throttle side of the carburetor (through a connecting tube) into the manifold. The valve closes as manifold pressure drops with the completion of deceleration. See Fig. 8.

Idle-Stop Solenoid

Because of a higher operating temperature when the mixture is leaned, as previously explained, together with a fast idle setting, an engine tends to continue running for a short time *after* the ignition switch is turned off (called "dieseling"). To prevent this, a solenoid, energized by the primary circuit of the ignition switch, is mounted on the carburetor and arranged to positvely return the throttle to a predetermined "off" position when the ignition switch is off.

Gas Engine Manual

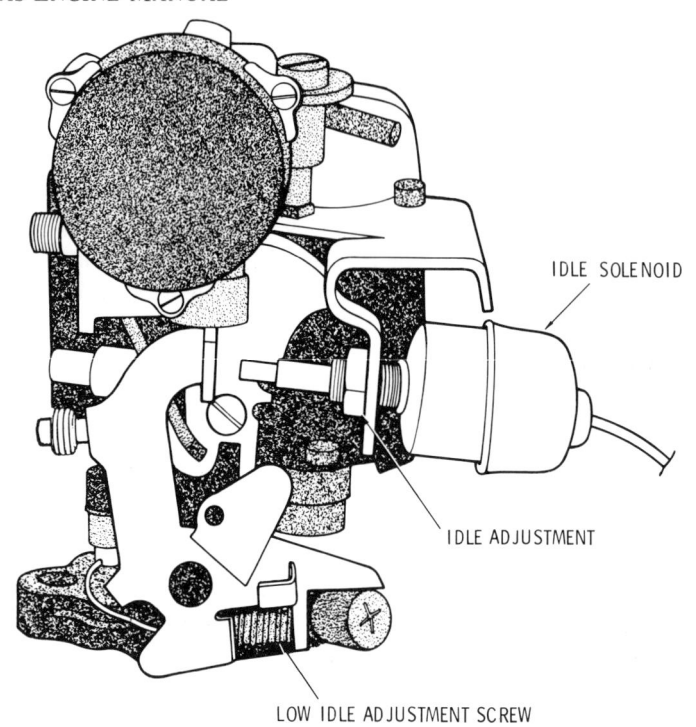

Fig. 8. An idle-stop solenoid installation.

When the ignition switch is on (or at start), the energized solenoid's plunger moves the throttle lever up to the (normal) idle-adjustment-screw setting; turning the ignition switch off deenerigizes the solenoid so that the spring-loaded plunger moves the throttle lever back against a low-idle adjustment screw (at which point the throttle is as nearly closed as it can be without danger of scuffing the throttle bore).

Distributor Solenoid

This is the simplest device for retarding ignition during idle. A solenoid is mounted so that movement of its plunger, when it is energized, will override the vacuum-advance mechanism and retard the spark timing. The solenoid is energized through electrical contacts on the carburetor throttle stop. Opening the throttle breaks the contacts to permit conventional vacuum distributor advance; therefore, the solenoid does not operate during part-open throttle starting—and other controls in the system with which this is used set the throttle at part open (fast idle) for cold starting.

EMISSION CONTROL SYSTEMS

A different type of distributor solenoid is used on some engines merely to override other (spark-retarding) controls so as to advance the spark during engine starting. This solenoid is activated when the ignition switch is at start. The solenoid valve is located in a vacuum line to the distributor which is open, by the valve, whenever the solenoid is energized.

Dual Diaphragm Distributor

Idle-speed spark retarding is also accomplished by a distributor having two diaphragm mechanisms. One diaphragm is operated by vacuum taken from the carburetor above the throttle—where the strong vacuum created during acceleration and cruising will serve to advance the spark in a conventional manner. The second diaphragm is operated by vacuum taken from the intake manifold, and serves to retard the spark during idle when this vacuum is high. See Fig. 9.

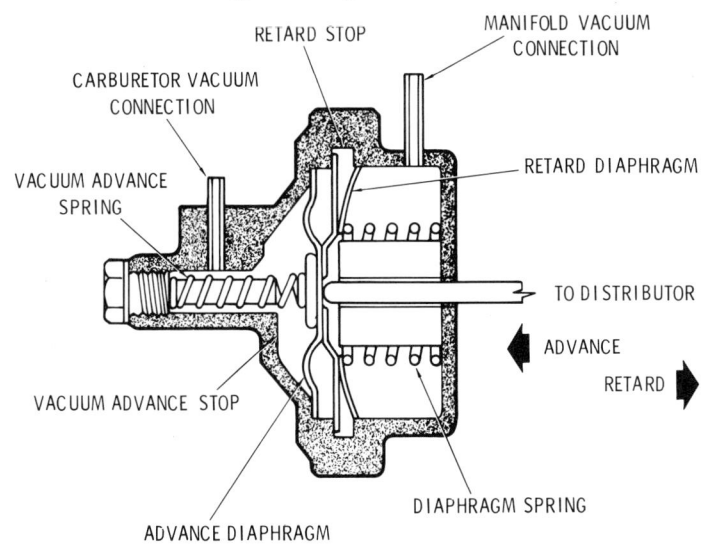

Fig. 9. Dual-diaphragm distributor mechanism.

This distributor mechanism is usually combined with a *vacuum control valve,* the purpose of which is to open a by-pass between the manifold vacuum and the conventional-advance distributor diaphragm. Installed where it can sense the engine-coolant temperature, this normally-closed valve opens whenever prolonged idling overheats the engine. When it is open, the manifold vacuum takes over to advance the spark and allow the engine to run cooler.

GAS ENGINE MANUAL

Vacuum-Advance Control Valve

A conventional distributor vacuum-advance mechanism advances the spark during idle and retards the spark during deceleration, due to the very low carburetor vacuum at such time. One way to assure the early spark needed for better combustion during deceleration is to use manifold vacuum (which, at this time, is strong enough) to overcome the mechanism spring and keep the spark advanced. See Fig. 10.

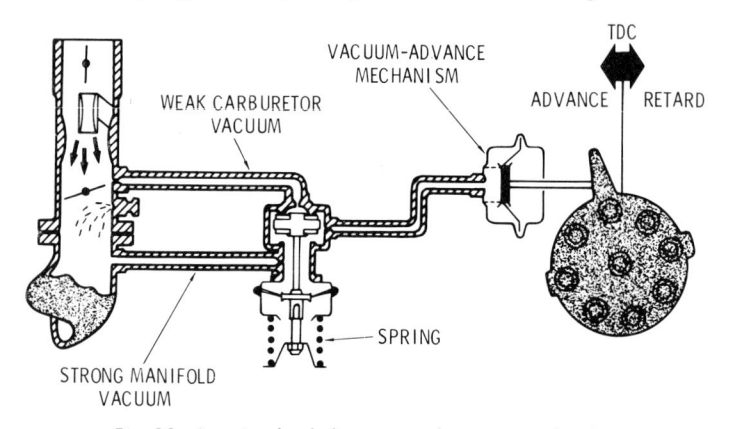

Fig. 10. A spring-loaded vacuum-advance control valve.

This is accomplished with a *vacuum-advance control valve,* which is used with a distributor specially designed to provide a retarded spark during idle. See Fig. 11. The spring-loaded valve is subjected both to carburetor vacuum and to manifold vacuum. During idle, manifold vacuum is too weak to overcome the spring; the carburetor vacuum is transmitted to the distributor, and is too weak (at this time) to advance the special distributor. Also during acceleration and cruise, the manifold vacuum is too weak and carburetor vacuum acts on the mechanism—but, since carburetor vacuum is now strong, the mechanism is advanced in a conventional manner. During deceleration, when manifold vacuum is high, the valve diaphragm moves its plunger to transmit this vacuum to the distributor, and the spark remains well advanced.

Instead of the spring-loaded valve, some systems use a solenoid valve. The solenoid is energized by operation of either an ambient-air or a coolant-temperature switch. At temperatures below a predetermined degree, the switch energizes the solenoid to position the valve so that carburetor vacuum is selected for spark-advance control; at higher temperatures the deenergized solenoid positions the plunger so that manifold

Emission Control Systems

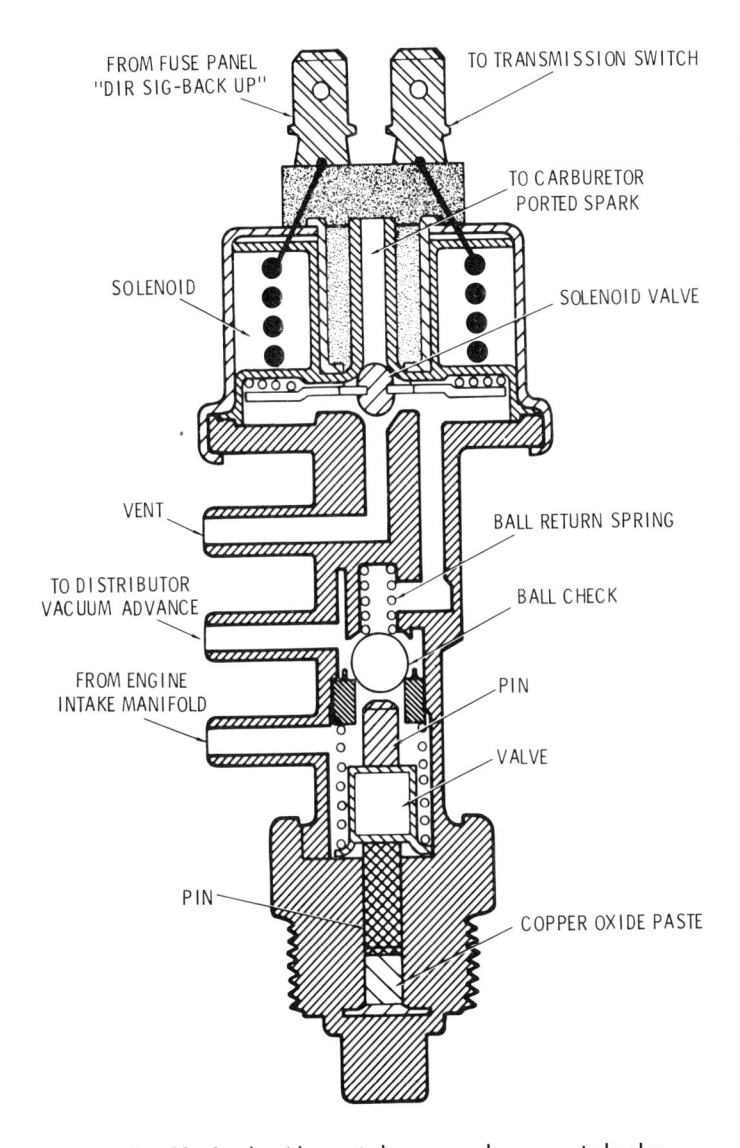

Fig. 11. A solenoid-operated vacuum-advance control valve.

vacuum is selected. Thus, selection of vacuum source becomes a function of the temperature rather than of the relative "strengths" of the two vacuum sources.

Spark-Delay Valve

This valve is used in a "system" which also includes a *one-way check valve*, a *coolant-temperature sensing switch* and a *solenoid vacuum valve*. The spark-delay valve is constructed to permit free flow of air in one direction (through an integral one-way check valve), but to delay flow in the opposite direction (there is a by-pass containing a sintered metal restrictor). See Fig. 12.

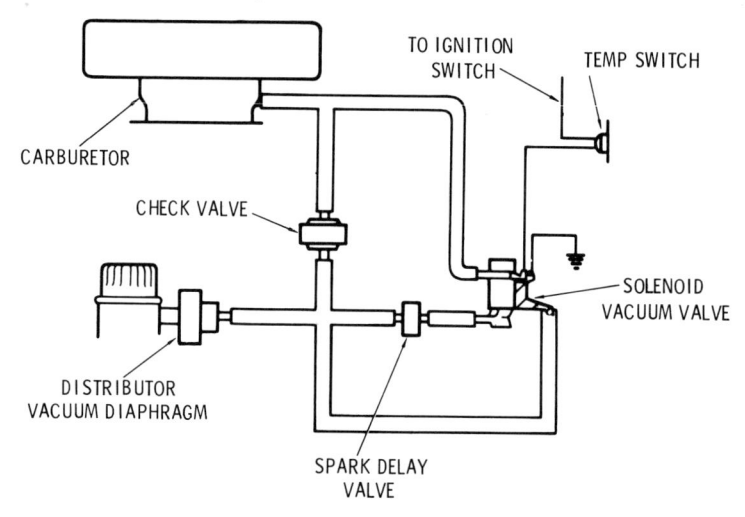

Fig. 12. A typical spark-delay valve application.

The one-way check valve is installed in the vacuum line between the carburetor and distributor so as to check the flow from the distributor to the carburetor, but to allow flow in the opposite direction. The spark-delay valve is in a by-pass line around the check valve, and restricted flow through the spark-delay valve is in the direction from the distributor to the carburetor.

During acceleration, the increasing carburetor vacuum creates a lower pressure in the line between the check valve and carburetor than between the distributor and check valve—and the check valve closes to stop the distributor vacuum advance. However, this lower pressure is also applied through the solenoid vacuum valve to the spark-delay valve, which permits air to bleed through its restrictor until the pressures at both sides of the check valve are equalized and the check valve again opens. Thus, vacuum advance of the spark is delayed by the amount of time required for the air to bleed through the spark-delay valve; and this time is

Emission Control Systems

proportional to the rapidity of the acceleration (which determines the amount of pressure differential to be dissipated).

During deceleration the pressure differential across the check valve and the spark-delay valve is reversed. The check valve therefore remains open (or instantly opens, if closed). Vacuum operation of the distributor remains normal.

The soleniod vacuum-valve is connected into the system so that, when open, it will by-pass both the check valve and the spark-delay valve, and provide an open line between the carburetor and distributor. When it is closed, air flow in the vacuum line around the check valve must pass through the spark-delay valve, as already explained. This valve is normally open; it is activated and closed only when the ignition switch is on and the temperature switch is closed. This occurs when ambient temperature is below a specified degree. Consequently, at temperatures above the design degree, the spark-delay system operates at temperatures below this design degree; the spark is advanced in a conventional manner.

A Coolant-Sensing Switch

Prolonged idling with a leaned mixture and retarded spark (as explained later) can overheat an engine. To avoid overheating, a thermal sensing device (called a "switch") is installed somewhere in the engine block or head where it will react to coolant temperature. This spring-loaded, thermal-expansion device moves a check ball to open or close vacuum lines which control distributor timing. See Fig. 13.

It functions to cut off manifold vacuum and all the "spark control" (explained later) to operate, unless coolant temperature exceeds a predetermined amount (approximately 220°F)—at that temperature it "takes over" by allowing the manifold vacuum to advance the distributor timing (thus permitting the engine to idle at a cooler temperature).

Electronic and Computer-Controlled Timing

Either an electronic ignition distributor or a modified conventional distributor may be used. The modified distributor has two separate spark-timing mechanisms offset from each other by the desired spark advance increment. Thus, one mechanism determines initial timing; the other, advanced timing — and an electronic control selects the one to be used in accordance with engine operating conditions.

Both systems employ a number of data feed-in sensors. Typically, these "report" engine condition (cold or hot) at starting, engine rpm (or transmission gear), intake manifold vacuum, throttle position, and car-

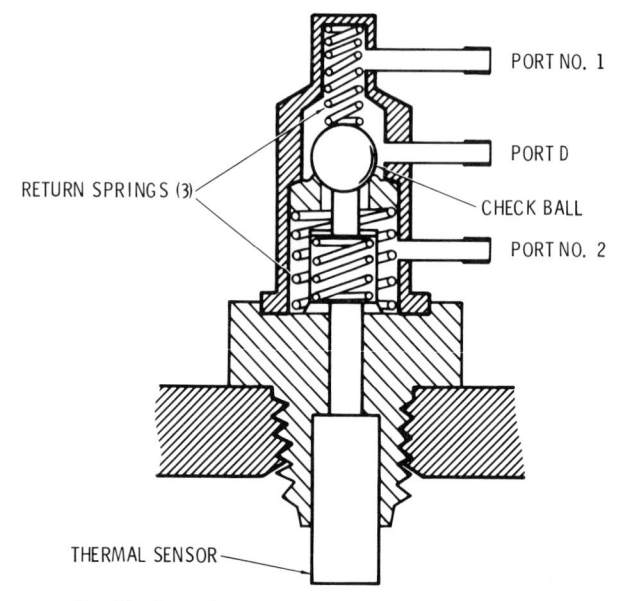

Fig. 13. Typical coolant sensing (thermo-vacuum) switch.

buretor air temperature and/or coolant temperature. Sensors generally are non-serviceable units that must be replaced if malfunctioning.

Unburned Exhaust Controls

Two different systems are used to reburn the unburned portions of exhaust in the exhaust manifold. One is an *Air-Injection System* that uses an air pump to feed fresh air (and, therefore, oxygen) into the manifold. The other, *Pulse Air-Injection Reactor System*, uses exhaust pressure pulsations to accomplish the same purpose. See Fig. 14.

Air-Injection System

A typical system contains an *air pump* (driven by the engine fan belt), a *diverter valve*, a *check valve(s)* (one valve is used, as illustrated, for an in-line engine; two valves—on each side, for a V-8 engine), and (not shown) one (or two) *air manifold(s)* with a separate *injection tube* to the exhaust manifold(s) at each of the exhaust port locations. The system pumps filtered under-hood air into the exhaust manifold at all the cylinders' exhaust-port locations, simultaneously. Excess air is dumped to the atmosphere through a pressure relief valve, either in the air pump or the diverter.

EMISSION CONTROL SYSTEMS

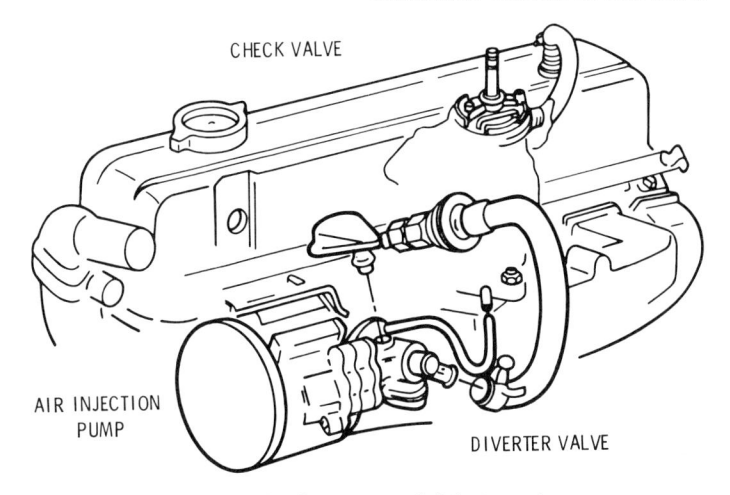

Fig. 14. Typical in-line engine air-injector system.

Fig. 15 is a rotary-type pump having a centrifugal air filter and may (or may not) be fitted with an integral pressure-relief valve. If this valve is not in the air pump there must be one located in the diverter, since the system has to have means of purging excess air flow created by the pump. Air

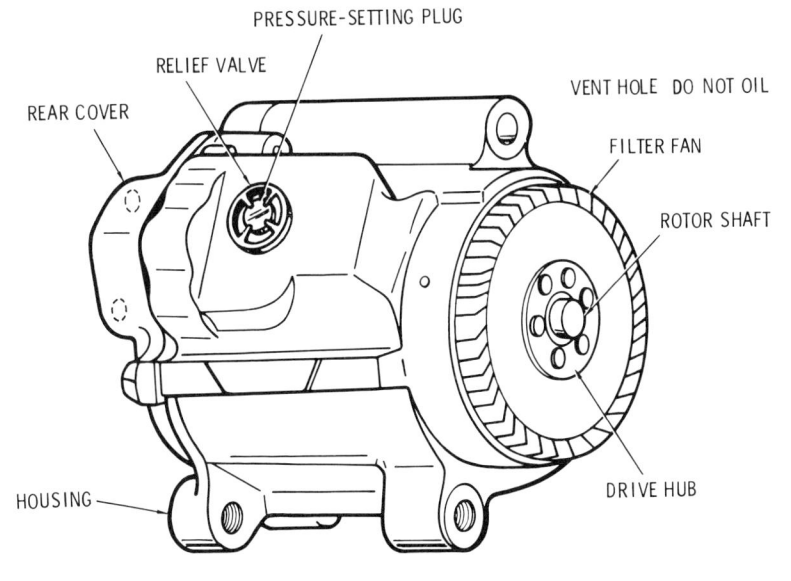

Fig. 15. An air pump with integral relief valve.

Gas Engine Manual

flow becomes excessive whenever the engine is operating at high speed and during deceleration when there is a sharp increase in intake manifold pressure. During engine overrun, the entire air supply is dumped by the relief valve.

Diverter (Air By-Pass) Valve

The diverter valve operated by vacuum from the intake manifold, which is a spring-loaded diaphragm-actuated valve that is normally open to allow air to flow through to the system. Any sudden increase in manifold vacuum causes the diaphragm to collapse against the spring, partially or wholy closing the valve. Whenever the valve is closed or there is an excess of air entering the inlet, air pressure opens the spring-loaded relief valve to allow air to escape from the system (or, as previously mentioned, the relief valve may be in the air pump, instead). See Fig. 16.

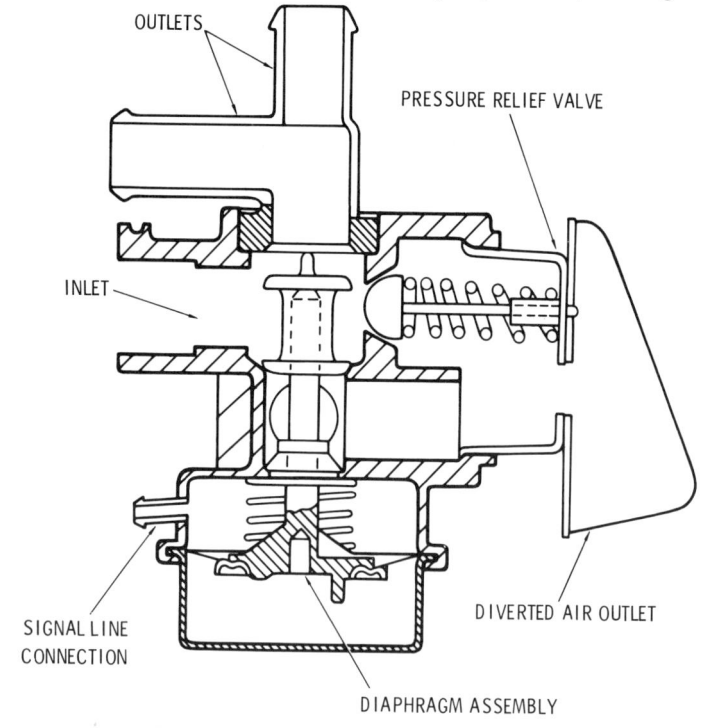

Fig. 16. A diverter with integral relief valve.

A simple ball-type check valve is used to prevent backflow of exhaust gasses through the system. Some systems also use a *mixture-control (or*

EMISSION CONTROL SYSTEMS

backfire by-pass) valve; other systems incorporate a *vacuum differential valve* together with (or without) a solenoid valve to regulate the diverter operation.

Backfire By-Pass Valve

Similar to the diverter valve, and also operated by intake-manifold vacuum, this normally-closed valve is opened by any sharp increase of manifold vacuum (as during deceleration). When open, it supplies fresh air to the intake manifold to offset the mixture enrichment—resulting from the high manifold vacuum—and to thus prevent backfire through the air-injection system. See Fig. 17.

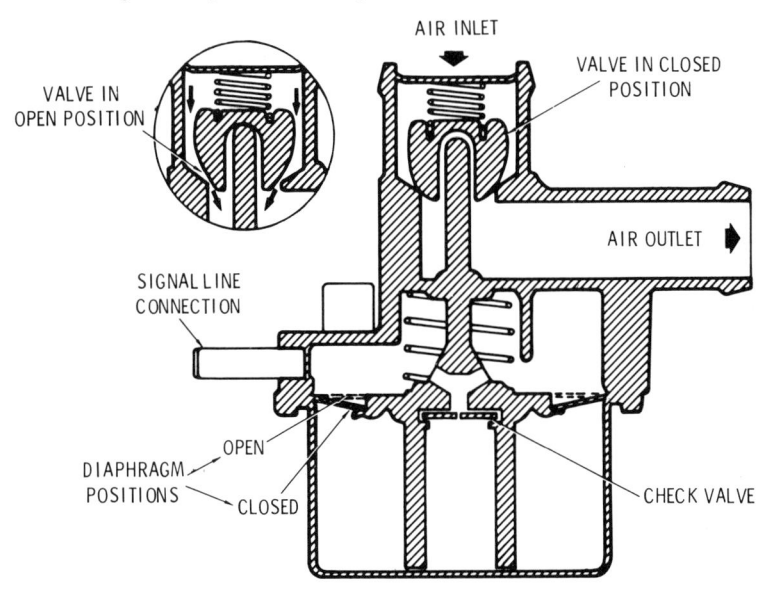

Fig. 17. A backfire by-pass valve.

Vacuum Differential Valve

This diaphragm-actuated valve is installed in the vacuum line to the diverter valve and is operated by the pressure differential at the two sides of its diaphragm. A small orifice between the two chambers keeps the pressures equalized—and the valve closed—except when the manifold vacuum is increasing (as during acceleration or deceleration), at which times the sudden excess of pressure in one chamber opens the valve. Opening of the valve admits air into the vacuum line thus "dumping" the vacuum and closing the diverter valve (which, in this system is a normally-closed valve). Shortly after opening, air bleed through the orifice will re-equalize the pressures and allow the valve to again close. Air injection by the system is thus wholy or partially interrupted in accordance with manifold vacuum.

Solenoid Valves

When used, this type of valve is installed in the vacuum line ahead of the vacuum differential valve, and also serves to open or close the vacuum line. The solenoid is electrically operated by a circuit containing a temperature sensor (switch). This sensor is located in the air cleaner (in some models, there is also one in the floor pan); and the system is designed to shut off the diverter valve (stop air injection) when the engine is cold and (some models) when it becomes overheated.

SUMMARY

Carburetors, distributors and the various system valves are specially calibrated for each engine application. Consequently, should any component malfunction, the entire engine operation is adversely affected, and the pollutant emmissions may be increased (rather than decreased).

Pulse Air-Injection Reactor System

This system contains one *check valve* for each cylinder and one *air shut-off valve* to serve the whole system—together with necessary air and vacuum lines. In operation, the system "sucks" air from the air cleaner into the exhaust manifold at each cylinder's exhaust port, doing so for each cylinder during the piston exhaust stroke. However, the system operates only when the engine is idling or cruising; will shut off at high engine rpm and whenever intake manifold vacuum suddenly increases (as during rapid deceleration). See Fig. 18.

Check Valves

These are one-way, disc-type valves installed in the lines so that each will open when its associated cylinder's exhaust valve opens (thus increasing manifold vacuum at this point), and will close when the exhaust valve closes. Due to inertia, however, valve opening is reduced as the engine speeds up (diminishing the flow of injected air), and it fails to open (shutting off the air) when engine rpm reaches the valve design limit.

EMISSION CONTROL SYSTEMS

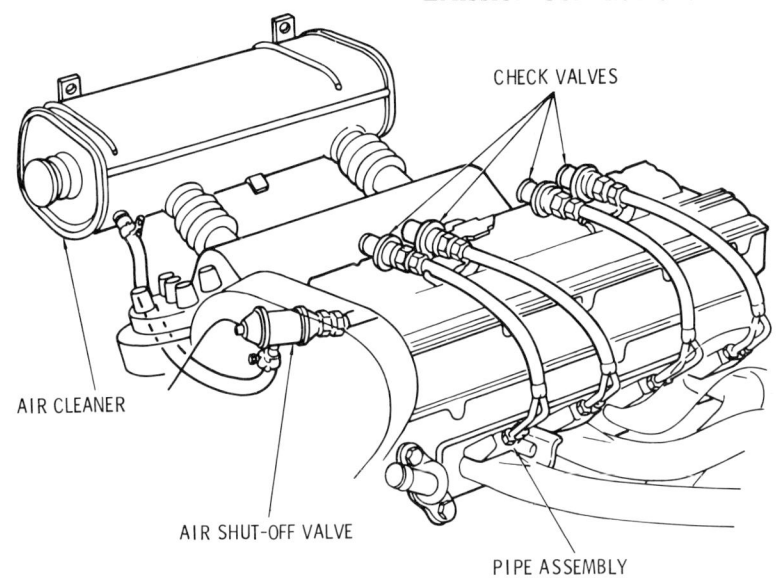

Fig. 18. A pulse air-injection reactor system.

Air Shut-Off Valve

This is a spring-loaded diaphragm valve operated by intake manifold vacuum and installed so as to shut off the air supply to all the check valves whenever intake manifold vacuum exceeds a predetermined value. The purpose is to prevent backfire, which could be caused by air induction into the exhaust during deceleration when the mixture is overly enriched.

NO_x Controls

Now used on all engines, an *exhaust-gas recirculation system (EGR)* reduces the formation of oxides of nitrogen by metering sufficient exhaust gas into the intake manifold to reduce the peak temperature of the burning mixture within the combustion chambers. Some engines also have an added spark-advance control which contributes to the reduction of NO_x by preventing vacuum spark advance at engine speeds below and at temperatures above predetermined amounts.

EGR Systems

A typical system, as illustrated, contains an *EGR Valve,* a *coolant temperature switch,* two vacuum signal modulators (not shown) —one for low temperature; the other for high—and connecting hoses. Other units included in some systems are an *exhaust back-pressure transducer,*

Gas Engine Manual

delay timer, and a *high-speed modulator.* Basically, a system introduces exhaust gases into the intake manifold at engine speeds other than idle (when the addition of exhaust gases would result in rough engine operation). This is accomplished by the EGR valve; all the other units are used, in one way or another, to modulate the flow of exhaust gases through the EGR valve to compensate for periods of deceleration or engine overload—or to close down the EGR system at temperatures below or above specified degrees. See Fig. 19.

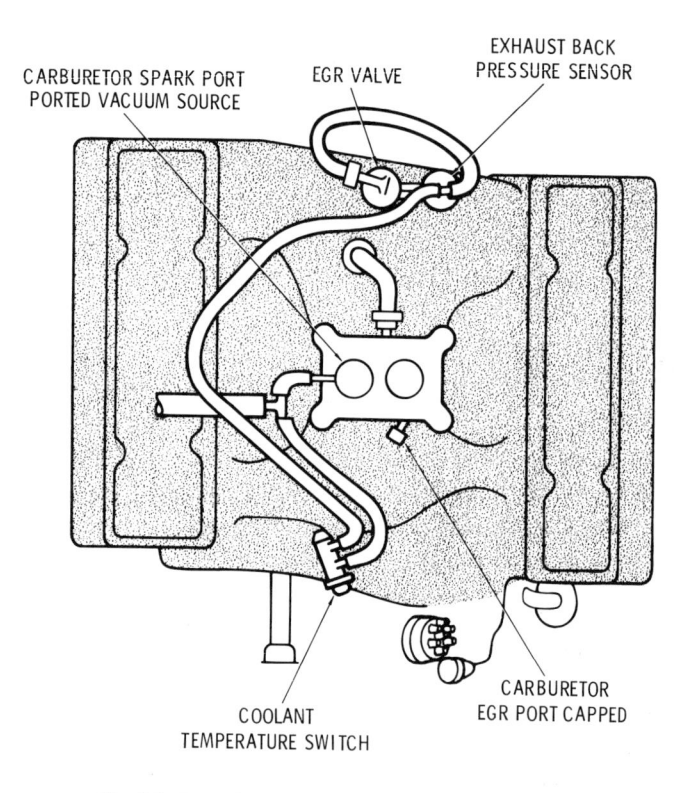

Fig. 19. Typical EGR system with a back-pressure sensor.

Except in certain modulated systems, the EGR valve is operated by vacuum taken from the carburetor EGR port (below the throttle); exhaust gases are taken from the exhaust crossover passage or near the bottom of the riser area, and are ported to the intake manifold floor below the carburetor.

EMISSION CONTROL SYSTEMS

EGR Valves

Two basic types are illustrated in Fig. 20. Both types are mounted on the intake manifold, at top or side. The single-diaphragm valve is operated by vacuum from the EGR port to start opening at a specified vacuum measure (in. hg) and to become—and remain—fully open at a higher specified measure. This is a spring-loaded, diaphragm-actuated valve that is normally closed at idle, when spring pressure exceeds the "start-to-open" vacuum suction—is partially open at slow speeds and during acceleration (when vacuum suction is low), and is fully open during cruising and deceleration. To offset the higher formation of NO_x during slow-speed operation and acceleration, by increasing exhaust gas recirculation at these times, systems using this type of valve generally included a back-pressure switch or transducer valve.

When the more flexible (in operation) dual-diaphragm valve is used, the EGR vacuum is taken from the carburetor spark port (above the throttle). An additional vacuum, taken from the intake manifold, is transmitted to the area between the two diaphragms (which are mechanically joined to move together). The upper diaphragm has a larger piston than the lower diaphragm. Therefore, suction created by manifold vacuum helps the spring to keep the valve closed or partially opened. It follows, that during cruising, deceleration or engine overload, when manifold vacuum is high, the valve is only partially open; but, at low speeds and during acceleration, when manifold vacuum is low, the valve becomes fully opened to recirculate a maximum amount of exhaust gases.

Exhaust Back-Pressure Transducer (Switch or Valve)

This unit serves to regulate EGR operation by permitting exhaust gas recirculation or stopping it, according to the back pressure in the exhaust manifold. The single-diaphragm EGR valve that is used takes its vacuum from the carburetor spark port (instead of the EGR port), and the transducer is connected into the vacuum line ahead of the EGR valve. Valve operation either opens a vent (in the valve) to dissipate the vacuum and keep the EGR valve closed, or to close the vent and allow normal operation of the EGR valve. See Fig. 21.

The spring-loaded, diaphragm-actuated valve normally holds the vent open. A probe (tube) connects the valve with the exhaust manifold crossover. Whenever exhaust back pressure is high (during acceleration and some cruise conditions), the diaphragm is moved to close the vent; at other times the vent remains open. The valve is precisely calibrated for its particular engine application.

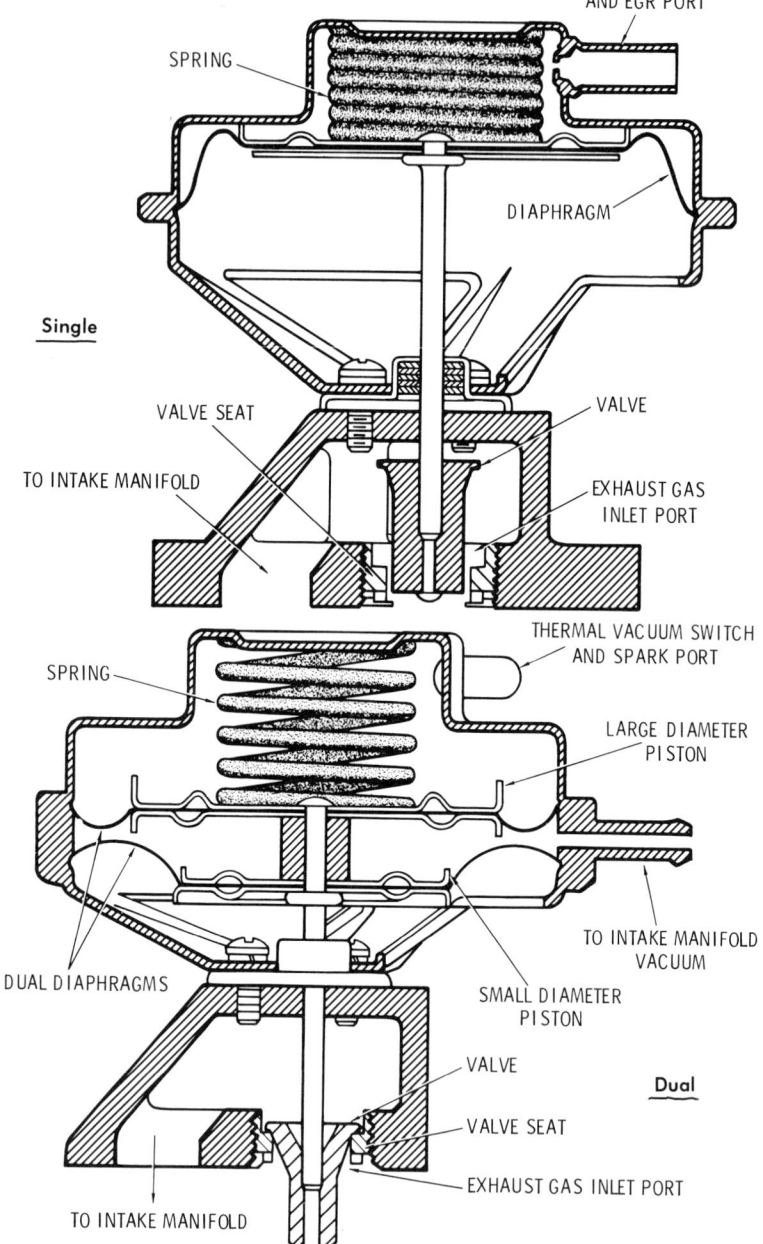

Fig. 20. Single and dual diaphragm EGR valve.

Emission Control Systems

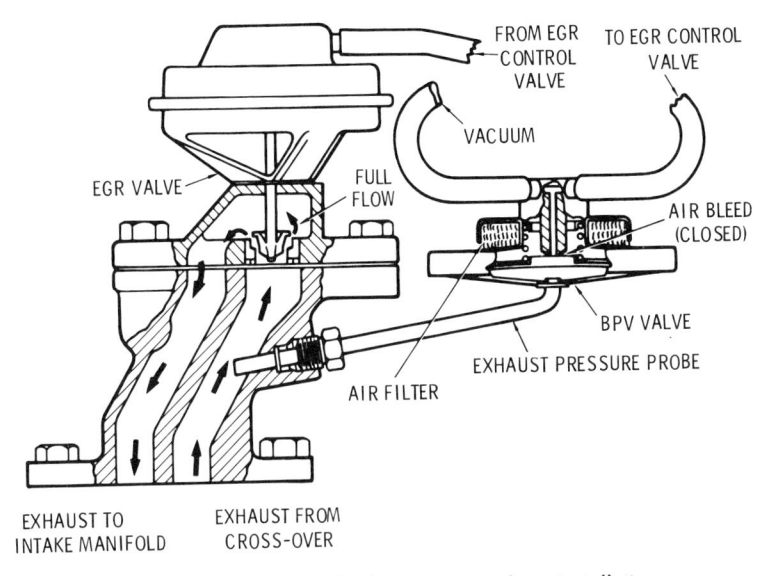

Fig. 21. Typical exhaust back-pressure tranducer installation.

Coolant Temperature Switch

This is the same as the coolant-sensing switch Fig. 13. Mounted where it can sense engine-coolant temperature and in the vacuum line ahead of the EGR valve, the switch serves to close off the EGR valve vacuum whenever temperature is below a predetermined degree. In short, it prohibits exhaust gas recirculation when the engine is cold. All systems include such a switch.

Vacuum Signal Modulators

These are electrically operated ambient-temperature switches installed, when used, in the vacuum line ahead of the coolant temperature switch. Operation is such as to partially close the vacuum line and weaken the signal to the EGR valve whenever either switch is activated. The low-temperature switch is usually mounted near the engine radiator and is activated at temperatures below a predetermined degree. The high-temperature switch is mounted at the rear of the engine compartment and is activated at temperatures above a predetermined degree.

Delay Timer

This is an electrically-operated timer connected with a solenoid valve in the EGR vacuum line ahead of the EGR valve. Energized when the ignition switch is turned on, the timer circuit holds the solenoid valve closed for a predetermined number of seconds (after engine starting) to prevent exhaust gas recirculation. Afterwards, the solenoid valve opens to allow normal EGR system operation.

High-Speed Modulator

This is a sub-system sometimes used to cut off exhaust gas recirculation at high vehicle speeds to improve performance. The system includes a normally open solenoid valve in the vacuum line ahead of the EGR valve, together with an electrical system to energize (and close) the solenoid valve. Electrical components include a speed sensor, driven by the speedometer cable, that produces a voltage proportioned to the vehicle speed, together with an electronic module that monitors this voltage and switches on the solenoid circuit according to design plan.

NO_x Spark-Advance Controls

NOx emissions are increased by an advanced spark during periods when the engine is already warmed-up and the gears are being up-shifted or the vehicle is traveling slowly (as in a crowded city area). During such periods, it is therefore desirable to prevent vacuum spark advance. Some manufacturers therefore install additional controls to prevent spark advance except under cruising conditions. There are two types.

The manual transmission control is a solenoid vacuum valve operated by a switch mounted on the transmission housing. The normally open valve is in the vacuum line to the distributor, and the switch is in ''off'' position whenever the transmission is in high gear. Shifting to any other gear turns the switch ''on'', activates the solenoid, and causes the valve to close the vacuum line and prevent vacuum spark advance.

The automatic-transmission control is also a solenoid vacuum valve that functions in a similar manner, but this valve is normally closed and is operated by a speed sensing switch similar to the one previously described *(refer to High-Speed Modulator, preceding)*. At speeds below the design mph the valve remains closed to prevent vacuum spark advance, but at higher speeds the switch activates the solenoid to open the valve and permit conventional spark advance.

Both control systems include an over-riding ambient-temperature thermal switch which serves to prevent closing of the solenoid valve when the temperature is below a design degree. Hence, during cold engine operations the spark advance is conventional.

EMISSION CONTROL SYSTEMS

CATALYTIC CONVERTERS

The converter is an integral, nonserviceable unit installed in the exhaust pipe between the exhaust manifold and the muffler. Converters are made in different shapes and sizes to suit their applications, but all operate on the same basic principal. A catalyst, consisting of noble metals (platinum, etc.) is used to combine HC and CO pollutants in the exhaust gases with oxygen to form, instead, harmless H_2O (water) and CO_2 (carbon dioxide). This catalyst is made permanent by being coated on an inert material formed either into pellets or into a single honeycombed shape, and does not become chemically changed or dissipated by the process. Consequently, a converter unit will last indefinitely if not abused.

A converter, however, cannot be used with leaded fuel; the lead will destroy the catalyst. For this reason, vehicles so equipped are fitted with

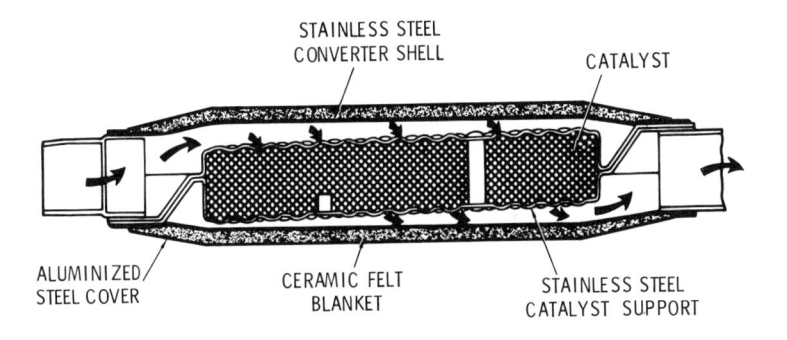

Fig. 22. A typical catalytic converter.

gas-tank filler, tubes too small in diameter to admit conventional gas-pump nozzles—a smaller nozzle, now universally adapted for unleaded fuel pumping, must be used. See Fig. 22.

A converter creates considerable chemically-generated heat. For this reason, most vehicles so equipped are also fitted with *heat shields*. These heat deflectors serve to protect the vehicle flooring and chassis components.

Some cars also have a *catalyst protection system* designed to prevent overheating of the converter during high-speed deceleration (when HC and CO pollutants are at maximum). The system contains an electronic engine-speed switch, or control unit, and a solenoid mounted so as to regulate throttle position when energized. At engine speeds above 2000 rpm the solenoid holds the throttle slightly open (a fast idle) so that during deceleration speed must drop below 2000 rpm before the throttle is allowed to close to normal idle position.

CHAPTER 16

Friction Clutches

Although friction clutches as such do not constitute a part of a gas engine, they are nevertheless used as a means of power transfer between the engine and its load in numerous instances. This is particularly true in automotive applications, and it is because of this fact that a brief explanation of its construction and operation has been included.

By definition, a clutch is a mechanical device for connecting or disconnecting two pieces of shafting in the same line, or a shaft and a wheel so that they revolve together or are free.

The function of a clutch therefore, is to connect the engine to the driven mechanism of the driven machinery or gear transmission system. Since the internal combustion engine does not develop a high starting torque it must be disconnected from the transmission system and allowed to operate without load until it develops enough torque to overcome the inertia of the load when starting.

CLUTCH PRINCIPLES

The transmission of power through the clutch is accomplished by bringing one or more rotating drive members secured to the crankshaft into gradual contact with one or more driven members secured to the unit being driven.

These members are either stationary or rotating at different speeds. Contact is established and maintained by strong spring pressure controlled manually through the clutch pedal and suitable linkage. Thus, as spring pressure increases, the friction increases; therefore, when the pressure is light the comparatively small amount of friction between the members permit a large amount of slippage.

As the spring pressure increases, less slippage occurs until when the full spring pressure is applied, the speed of the drive and driven members is the same.

Clutch Elements

In any clutch construction there are two principal elements:
1. A drive member,
2. A driven member.

The drive member is attached to the engine and turns with it; while the *driven* member is attached to the transmission and operates in a similar manner. The operating members include the spring or springs and the linkage required to apply and release the pressure which holds the drive and driven members in contact with each other.

The drive members of a clutch usually consists of two cast-iron plates or flat surfaces machined and ground to a smooth finish. Cast iron is desirable because it contains enough graphite to provide some lubrication when the drive member is slipping during engagement. One of these surfaces is usually the rear face of the engine flywheel, and the other is a comparatively heavy flat ring with one side machined and surfaced. This part is known as the pressure plate, and is fitted into a steel cover, which also contains some of the operating members, and is bolted to the flywheel.

The driven member is a disk with a splined hub which is free to slide lengthwise along the splines of the clutch shaft, but which drives the shaft through the same splines. The clutch disk is usually made of spring steel in the shape of a single flat disk of a number of flat segments. Suitable frictional facings are attached to each side of the disk by means of copper rivets.

These facings must be heat resistant since friction produces heat. The most commonly used facings are made of cotton and asbestos fibers woven or molded together and impregnated with resins or similar binding agents.

Clutch Operation

The operation of an elementary clutch such as that shown in Fig. 1, is essentially as follows: The drive member consists of a plate **A**, attached to or forming a part of the flywheel. The driven member consists of a friction plate **B**, attached to a sleeve which engages with a splined section of the transmission shaft so that while free to slide along the shaft, must always turn with it.

Friction Clutches

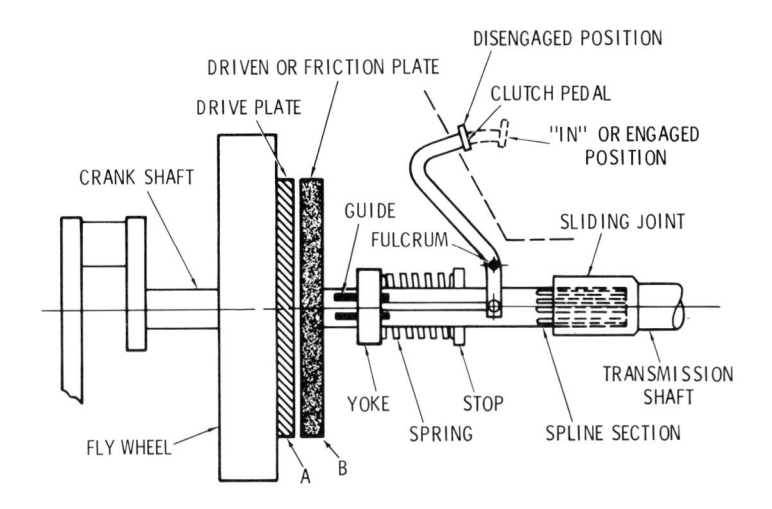

Fig. 1. Elementary plate clutch showing drive and driven members and releasing gear.

The driven plate is normally held against the drive plate by a spring unless released or placed in the disengaged position by pressing down the clutch pedal. Movement of the clutch pedal is transmitted to the friction plate through a yoke and collar.

In operation, when the clutch is engaged or "let in", the engine shaft is connected frictionally to the transmission shaft and if the transmission gear is in mesh, the entire drive is connected.

When the clutch is disengaged, that is in the position shown in Fig. 1, there is no connection between the engine and the drive and the engine is free to turn without movement of the transmission shaft.

Heavy Duty Clutches

In the elementary clutch arrangement shown in Fig. 1, the driven plate makes frictional contact on only one side. Evidently by providing another drive plate with construction such that the two drive plates can engage the driven plate with construction such that the two drive plates can engage the driven plate in frictional contact, the capacity of the clutch is greatly increased.

Such arrangement should be called a three plate clutch, but it is known as a single plate clutch. It should also be noted that multiple plate clutches are referred to by the number of driven or friction plates.

Gas Engine Manual

In the elementary so called one plate clutch, shown in Fig. 2, the driven friction plate **B,** is attached to the shaft **C,** which has a splined section as shown. The latter meshes into a splined sleeve on the end of the transmission shaft to provide a sliding joint and positive engagement between shaft **C,** and the transmission shaft. The dowel pins on the flywheel provide a similar sliding joint for the drive plate **A',** and also forces it to turn with the flywheel.

In operation, releasing the clutch pedal brings drive plate **A',** into contact with plate **B,** and pushes it over until it contacts with drive plate **A.** The pressure due to the spring holds the three plates into firm frictional contact.

For extra heavy duty service, so-called two plate clutches are used consisting of two driven or friction plates and three drive plates. The term multiple plate clutches should not be confused with multiple disk clutches.

Driven or Friction Plate

This part of the clutch is sometimes called a disk. It is made in various ways, a typical construction consisting essentially of a thin metal disk to which is attached friction lining or facing.

At the center is a hub having internal teeth which mesh with spline teeth on the transmission shaft. The essentials are shown in Fig. 3. Note the internal teeth on hub, which engage the external teeth of the transmis-

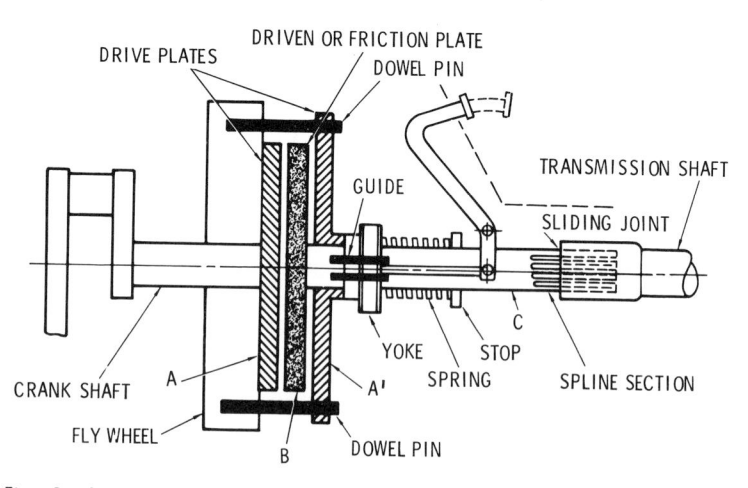

Fig. 2. Elementary so-called one plate clutch in which the driven plate is engaged by two drive plates.

FRICTION CLUTCHES

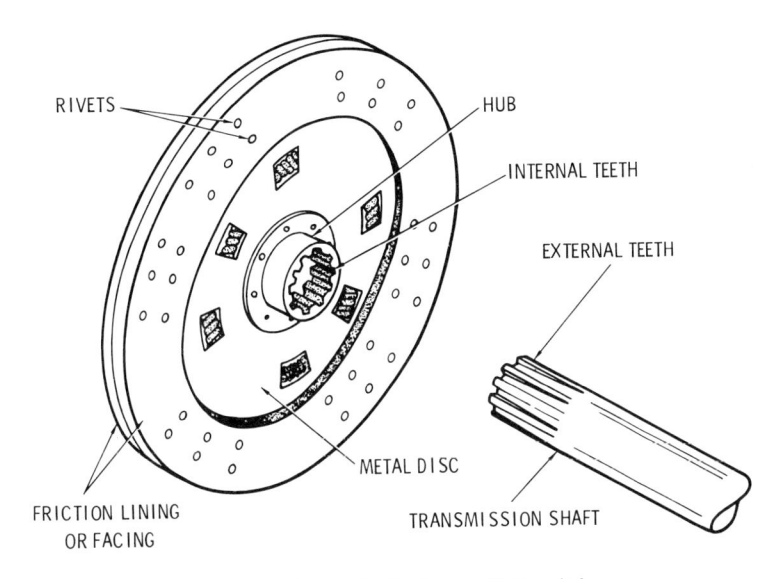

Fig. 3. Plain driven clutch plate or friction disk.

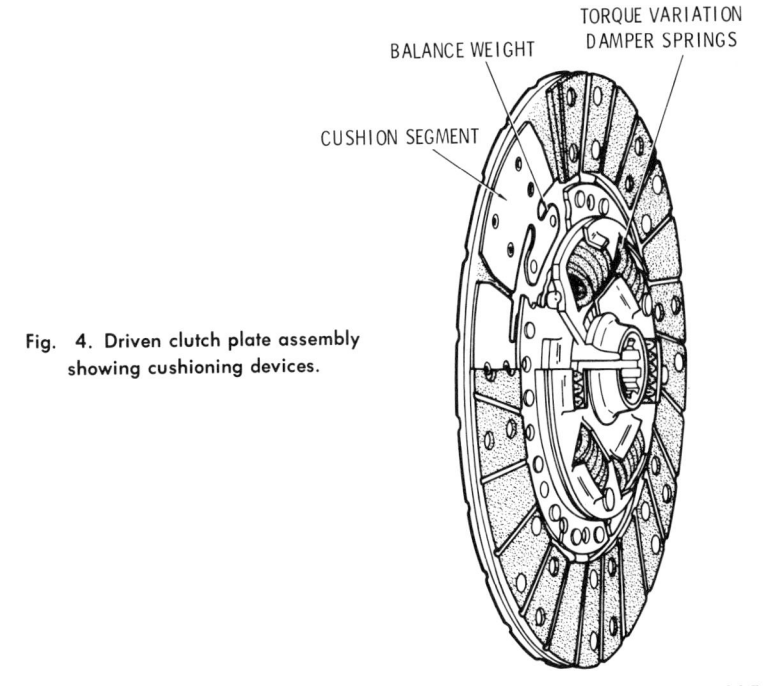

Fig. 4. Driven clutch plate assembly showing cushioning devices.

205

sion shaft, thus providing the sliding and engaging joint. The friction liner or facing is attached to the disk with rivets. These linings are easily renewable when worn.

Cushioning Devices

In order to make clutch engagement as smooth as possible and eliminate "chatter", various cushioning devices are employed, such as crimped spring segments, coiled springs, etc. These devices are shown in Fig. 4.

As noted in the illustration, a number of crimped spring steel cushion segments are placed between the rear lining and disk. These segments constitute an independent cushioning means for the lining. The objects of these segments is to eliminate any possibility of clutch chatter or jerk when engaging the clutch. These segments also aid in prolonging the life of the lining, as they assist in giving a nearer uniform wear over the entire contacting surface.

Fig. 4, shows six coiled springs mounted concentric with the disk hub. These are cushion springs and form the only means of contact between the hub and the disk carrying the friction facings. By this method, torque impulses of the engine are cushioned out before being transmitted to the spline shaft of the transmission.

CHAPTER 17

Horsepower Measurement

By definition, one horsepower is a unit of power numerically equal to a rate of 33,000 foot-pounds of work per minute, or 550 foot-pounds per second. The horsepower of an engine may be expressed in several ways, depending upon how it is measured.

According to definitions and the manner in which it is measured, horsepower may be classed as:

1. Brake horsepower,
2. Indicated horsepower,
3. SAE horsepower, etc.

Other types of horsepower used in the various branches of engineering are: *boiler horsepower, hydraulic horsepower, electrical horsepower,* etc.

One of the simplest means to determine useable horsepower of small engines is by means of the Prony brake method which consists in causing the engine flywheel to rotate against a braking device.

PRONY BRAKE

The output of small and medium size engines is frequently determined by means of a Prony brake, the method of which is shown in Figs. 1 and 2.

It consists essentially of two blocks of wood shaped to fit around the flywheel of the engine. The pressure of the blocks may be adjusted by means of tightening or loosening of the bolts holding the blocks in place.

The two bars fastened at the top and bottom of the brake form an arm which is attached to a scale. When the engine turns, the friction between the flywheel and the wooden blocks converts the power supplied by the engine into heat.

Gas Engine Manual

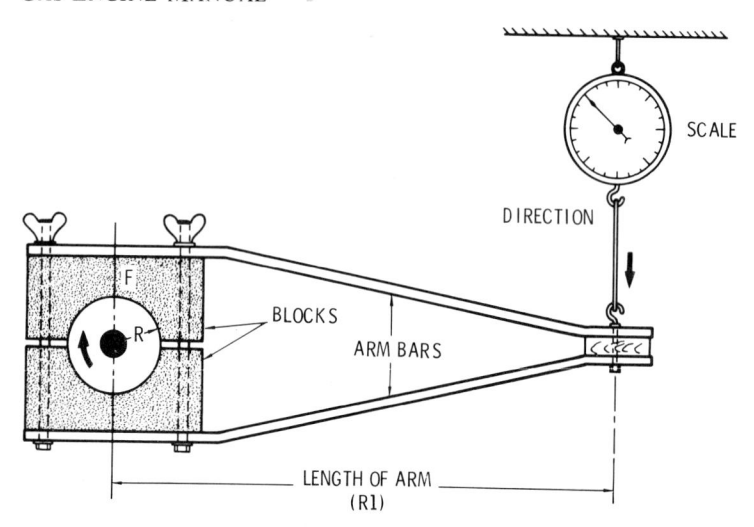

Fig. 1. Typical Prony brake arrangement with suspended spring balance.

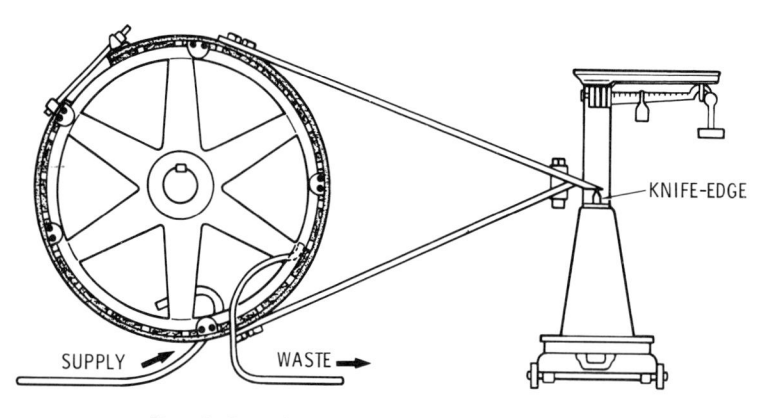

Fig. 2. Prony brake arrangement with platform scale.

The heat generated by the flywheel as it turns under the pressure of the blocks is absorbed by a stream of water applied to the flywheel rim.

In developing the expression for the horsepower converted into heat, let **R**, represent the radius of the flywheel in feet, and **F**, the force supplied at the surface of the flywheel. Then when the flywheel completes one revolution, the force **F**, moves a distance of **2πR** feet, and therefore **F2πR** foot pound of work is done.

HORSEPOWER MEASUREMENT

When the flywheel rotates at **N**, revolutions per minute **FN2πR** foot pound work is being done.

Since one horsepower is developed for every 33,000 foot pounds of work done per minute, we may write:

$$\text{Horsepower} = \frac{FN2\pi R}{33,000}$$

which after reduction of terms, may be written:

$$\text{Horsepower} = \frac{FNR}{5252}$$

Example—It is desired to measure the horsepower output of a gas engine by means of a Prony brake attached to the engine flywheel. Compute the horsepower output with the engine running, at its rated speed of 750 rpm, lever length 36 inches, and scale deflection 50 pounds.

Solution—The power output may easily be computed by substituting numerical values in our formula for horsepower, or

$$\text{Horsepower} = \frac{FNR_1}{5252} = \frac{750 \times 3 \times 50}{52520} = 21.4 \text{ Ans.}$$

ROPE BRAKE

For comparatively small engines, various forms of rope brakes are often employed. In such cases a weight is hung to one end of the rope and a spring scale to the other as shown in Fig. 3. In order to compute the

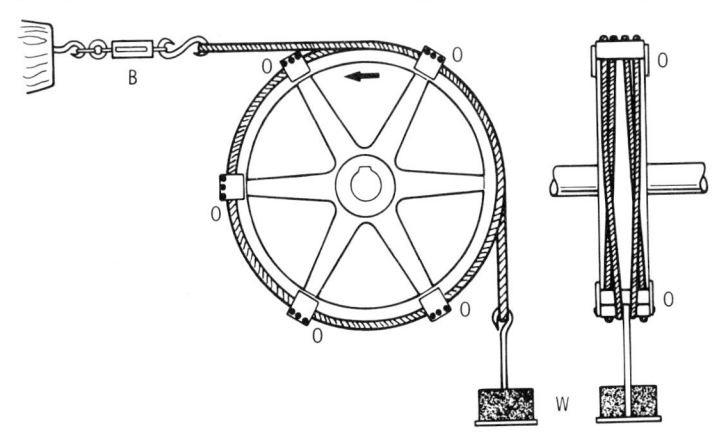

Fig. 3. Side and view of typical rope brake.

horsepower in brake systems of this sort, multiply the difference between the weight **W** and the weight registered on the spring balance **B,** by the number of revolutions of the flywheel per minute, and by the circumference of a circle passing through the center of gravity of the rope expressed in feet, finally divide the product by 33,000.

This type of brake is easily constructed of material at hand and being self-adjusting needs no accurate fitting. For large powers the number of ropes may be increased. It is considered a most convenient and reliable brake. In Fig. 3 the spring balance, **B,** is shown in a horizontal position.

This is not necessary; if convenient the vertical position may be used. The ropes are held to the pulley or flywheel face by blocks of wood **O.** The weight **W,** may be replaced by a spring balance is desirable. To calculate the brake horsepower, subtract the pull registered by the spring balance, **B,** from the weight **W.** The lever arm is the radius of the pulley plus one-half the diameter of the rope. The formula is,

$$\text{B.H.P.} = \frac{2\pi RN(W-B)}{33,000} = \frac{RN(W-B)}{5252}$$

In the formula **R** = radius from center of shaft to center of rope; **N** = *revolutions* per minute; **W** = weight; **B** spring balance.

Torque

The torque of an engine is usually expressed in pound-feet, and is the measurement of the turning effort exerted by the flywheel at a radius of one foot from the center of rotation.

The turning effect produced by an engine does not depend only upon the magnitude but also upon the radial distance through which the force acts. If torque represented by **T,** and is expressed in pound feet, it may be written:

$$T = FR$$

where: F = force in pounds,
 R = radial distance through which the force acts.

We also have, $\text{HP} = \dfrac{FNR}{5252}$

which after substitution may be written as:

$$\text{HP} = \frac{TN}{5252}$$

From the foregoing it will be noted that since the horsepower is a product of torque and rotation speed, the horsepower at any given rpm

Horsepower Measurement

will vary directly with the torque. Thus, for example, if at a constant engine speed the torque is doubled, the horsepower output will also be doubled.

Actually, however, the torque does not remain constant throughout the various speed of an engine, but after a short rise begins to fall off. This falling off of torque as the speed increases above a certain point is caused by the inability of the engine to obtain as full a charge of gas as at a lower speed. See Fig. 4.

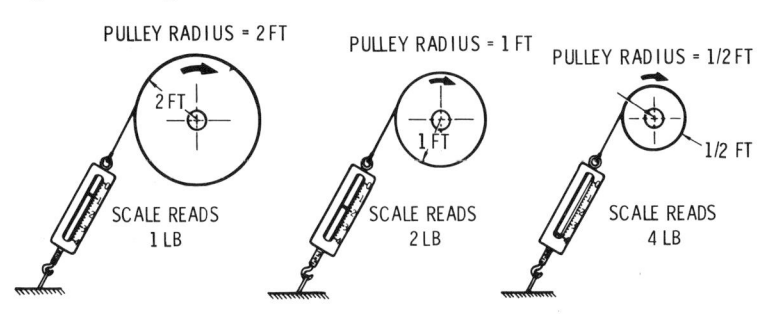

Fig. 4. Three examples of equal torque. If a cord is wound around a pulley as shown and the force produced in the cord measured on a spring balance, the torque produced in each instance will be equal or 2 foot-pounds, although the pulley radius in each case may differ.

DYNAMOMETER

Another device for measuring the developed horsepower of an engine is commonly known as a dynamometer. This is a very convenient method of power measurement and consists in connecting an electric generator to the main shaft of the engine to be tested.

If the efficiency of the generator is known at the particular speed and output at which it is to be operated, a very accurate method of measuring the power of the engine becomes available.

Since the amount of electrical power delivered will vary according to the speed at which the armature revolves, and this in turn is dependent upon the power of the engine under test, it is evident that the power delivered to the engine can be read directly from the instrument recordings.

For direct current generators, brake horsepower is equal to the products of amperes multiplied by volts and divided by the products of 746 times the efficiency of the generator. By formula,

GAS ENGINE MANUAL

$$HP = \frac{Current \times Volt}{746 \times Efficiency}$$

For dependable test results the two machines should be of comparative size, and the generator loaded accordingly by suitable power consuming devices.

Another means employed to measure the power of an engine is by means of a so called hydraulic brake. In this system the engine shaft is delivering power to a paddle wheel revolving in a fluid which is designed to offer enough resistance to absorb all the power furnished by the engine.

INDICATED HORSEPOWER

This method of measuring engine horsepower is used mainly for experimental and laboratory purposes, and is a measurement of the force delivered by the expanding gas to the piston inside the cylinder.

Briefly, indicated horsepower is based upon the pressure exerted on the piston during the power stroke obtained from indicator diagram, area of the piston head, length of stroke, and the number of power strokes in a given period.

To express the indicated horsepower of an engine the following fomula is used:

$$I.H.P. = \frac{PLAN}{33,000} \times K$$

where: P = mean effective pressure in lbs. per sq. in. acting on the piston (as shown by indicator diagram)
L = length of stroke in feet
A = area of piston in square inches
N = number of working strokes per min.
K = co-efficient equal to ½ times number of cylinders in gas engines and in double acting steam engines 2 times the number of revolutions.

SAE HORSEPOWER

The expression "SAE" horsepower formula is developed by the Society of Automotive Engineers, and is used in certain states to determine the tax for auto license plates.

The assumption upon which this formula for brake horsepower is based is that all gas engines will, or should deliver their rated horsepower at a

HORSEPOWER MEASUREMENT

piston speed of 1000 feet per minute, that the mean effective pressure in such engine cylinders will average 90 lbs. per sq. in. and that the mechanical efficiency will average 75 percent. The formula as presently employed is written:

$$BHP = \frac{D^2N}{2.5}$$

where: D = cylinder bore diameter in inches.
N = number of cylinders.

It should be noted that this formula while being of value in the early days of the automobile with its inefficient engine, is today very misleading, since modern engines develop about four times the horsepower computed by using this formula. A more suitable method of determining the auto license tax is by weight as has been adopted by most states, instead of so called horsepower. See Fig. 5.

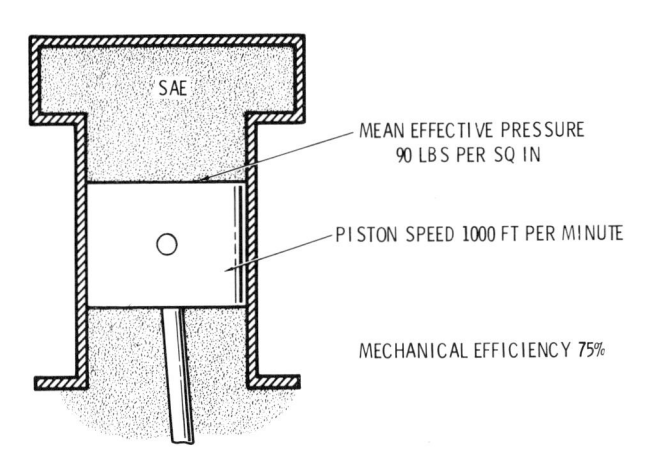

Fig. 5. Factors involved in SAE horsepower rating formula.

EFFICIENCY

By the term "efficiency" as applied to an engine, or any other power producing machine, is generally meant the relationship between the output and input, measured in the same units. It is commonly written

$$\text{Efficiency} = \frac{\text{Power output}}{\text{Power input}}$$

GAS ENGINE MANUAL

Since the power output of any power producing machine is always smaller that its input, it follows that the quotient in the foregoing equation will always be less than one. Efficiencies of machinery are commonly given as a percentage factor of one, such as 28 percent, 52 percent, 92 percent etc., and while some power conversion methods, result in comparatively high efficiencies, others, particularly all forms of thermal power conversion machinery, have a low efficiency.

THERMAL EFFICIENCY

The overall thermal efficiency of a gas engine is the relationship between the fuel input and the power output. This relationship is commonly expressed in "heat units" or "British thermal units" (abbreviated Btu.). It has been found experimentally, that one Btu. equals 778 foot pounds of work, it is therefore a simple matter to convert the horsepower output of an engine into Btu.

The thermal efficiency of an engine is written:

$$\text{Thermal efficiency} = \frac{\text{output in Btu.}}{\text{input in Btu.}}$$

To compute the thermal efficiency of engines it is necessary that identical units be used throughout.

Example—An engine delivers 50 brake horsepower for a period of one hour, and in that time consumes 30 lbs. (approximately 5 gallons) of gasoline. Assuming that the gasoline has a heat value of 18,500 Btu. per pound. Calculate the thermal efficiency of the engine.

Solution—From the foregoing data, the energy developed by the engine at 50 horsepower for one hour equals 50 horsepower-hours.

Btu per hp hour = 33,000 × 60 ÷ 778 = 2,545 approximately
Btu output per hour = 50 × 2,545 = 127,250
Btu input per hour = 30 × 18,500 = 555,000

$$\text{Thermal efficiency} = \frac{\text{Power output}}{\text{Power input}} = \frac{127,250}{555,000}$$
$$= 0.23 \text{ nearly}$$
$$\text{or } 23\%. \text{ Ans.}$$

HORSEPOWER MEASUREMENT

MECHANICAL EFFICIENCY

The mechanical efficiency of an engine is the ratio of brake horsepower to indicated horsepower. It is written:

$$\text{Mechanical Efficiency} = \frac{\text{Brake horsepower}}{\text{Indicated horsepower}}$$

The factor which has the greatest effect upon mechanical efficiency is the friction between moving engine parts. Since the friction between the moving parts of an engine remains practically constant throughout its speed range, it follows that the mechanical efficiency will be highest when the engine is running at a speed which develops the maximum horsepower.

VOLUMETRIC EFFICIENCY

The quantity of charge drawn into the cylinder of an internal combustion engine is always less than the theoretical quantity of charge which would fill the working volume of the cylinder at atmospheric pressure and temperature. The ratio of the actual to the theoretical quantity is termed volumetric efficiency.

PISTON DISPLACEMENT

Piston displacement generally means the number of cubic inches of cylinder space displaced by the pistons during a single stroke. Since the length of the stroke, and the cylinder bore diameter are both measurable, it is a comparatively simple matter to obtain this piston displacement.

By means of a simple formula we obtain

$$\text{Piston displacement} = \frac{\pi D^2}{4} \times L \times N$$

where: D = cylinder bore diameter in inches
L = length of stroke in inches
N = number of cylinders

Example—An eight cylinder, 250 Horsepower automotive engine, according to specifications has a 4.125 × 3.4 inch bore and stroke. What is the piston displacement?

Solution—Substituting values in our previously developed formula, we may write:

Piston displacement $= \dfrac{\pi(4.125)^2}{4} \times 3.4 \times 8 = 363.5$ cu. in. Ans.

Example—A four cylinder gas engine with a 104 *cubic inch capacity has a stroke length of* 3.13 *inches. Calculate the cylinder bore diameter.*

Solution—Using the previously derived formula, we have:

Piston displacement $= 0.7854$ $D^2 \times L \times N = 104$

or $104 = 0.7854$ $D^2 \times 3.13 \times 4$

That is, $D = \sqrt{\dfrac{104}{0.7854 \times 3.13 \times 4}} = \sqrt{10.58} = 3.25$ in. Ans.

The foregoing 104 cubic inch engine accordingly has a 3.25 × 3.13 inch bore and stroke respectively.

SUPERCHARGERS

The principal purpose of a supercharger is to put additional pressure on the air fuel mixture so that the engine will be supplied with a greater weight of fuel mixture than would be normally induced at the prevailing pressure.

Superchargers or distributor impellers, as they are sometimes called, are simple air blowers whose function it is to force air or gas into the engine cylinders. This additional pressure increases the volumetric efficiency and power output of the engine. They have been used extensively on racing engines, radial type aircraft engines and to a limited extent on passenger cars and trucks.

The supercharger (when used) is located in the engine manifold between the carburetor and the engine. It is driven at high speed (usually about 6 to 1 relative to the crankshaft speed) by a gear train driven by the crankshaft gear.

The elements of a supercharger is shown in Fig. 6. *In operation,* the mixture is drawn in at **A**, by the rotor of the blower and builds up a pressure above that of the atmosphere at **B**. Accordingly, when the inlet valve opens on the admission stroke, the mixture is forced past the valve so that even during the pre-admission period before the piston has quite reached the top dead center, the mixture is flowing into the combustion chamber and cylinder, the motive force to move the mixture being much greater than that due to inductive exhaust during the post-exhaust period of exhaust valves timed for late closing.

The result is that as the piston traverses its admission stroke the incoming mixture is able to follow it at a pressure greater than atmos-

HORSEPOWER MEASUREMENT

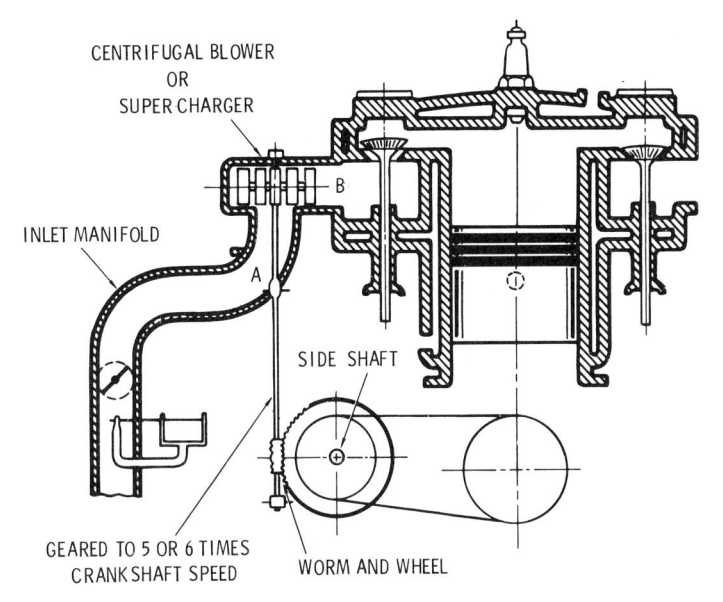

Fig. 6. Schematic illustration of centrifugal type supercharger.

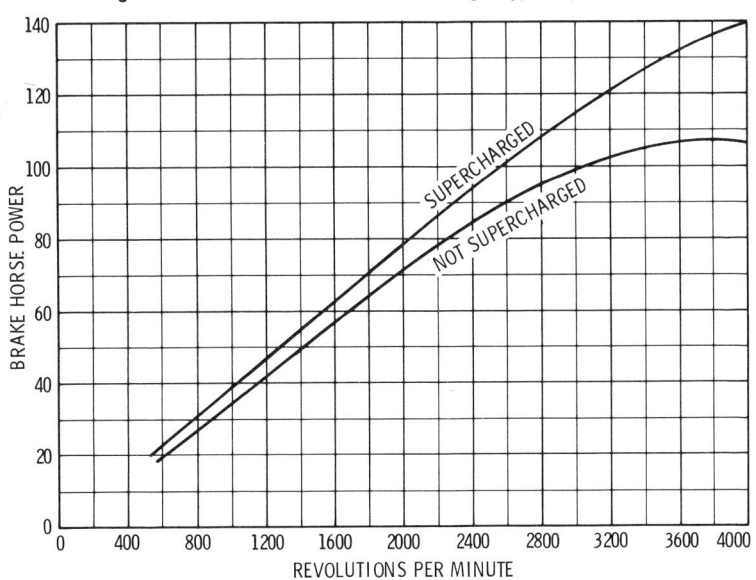

Fig. 7. Typical engine performance characteristics for eight cylinder (300 cu. ins. displacement) engine with and without super-charger. Note increase in brake horsepower in the higher speed range when supercharger is used.

pheric at highest engine speed. It must be evident therefore, that with a supercharge, that is, one whose pressure is greater than atmospheric, that the charge on the next stroke is compressed to a higher pressure than for ordinary operation without a supercharger.

It should be noted that the apparatus described adds considerably to the weight and cost of the engine. Also, the additional power obtained by installation of the supercharger will be reduced somewhat due to the extra engine power required to drive the air blower. See Chart in Fig. 7.

CHAPTER 18

Fundamental Electricity

A knowledge of electricity is of great importance to servicemen because it enables them to grasp more clearly the operating principles of the various units comprising the electric starting, lighting and ignition system associated with internal combustion engines.

Many different theories have been advanced as to the true nature of electricity. Presently however, the *electron theory* is accepted amongst scientists as the only means by which the nature of electricity can be satisfactorily explained.

In general, the electron theory explains how electrical effects are caused by the movement of extremely small particles of electricity known as *electrons*. Thus, it is actually the movement of electrons through a wire or conductor which constitute the current flow.

A mechanic is more concerned about the applications and uses of electricity than he is about the theoretical principles involved. A general understanding of the behavior of electric currents, however, is considered necessary for satisfactory work and troubleshooting or repairs of electric units on internal combustion engines.

OHM'S LAW

Briefly stated, Ohm's law expresses the fixed relationship between the current, pressure and resistance which always exists in an electric circuit. It states that the electrical current in *amperes* passing through a conductor, equals the pressure in *volts* divided by the resistance in *ohms*. It is written:

$$\text{Amperes} = \frac{\text{Volts}}{\text{Ohms}}$$

which is usually written:

$$I = \frac{E}{R}$$

or $\quad R = \dfrac{E}{I}$

Similarly $\quad E = I \times R$

The foregoing formulas are of the utmost importance to any electrical system, since they contain the means whereby *current, voltage* and *resistance* in a circuit may be determined. Ohms law also shows that for a given voltage, the lower the resistance the larger will be the current, and the higher the resistance the smaller the current.

Ques. What is a volt?

Ans. The unit of electric pressure called electromotive force. It is *that electric pressure which produces a current flow of* **one ampere** *through a resistance of* **one ohm.** See Fig. 1.

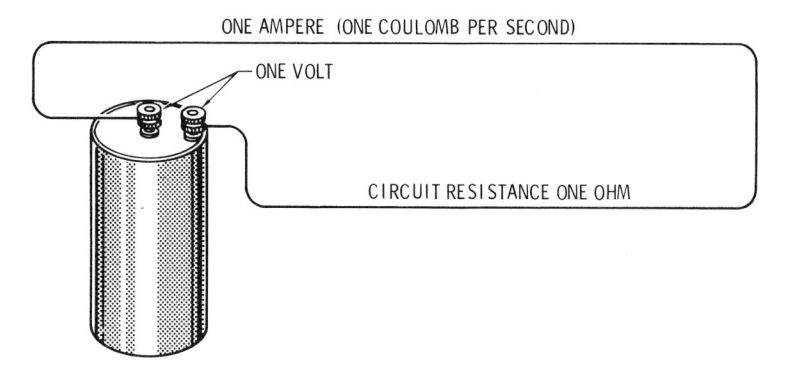

Fig. 1. Simple circuit showing relation between volts, amperes, coulombs and ohms.

Ques. What is an ampere?

Ans. The unit of electric current. It is *that current caused to flow through a resistance of one ohm by a pressure of one volt.*

Ques. What is an ohm?

Ans. The resistance offered to an unvarying current by a column of mercury at 32° Fahr. 14.4521 grams in mass of a constant cross sectional area, of one square millimeter and 106.3 centimeters long. See Fig. 2.

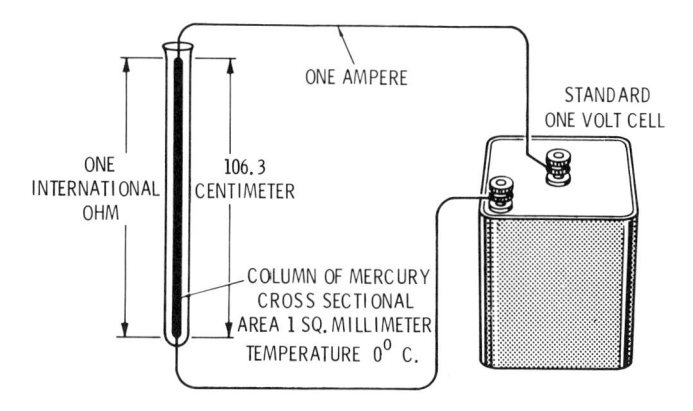

Fig. 2. The International ohm.

Ques. What is resistance?
Ans. That property of a substance that opposes the flow of an electric current. Unit, the *ohm*.

Ques. What is an insulator?
Ans. A substance which offers tremendous resistance to the passage of an electric current.

Ques. Name two kinds of wire conductors.
Ans. Copper wire (low resistance); resistance wire (high resistance.)

KINDS OF CURRENT

The electric current is said to be:
1. *Direct* when it is of unvarying direction.
2. *Alternating* when it flows rapidly to and fro in opposite directions. See Fig. 3.
3. *Primary* when it comes directly from the source.
4. *Secondary* when the voltage and amperage of a primary current have been changed by an *induction coil*.

Ques. What names are given to low and high voltage currents?
Ans. *Low tension* and *high tension* currents respectively.

Ques. What is an insulated circuit and why?
Ans. One in which the wires are covered with insulating material to prevent leakage.

Ques. What is a short circuit?
Ans. One in which the current *leaks* and so returns to the source without doing its work.

GAS ENGINE MANUAL

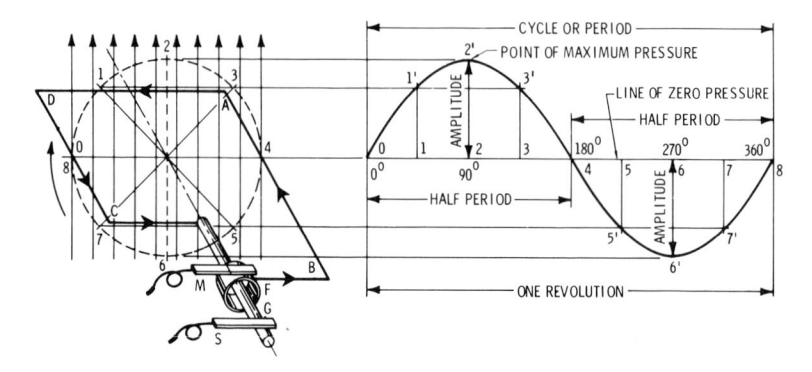

Fig. 3. Alternating current represented by the *sine curve. In principle,* as the elementary alternator rotates, the induced electric pressure (voltage) will vary in such a manner that *its intensity at any point of the rotation is proportional to the sine of the angle corresponding to that point.* The curve lies above the horizontal axis during the first half of the revolution and below it during the second half, *which indicates* that the current flows in one direction for a half revolution, and in the opposite direction during the remainder of the revolution.

Ques. Why?
Ans. Because the current will always take the path of least resistance.

MAGNETISM

By definition: *The property some bodies have to attract iron and steel, and those bodies having the property are called magnets.* Magnets have two opposite kinds of magnetism or *magnetic poles*. One of these poles tends to move toward the North and the other toward the South. They are accordingly called the *North* and *South* poles. See Fig. 4.

Ques. What are the two laws relating to the poles?
Ans. *Unlike* poles *attract* each other; *like* poles *repel* each other.

Ques. What is a permanent magnet?
Ans. One made of hard steel which holds its magnetism almost indefinitely.

Ques. What is an electromagnet?
Ans. One made up of an iron core over which is wound a number of turns of insulated wire. See Fig. 5.

Ques. What happens when an electric current flows through the winding?
Ans. The core becomes magnetized and strong *poles* are produced.

Ques. What is a solenoid?

Fundamental Electricity

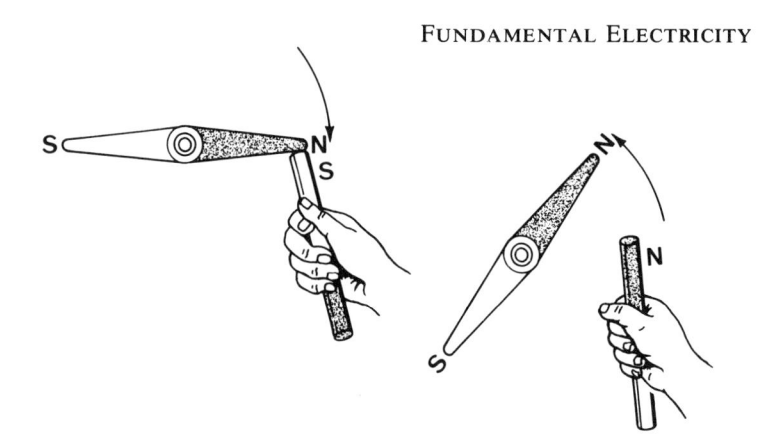

Fig. 4. Effect of unlike and like poles. Unlike poles attract each other; like poles repel each other.

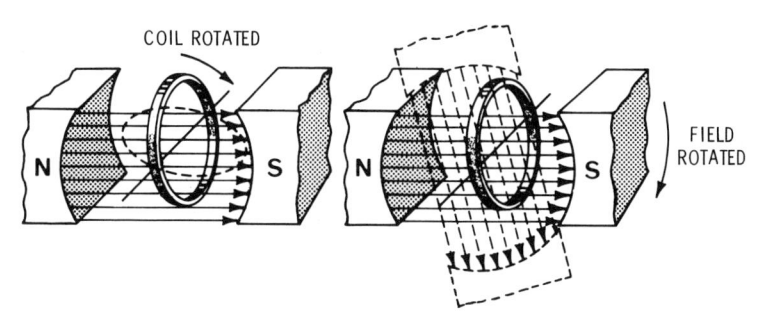

Fig. 5. Examples of electromagnetic induction. *Either the coil or the field may be rotated to cut lines of force.*

Ans. A spiral conductor of numerous turns with or without an iron core which forms a magnet when current passes through the coil.

The core greatly increases the strength of the solenoid.

Ques. What is a magnetic field?

Ans. The region surrounding a magnet in which magnetic force acts.

ELECTROMAGNETIC INDUCTION

By definition: *The tendency of electric currents to flow in a conductor when it is moved in a magnetic field so as to **cut** lines of magnetic force.* All generators of whatever form are based on this discovery made by Faraday.

Ques. What do you understand by the term *"cut lines of force"*?

Ans. *A conductor forming part of an electric circuit **cuts** lines of force when it is moved across a magnetic field in such manner as to **alter** the number of magnetic lines of force embraced by the circuit.* The current is called the *induced current* and that part of the wire moved in the magnetic field the *inductor;* ignorantly called conductor.

CELLS

By definition, a cell is *a device for producing electricity by placing two dissimilar metal plates in a material called the electrolyte.* The two dissimilar plates are called the *elements*.

Ques. What is a battery?

Ans. Two or more cells joined together so as to form a single unit.

Ques. What is the difference between *primary* **cells and** *secondary* **cells?**

Ans. Cells are said to be *primary* when they produce a current of themselves; *secondary* when they first require to be charged from an external source, storing up a current supply which is afterwards yielded in the reverse direction to that of the charging current. Accordingly an assembly of secondary cells joined together is known as a storage battery.

CELL CIRCUITS

There are three methods of connecting cells; 1, In *series* (Fig. 6), 2, in *parallel (Fig. 7.); and 3, in series-parallel,* as shown in Fig. 8.

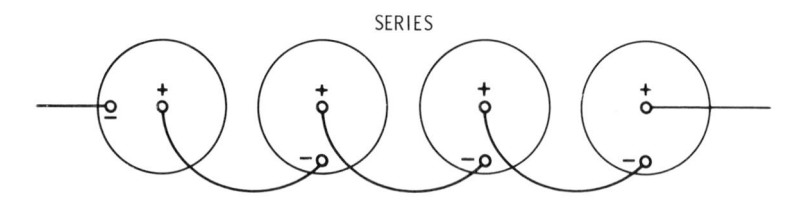

Fig. 6. Series battery connection. Voltage of battery equals product of the voltage of a single cell and multiplied by the number of cells.

FUNDAMENTAL ELECTRICITY

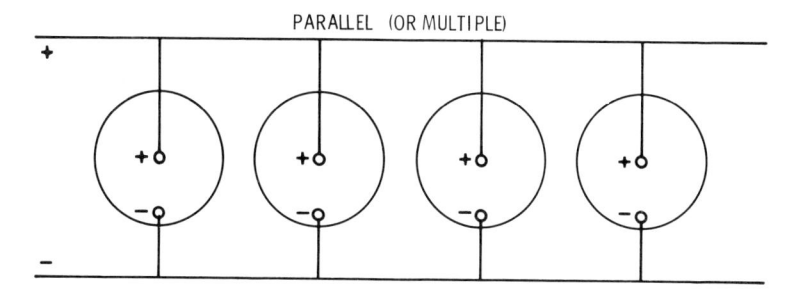

PARALLEL (OR MULTIPLE)

Fig. 7. Parallel battery connection. Voltage of battery equals voltage of a single cell, but the current is equal to the amperage of a single cell multiplied by the number of cells.

SERIES-PARALLEL

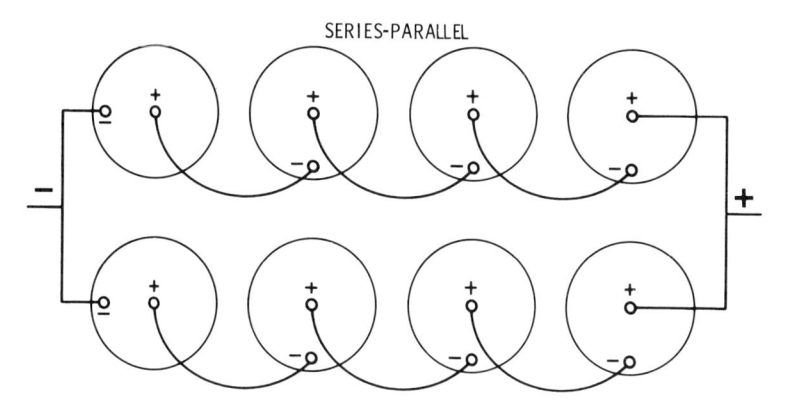

Fig. 8. Series-parallel battery connection.

PRIMARY INDUCTION COILS

This type of coil consists of *a long iron core wound with a considerable length of a **low** resistance insulated copper wire*. Its operation is due to *self induction*. Fig. 9.

Ques. What is self induction?

Ans. The property of an electric current by virtue of which *it tends to resist any change in its rate of flow*. Self induction becomes especially marked when the current passes through a primary coil.

Ques. How is the self induction brought into action?

Ans. By making and breaking the circuit connected to the source.

225

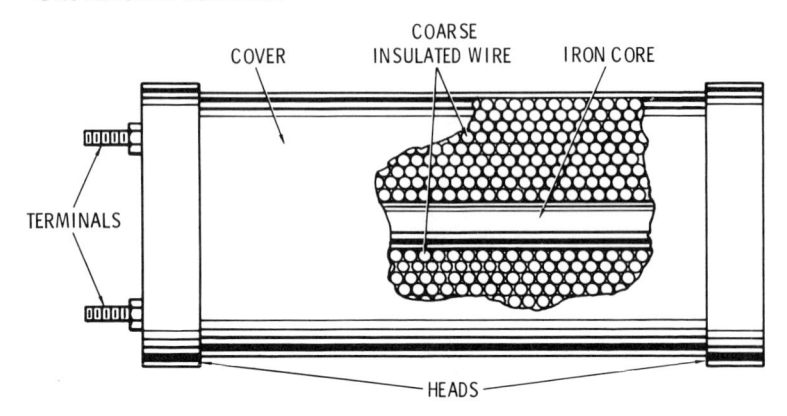

Fig. 9. Primary induction coil as used for low tension ignition. Coils of this type are made in a great variety of form and size. Ordinarily the winding consists of about six layers of No. 14 copper wire. The winding is usually covered and the ends capped with ebonite so that the core and wires are not exposed.

Ques. What is the application of the coil and how does it work?

Ans. It is used in *low tension* or *make and break ignition*.

In an ignition hook up, *the spark occurs at the instant of breaking the circuit, not at the instant of making,* because when the current is flowing it cannot be stopped instantly on account of self induction, that is, it acts as though it possessed weight.

SECONDARY INDUCTION COILS

This type coil consists of *a long iron wire core upon which is wound a primary and a secondary winding.* Its operation is due to *mutual induction.* Fig. 10.

Ques. What is mutual induction?

Ans. That particular case of electromagnetic induction in which *the magnetic field producing an electric pressure in a circuit is due to the current in a neighboring circuit.*

The circuit to which the current is applied is called the *primary circuit* and the circuit in which a current is induced, the *secondary circuit*. In the actual coil, the primary and secondary circuits are made up of *heavy* and *fine* insulated wire respectively.

Ques. What is the property of a secondary coil that makes it of great value for most purposes?

Ans. That property of mutual induction by which *the voltage of the induced current may be increased or diminished to any extent depending*

FUNDAMENTAL ELECTRICITY

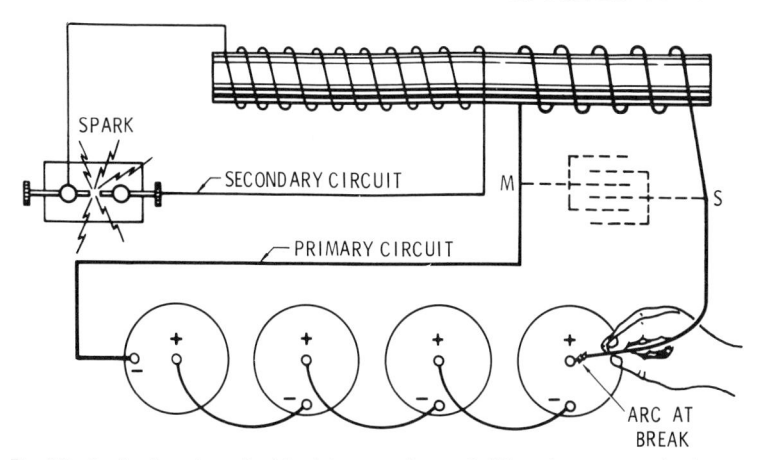

Fig. 10. Production of spark with plain secondary coil. When the contact is broken at R, there will be a pronounced spark at R and a weak spark across the gap in the secondary circuit. To avoid a spark at R and obtain a good spark at the gap, place a condenser across the primary circuit as at M and S, shown by dotted lines.

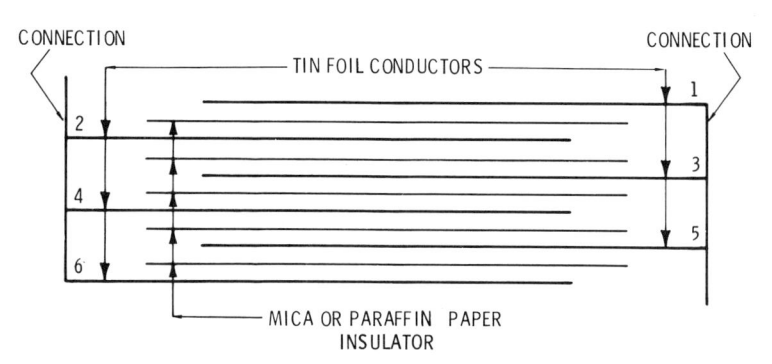

Fig. 11. Construction of one type condenser for an induction coil. Numbering the successive sheets of tin foil serially, *connect all even sheets together and all odd sheets together*, these connections forming the terminals of the condenser.

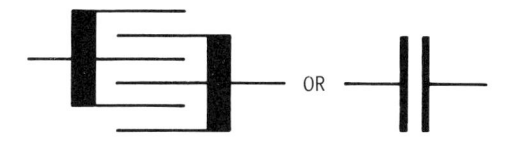

Fig. 12. Conventional symbols for a condenser.

upon the relation between the number of turns in the primary and secondary winding.

Ques. What is the rule?

Ans. *The voltage of the induced current is (approximately) to the voltage of the primary current as the number of turns of the secondary winding is to the number of turns of the primary winding.*

This makes it possible to get the *enormous* voltage of the induced current required for high tension or jump spark ignition.

Fig. 13. Condenser for induction coil.

Ques. What is necessary besides the secondary coil in a jump spark ignition system to make it work and why?

Ans. A condenser is used to absorb the induced current of the primary winding and thus prevent it opposing the rapid fall of the primary current.

CHAPTER 19

Ignition Systems

The function of an ignition system is to produce the sparks that ignite the combustible mixture and to do this so that each engine cylinder will be "fired" in correct firing sequence. There are two general types of ignition systems; *battery system* and a *magneto system*.

Battery ignition is used exclusively for motor-vehicle engines; magneto ignition is generally preferred for industrial and small gasoline engines (as for lawnmowers, etc.). Each type of system produces a very high-voltage (15,000 to 20,000 or more volts) current capable of jumping the gap between the two electrodes of a spark plug to create a "hot" (large) spark and efficiently ignite the combustible mixture. In a battery system, this high-voltage current is created by "stepping-up" (transforming) the low battery voltage. A magneto is designed to produce the required high-voltage current by simultaneously generating and stepping-up a lower-voltage current.

Basically, there are two kinds of battery-ignition systems which we shall refer to as; an *electro-mechanical system* and an *electronic-ignition system*. These will first be discussed separately, following which is a discussion of magneto systems.

ELECTRO-MECHANICAL BATTERY IGNITION

This type of system contains two separate electrical circuits known as the *primary circuit* and the *secondary circuit*. The primary is the low-voltage circuit with the battery as its source; the secondary is the high-voltage circuit with an *ignition coil* (a type of transformer) as its source.

Either a 6-volt battery or a 12-volt battery may be used. Most modern automotive engines use a 12-volt battery because 12 volts produce twice the power for engine starting, vehicle lighting, etc. that is produced by 6 volts *with the same amperes of current flow.*

The component parts of a 6-volt primary (low-voltage) circuit (Fig. 1) are the battery, an ammeter (when used), the ignition switch, the primary winding of the ignition coil, and the cam and breaker points of the distributor. A 12-volt primary circuit is the same except that a coil resistor is added into the circuit between the coil and the breaker points.

In either case, the secondary (high-voltage) circuit is composed of the secondary winding of the ignition coil, the cap and rotor of the distributor, the spark-plug leads, and the spark plugs.

In any battery-ignition circuit the battery serves as the source of electrical energy only until the engine is running; afterwards, the generator (or alternator) takes over.

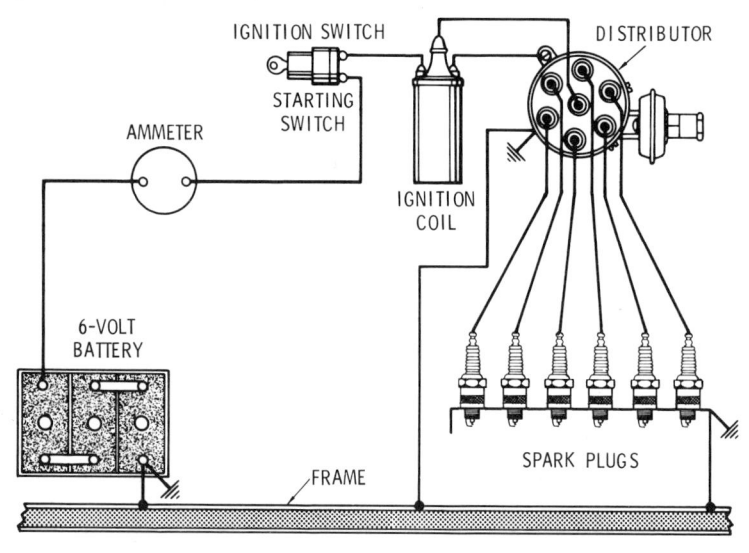

Fig. 1. Illustrating wiring diagram of a typical six-cylinder battery ignition system.

Ammeter

An ammeter, although not essential to the operation of the primary circuit, is sometimes used to indicate the condition of the battery and proper functioning of the battery-charging circuit (between the battery and the generator or alternator). In most automotive applications a dashboard light, which flashes on at an appropriate time, replaces the ammeter to indicate that the charging circuit is operative.

Ignition Systems

Ignition Switch

Fundamentally, the ignition switch is simply a manual circuit breaker used to open or close the primary circuit. In most automotive applications, however, it is a much more complicated device that serves, in one position, to energize the engine-starting and ignition-primary circuits-then, in a second position (to which it is returned by a spring), to energize the ignition-primary circuit and all other vehicle circuits except the starting circuit. There also may be a third position (manually selected) in which this switch energizes only selected vehicle circuits (such as for the lights radio, etc.).

NOTE: Because an engine starter draws a considerable amount of current (amperes) which tend to burn-up ordinary switch contacts, the starter circuit generally is not energized through the ignition switch. Instead, this switch when in starting position, energizes a *starter solenoid* (switch) which, when energized, closes heavy-duty (high-amperage) contacts through which the starter current is passed. See Fig. 2.

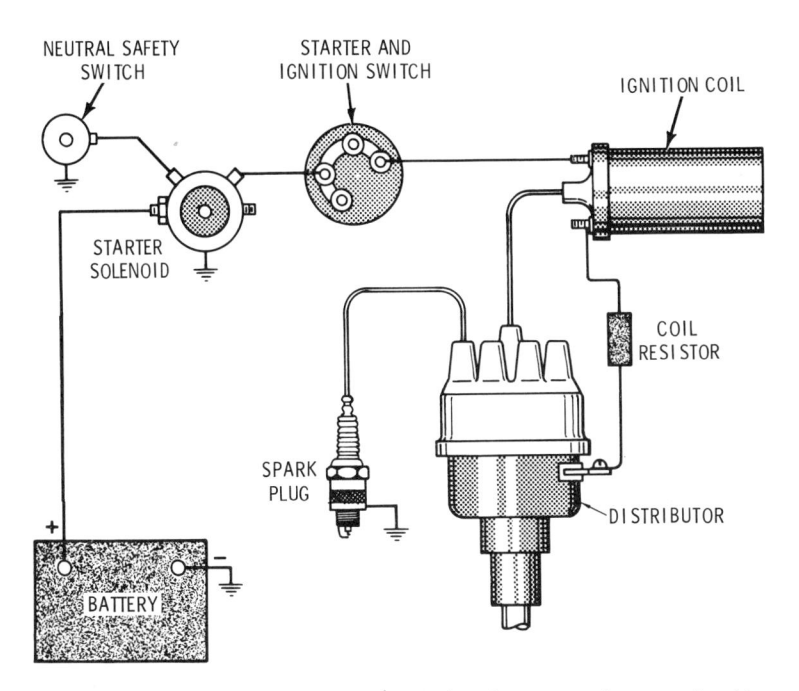

Fig. 2. Typical wiring for a 12-volt system with coil resistor and starter solenoid.

Ignition Coil

The primary winding of an ignition coil is made up of a few hundred turns of relatively large diameter wire (to carry the low-voltage, high-amperage current of the battery — or generator or alternator — circuit). The secondary winding consists of several thousand turns of relatively small-diameter wire (capable of carrying the high-voltage but low amperage secondary-circuit current). The primary winding is on the outside, the secondary winding is inside, and at the center there is a coil core made up of thin layers of soft iron laminations (which serve to increase the coil efficiency). This assembly is surrounded by a layer of insulating compound then a shell of soft iron (which further increases the coil efficiency by completing the magnetic circuit). Around this entire assembly there is a protective housing filled with insulating compound.

When the primary circuit is closed, the flow of current through the coil's primary winding builds up a magnetic field; when the primary circuit is opened, the cessation of current flow causes this magnetic field to suddenly collapse. Hence, as the distributor breaker points continuously close and open the primary circuit, the magnetic field continuously builds-up and collapses. This magnetic activity induces a corresponding current in the coil's secondary circuit — but, because of the difference between the numbers of turns of wire in the primary and secondary, the current in the secondary is of much greater voltage (more force with which to "jump" the spark plug's gaps).

Ignition Resistors

Ignition coils generally are designed to operate on a 6-volt primary circuit. Therefore, in a 12-volt engine system it is necessary to reduce the primary voltage from 12 to 6 volts. This is done by using a resistor (which "consumes" the excess voltage). In addition to protecting the coil from too-high a voltage, the resistor also protects the distributor breaker points (which, also, are intended only for 6 volts).

A resistor is simply a predetermined amount of resistance (to current flow) which, by resisting the current flow, lowers its force (voltage) by converting part of the force into heat. Some materials conduct electricity with very little resistance (heat loss); others will conduct electricity, but do so poorly so that the heat loss (and consequent reduction of voltage) is considerable. There are two types of "poor" conductors used to reduce the voltage from 12 to 6 volts.

1. *Block-type ballast resistor*. This is a compact unit consisting of a high-resistance material in a housing, which is connected in

IGNITION SYSTEMS

series with the primary circuit. The material used has a much higher resistance after it is heated (by the passage of current through it, which creates the heat) than when it is cold. Therefore, in the beginning (when it is cold) it allows almost the full 12-volts of the system to energize the coil's primary winding to assist starting — but, as current flows and it heats up, it reduces primary-circuit current to the desired operating force of 6-volts.

2. *Wire-type resistor.* Resistance is provided simply by an extra-long special-material wire (5 to 6 ft. in length) in the coil's primary circuit. As with the preceding, resistance (and voltage reduction) is minimal in the beginning (before the wire becomes heated), but increases to the desired amount after the engine is started and current flow heats the wire.

Spark Action

When the primary contacts (cam and breaker points) in the distributor are closed, current flows from the battery through the primary winding and primary contacts to ground. This flow of current produces a magnetic field, which collapses very rapidly when the circuit is opened. This collapsing of the field intersects the coil windings and induces the necessary high voltage in the secondary winding of the ignition coil. The high voltage surge is great enough to overcome the resistance of the spark plug gap, and to produce a spark which ignites the fuel mixture. The induced voltage ratio between the secondary and primary winding is commonly about 20,000/200 that is the voltage of the secondary is approximately 100 times that of the primary.

Distributor

The distributor is essentially a switching device consisting of a combination of switches working in unison with each other; one makes and breaks the primary circuit, and the other makes and breaks the secondary circuit while distributing secondary current to the spark plugs in correct firing order. There are two basic types of distributors: the *conventional* (electro-mechanical) type, and the more recently developed *electronic type* (referred to as an *electronic ignition system*).

The cam and breaker points, Fig. 3, serve to interrupt the primary circuit at certain definite intervals. The cam is located on the distributor shaft, which is driven indirectly by the engine crankshaft.

Ordinarily, the cam will have as many lobes as engine has cylinders and is rotated half as fast as the crankshaft. As the distributor shaft

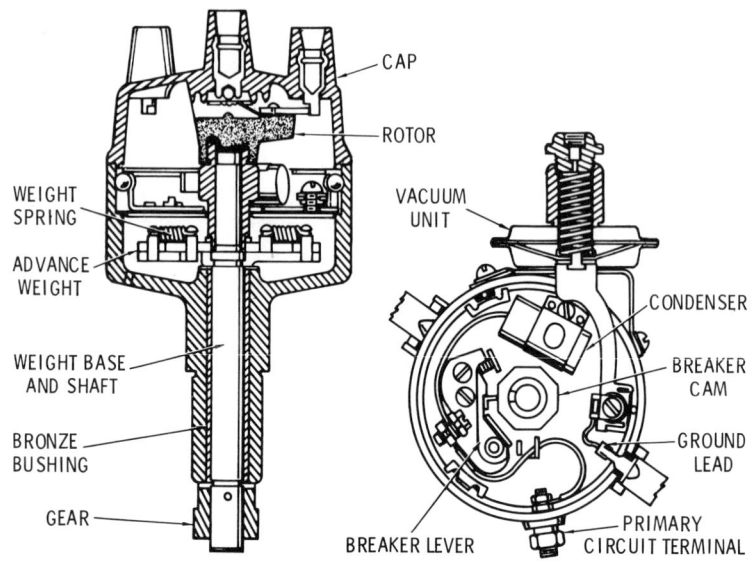

Fig. 3. Details of a typical *Delco-Remy* distributor.

rotates, the cam opens the breaker points and interrupts the primary circuit. Within the time necessary for one revolution of the cam, the four-stroke cycle has been completed within each cylinder and the primary circuit has been interrupted once for each cylinder. Thus, the cam and breaker points, located in the distributor is actually a precisely set timing device operating in exact synchronism with the engine.

The second function of the distributor is to conduct the high voltage surges from the secondary winding of the ignition coil to the proper spark plug at the correct time.

This is performed by the distributor cap and rotor. As will be noted in the circuit diagram, Fig. 1, the circuit continues through the cap to the rotor, which is a revolving arm. As it rotates, the outside edge of the arm lines up with the electrodes connected up through the cap to the terminals in which the spark plug leads are inserted. These terminals are located near the outside edge of the cap.

Since the rotor must line up with the lead to a spark plug each time the high voltage surge is produced, and since a surge occurs each time the cam interrupts the primary circuit, the cam and the rotor must be driven by the distributor shaft to synchronize their operation.

Timing is accomplished by adjusting the position of the distributor so that the breaker points open and the rotor is aligned with the spark plug lead at the proper moment for igniting the fuel mixture in the cylinder.

IGNITION SYSTEMS

The leads to the spark plugs must be well insulated to withstand the high voltage to which they are subjected. One end of the leads are inserted into the terminals of the distributor cap and the other end is connected to the spark plug.

Spark Plugs

The spark plug serves to produce the spark necessary to ignite the compressed charge of air and fuel in the engine cylinder. Ignition occurs at the gap of the plug across which high voltage from the ignition coil jumps in the form of a spark. The second electrode of the spark plug is attached to the steel shell part of the plug and is grounded to the frame through the engine block assembly. See Fig. 4.

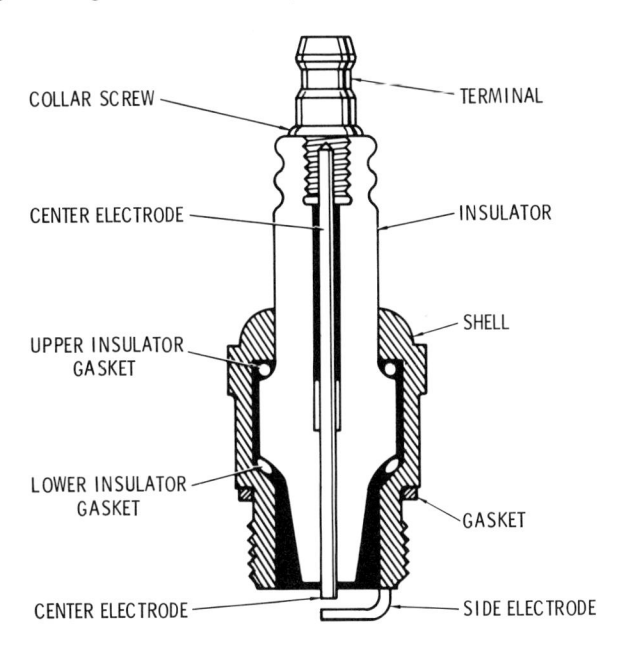

Fig. 4. Section of typical spark plug with names of parts.

Condensers

The condenser as employed in battery ignition systems has a double duty in that it (1) prevents arcing at the primary contacts and (2) speeds up the collapse of the magnetic field by reversing the primary voltage surge.

The collapse of the magnetic field which induces a high voltage in the secondary circuit simultaneously induces a fairly high voltage in the primary circuit. This voltage would produce a strong spark across the small gap established as the breaker points open to interrupt the circuit. Without the condenser action, the contacts would be severely burned, and most of the energy stored in the coil would be lost.

With the condenser connected across the primary contacts, an additional path for the current flow during the first instant the contacts begin to separate is established. The current thus flows into the condenser instead of arcing across the contact points. This action stops the flow of primary current, charges the condenser, and hastens the collapse of the magnetic field. Very soon after the high-tension spark appears at the spark-plug electrode, the current stored in the condenser discharges back through the primary circuit. This process is repeated for every power stroke of the engine.

The capacity of a condenser is measured in picofarads and is determined by the area of the aluminum or lead foil and the thickness of the insulating sheets. Thus, the thinner the insulation for a given area of foil the greater the capacity. Ignition condensers are usually manufactured in the range of from 0.15 to 0.30 uF. See Fig. 5.

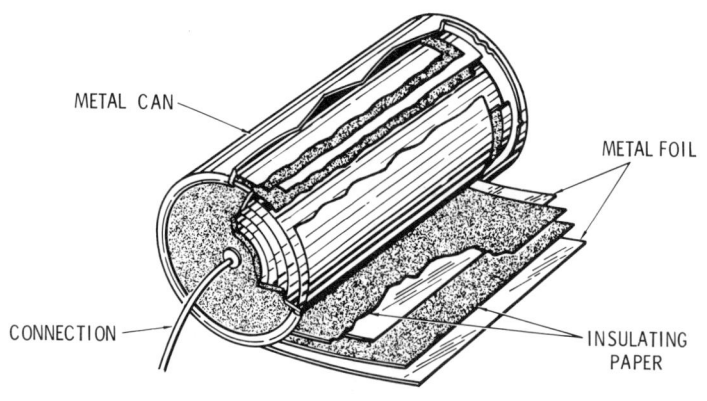

Fig. 5. Details of typical ignition condenser. One of the metal foil strips is grounded to the metal case, whereas the other is connected to the live breaker contacts, fig. 3.

Spark Control

In order to obtain efficient operation of an internal combustion engine, throughout the speed range under various operating conditions, it is essential that the spark occur at the correct instant. This exact instant will vary according to engine load and speed. Various mechanisms have been provided to automatically provide the advance or retard of the spark as

IGNITION SYSTEMS

load and speed conditions require. The two most common methods employed for spark control are:
1. By the centrifugal force method,
2. By the engine vacuum method.

On motor vehicles spark control may be obtained by either method, or by a combination of both.

Centrifugal Force Method

The centrifugal force method of spark control as shown in Figs. 6 and 7, consists essentially of a centrifugal governor having two weights that

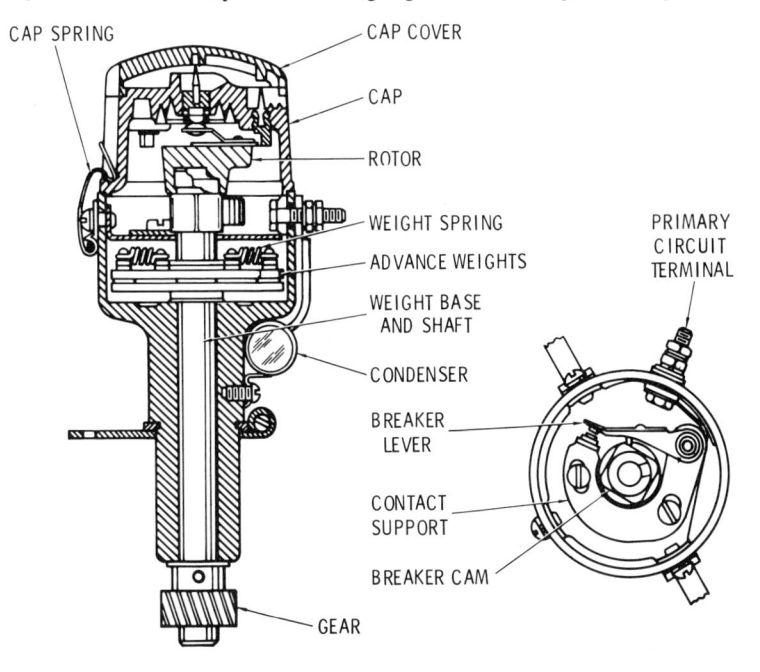

Fig. 6. Automatic spark advance type distributor having centrifugal weights built into the distributor mechanism, which automatically advances the spark after the engine has been started.

swing out against spring tension as the engine speed increases. The centrifugal governor is mounted on the distributor shaft, beneath the breaker plate in the distributor and is linked through the shaft to the cam which is mounted above the breaker plate and which operates the breaker points.

237

Gas Engine Manual

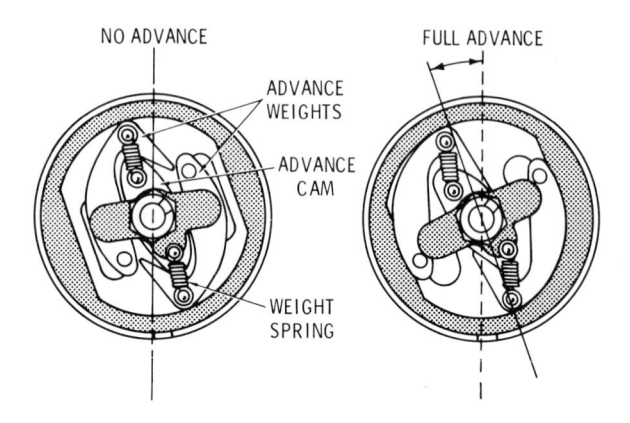

Fig. 7. Illustrating typical spark advance mechanism. As engine speed increases the weights move out to the dotted line position and advance the breaker arm.

As the engine speed increases, the advance weights of the governor fly outward due to the centrifugal force, against action of the weight springs, and cause the breaker arm to shift forward in relation to the distributor shaft. The shifting of the cam will make the breaker points open earlier and thus advance the spark automatically to the correct position in relation to the engine speeds.

As the speed decreases, the weights are gradually returning to their slow speed position by the springs. This shifts the cam in opposite direction, thus making the points open later and retarding the spark. The centrifugal governor reacts only to engine speed.

When the throttle is only partly open, the engine cylinders take in only part of the full charge during each intake stroke, resulting in low compression pressures and slow burning of the charge. For improved performance and full economy during such operation, intake manifold vacuum is used to obtain greater spark advance than is possible with the centrifugal governor alone.

Engine Vacuum Method

In the engine vacuum method of spark control, also termed, "the vacuum advance spark control," a vacuum diaphragm is usually mounted on the distributor housing and linked to the breaker plate, Fig. 8. A vacuum chamber next to the diaphragm is connected to the intake manifold so that changes in the manifold vacuum will operate the diaphragm.

Ignition Systems

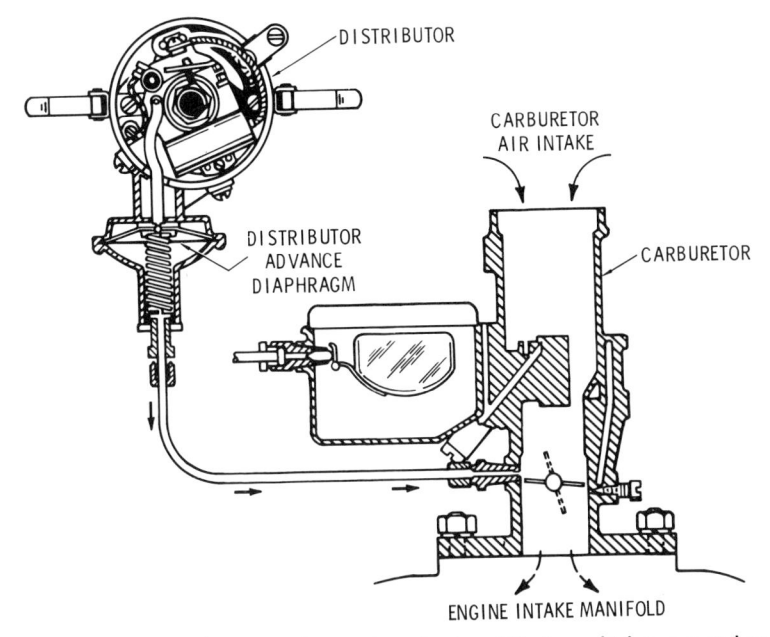

Fig. 8. Vacuum advance spark control mechanism of the type which is mounted on the side of the distributor. Breaker plate is supported on a ball bearing, and the breaker plate alone rotates in the housing as vacuum conditions change.

When the engine is not running, the breaker plate is in the retard position with the diaphragm held in the normal position by the diaphragm spring. A vacuum created in the intake manifold actuates the diaphragm against the spring, shifting the breaker plate to an advanced position, thereby giving more spark advance than that obtained with the centrifugal governor alone. On quick acceleration or open throttle condition, the vacuum in the manifold decreases, allowing the compressed diaphragm spring to return the breaker plate to the retard position.

In another method of vacuum advance, intake manifold vacuum is utilized to actuate a piston brake against the advance plate of the governor. In this method, a high vacuum draws the piston brake away from the advance plate, compressing the spring and permitting the governor to control the spark advance.

If the engine is suddenly accelerated or operated under heavy load, the intake manifold vacuum decreases, allowing the compressed spring to actuate the brake against the advance plate. The friction imposed by the brake upon the advance plate exerts a force counter to the centrifugal

Gas Engine Manual

force of the weights so that the weights move inward, retarding the cam. As soon as the intake manifold vacuum increases, the piston brake is pulled back, permitting the weights to fly outward and advance the timing.

ELECTRONIC BATTERY IGNITION

The essential difference between this system and an electromechanical system is the elimination of the (mechanically operated) breaker points and cam. Breaker point pitting and cam wear are dispensed with, thus making it unnecessary to service the distributor for faulty timing and dwell of the ignition. In some systems a conventional type ignition coil (discussed preceding) is used; in other systems a different type, but separate, ignition coil is used; and in still other systems, the ignition coil is an integral part of the distributor assembly.

In addition, there are several types of electronic ignition systems, some requiring a (separate) control unit, and others that require a (separate) amplifier unit. These systems may use 6-volt batteries though, generally, 12-volt batteries are used. Each system also incorporates an ignition switch, either an ammeter or a dashboard light, a starter and a starter solenoid, as previously discussed.

A conventional ignition coil is used, and the distributor is conventional except that a stationary *sensor* and rotating *trigger wheel* replace the cam, breaker points and condensor. See Fig. 9. There is a solid state, modular, electronic *control unit* connected into the circuits between the distributor and coil primary, and the coil primary and the battery (through the ignition switch). This unit contains a current regulator and a power transistor which control battery current in the coil's primary winding. It also contains an oscillator that furnishes current to excite the sensor which, in turn, develops an electromagnetic field through which the trigger wheel rotates. Each time one of the trigger wheel teeth enters this field the oscillator reacts upon the transistor causing it to momentarily cut off current flow through the ignition coil primary, thus creating the high-voltage secondary-circuit current as previously described.

In similar systems, the sensor is called a *pick-up coil assembly* and the trigger wheel is called a *reluctor*. Also, in place of a current regulator in the control unit a (separate) *ballast resistor* may be connected so as to control voltage to the coils primary after the ignition switch is turned (from start) to the on position.

IGNITION SYSTEMS

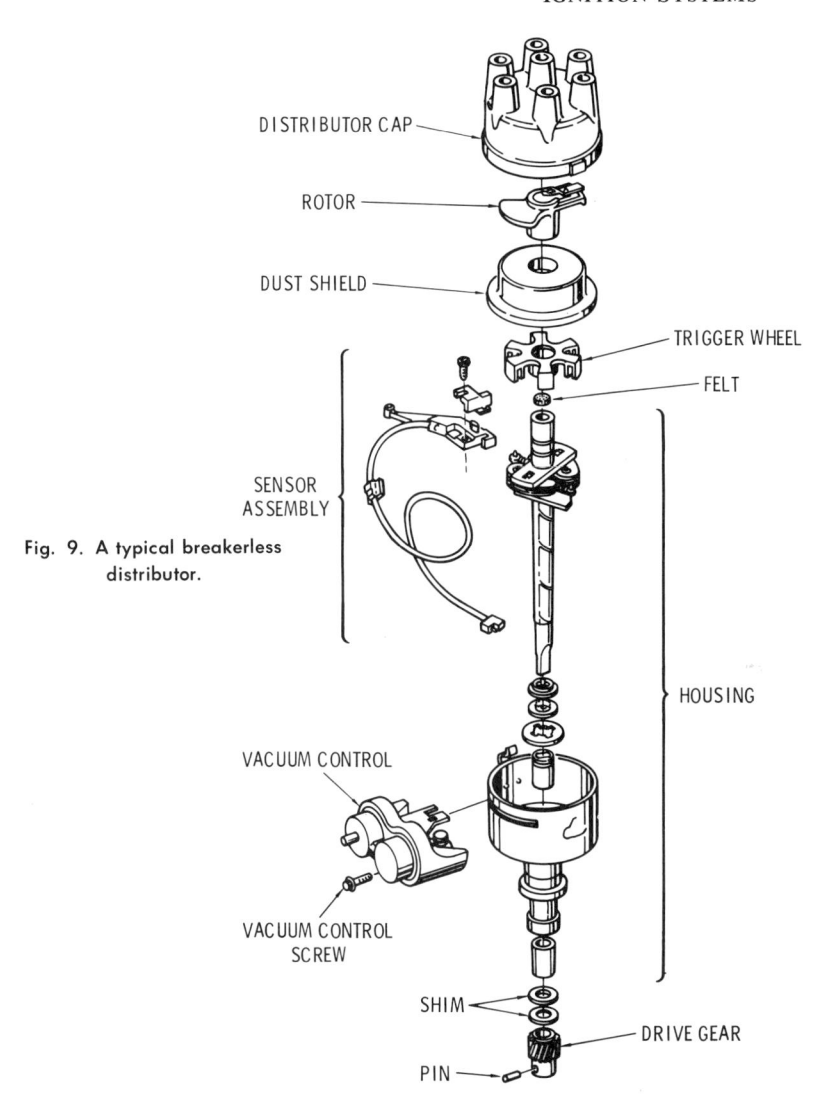

Fig. 9. A typical breakerless distributor.

- DISTRIBUTOR CAP
- ROTOR
- DUST SHIELD
- TRIGGER WHEEL
- FELT
- SENSOR ASSEMBLY
- HOUSING
- VACUUM CONTROL
- VACUUM CONTROL SCREW
- SHIM
- DRIVE GEAR
- PIN

Capacitor-Discharge (CD) System

This system has a *magnetic-impulse distributor*, an *amplifier unit,* a *high-capacity condenser* and a standard ignition coil. The secondary (high-voltage) circuit is conventional (distributor has a cap and rotor), but the primary circuit is electronically timed. This is accomplished by a

241

timer core and a *magnetic pick-up assembly* in the distributor. See Fig. 10.

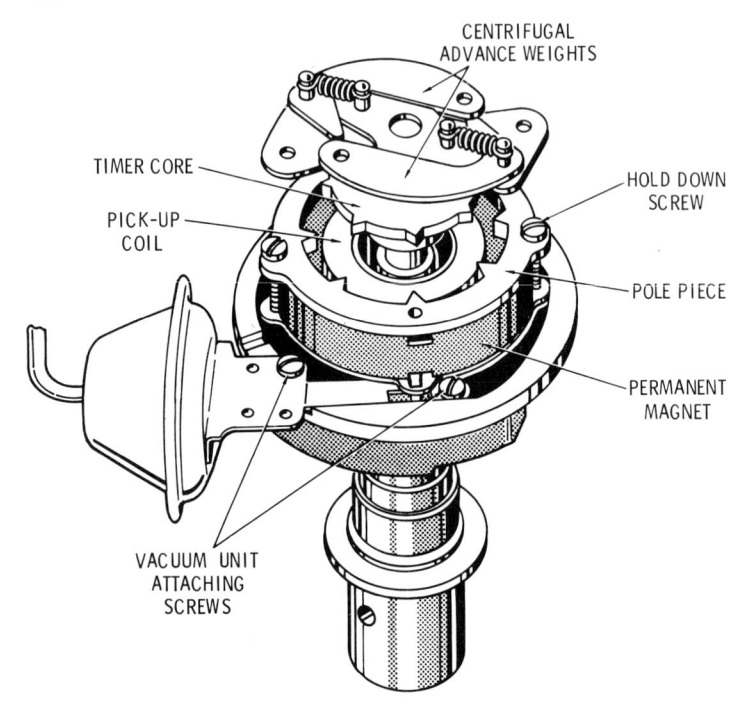

Fig. 10. Primary-circuit components of a magnetic-impulse distributor.

Battery voltage is applied to the high-capacity condenser (rather than to the ignition-coil primary), which builds up a high (300 volt) charge in the intervals between discharge. The iron timer core of the distributor has projections, one for each cylinder, and the pick-up assembly consists of a permanent magnet, a coil and a pole piece with internal "teeth" to match the timer-core projections. At the proper intervals, the projections line up with the teeth to establish a magnetic path through the coil, which is energized through the system amplifier and thereupon permits a current to flow through a circuit in the amplifier. The voltage of this current is amplified to activate a thyristor (in the amplifier) which then "turns on" the primary circuit.

"Turning-on" the primary circuit allows the high-capacity condenser to discharge through the amplifier and the ignition-coil primary, thus creating a momentary current which, in turn, is transformed by the coil

IGNITION SYSTEMS

into the required very-high-voltage secondary current. In short, this system creates the secondary current by momentarily sending a relatively high-voltage current through the primary, rather than by momentarily interrupting primary current. The amplifier unit is a solid-state circuit which has a modular unit containing transistors, diodes, resistors, a thyristor, transformer and printed circuits. See Fig. 11.

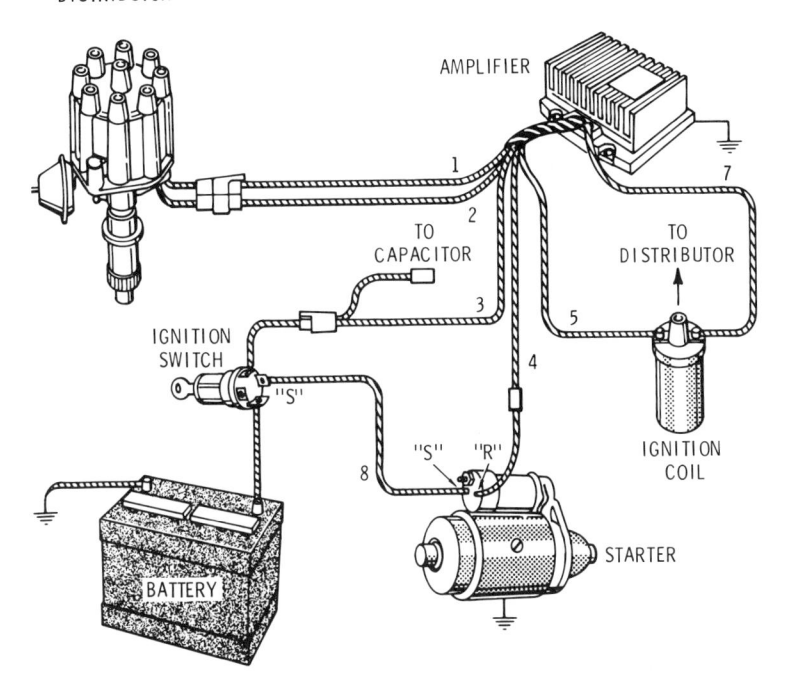

Fig. 11. A typical capacitor-discharge (CD) system.

Variations of this system (referred to as "unit ignition" and "high-energy" systems) differ, principally, in the elimination of the standard ignition coil. Instead, a type of transformer coil is incorporated within the distributor assembly. Operation is similar to the foregoing except that (as in a conventional system) secondary-circuit current is induced by interrupting primary-circuit current. An *electronic module* (in place of a pick-up), a *pole piece* (as before), a *rotor* (in place of a timer core), and an integral *capacitor* accomplish this purpose.

Breakerless (BL) System

This system functions much the same as a CD system except that interruption of the primary circuit causes the secondary-circuit current to flow. Instead of a control unit, an all-electronic *module* is used — and a special ignition coil *(non-*standard) is required.

MAGNETO IGNITION

Magneto ignition, as the name implies, is provided by a magneto which is a compact combination of a generator, ignition coil and distributor. Magneto ignition is usually provided on internal combustion engines, which do not require a lighting system or other current consuming devices in connection with their operation. In some applications, however, such as on outboard motors, both magneto and battery operated ignition systems are available.

In ignition systems suitable for magneto use there are certain advantages over battery ignition in that it is generally more reliable, requires little maintenance, and perhaps the most important that it does not have a battery to run down or wear out.

Operation Principles

Electrical energy in magneto ignition is obtained by a generator utilizing the principle of electromagnetic induction to produce electricity. It should be clearly understood that in order to generate electricity, three things are necessary: 1. An electrical conductor, 2. A magnetic field, 3. Relative motion between the field and the conductor.

In the magneto, a permanent magnet supplies the magnetic field, a wire coil serves as conductor and the engine provides mechanical energy for the motion between the field and the conductor. There are two types of magnetos in general use. They are:
  1. Magnetos with rotating conductors,
  2. Magnetos with rotating magnets.

Both principles, the moving conductor and the moving magnet have been utilized in providing ignition. Due, however, to improved magnetic materials, the rotating magnet principle is now commonly used.

Rotating Conductor Magnetos

Fig. 12 illustrates schematically the circuit arrangement of a rotating conductor magneto. The primary circuit consists of a primary winding on a rotating armature, a condenser and interrupter. One end of the primary

IGNITION SYSTEMS

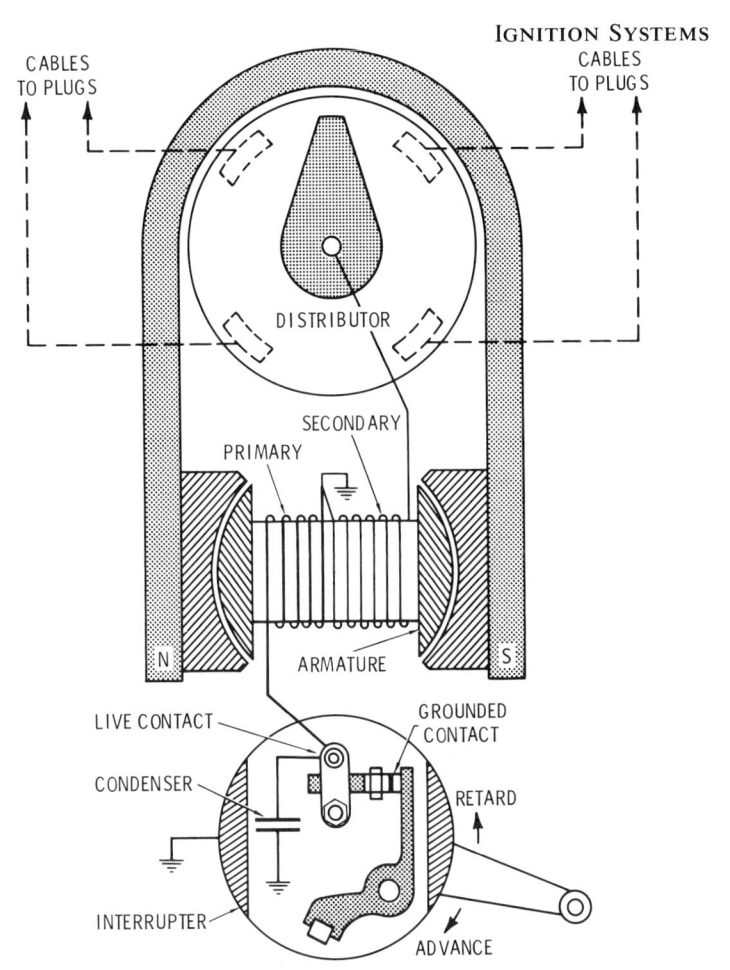

Fig. 12. Schematic circuit diagram of rotating conductor magneto showing relationship of the various parts and circuits.

winding is grounded through a grounding brush, while the other end is brought out to the interrupter through a slip ring or contact button.

The interrupter, or breaker mechanism is shown below the magneto (for clearness) but in reality the interrupter is mounted on an extension of the armature shaft. When the contact points are closed, the primary current passes to ground. The condenser is connected across the contact points. The ground terminal is electrically connected to the insulated contact point. A wire is connected between the ground terminal and the switch. With the switch in the off position, this wire provides a direct path

245

Gas Engine Manual

to ground for the primary current. This prevents the production of too high voltage in the secondary winding.

The secondary circuit consists of a secondary winding on the armature and the distributor. One end of the secondary circuit is grounded to the primary and the other end to the rotor in the distributor, which conveys the secondary current in the proper sequence to the spark plug electrodes of the distributor.

The high tension current produced in the secondary winding passes to the central insert of the distributor finger by means of a carbon brush. From this point the secondary current is conducted to the high tension segment of the distributor finger and across a small air gap to the electrodes of the distributor block. See Fig. 13.

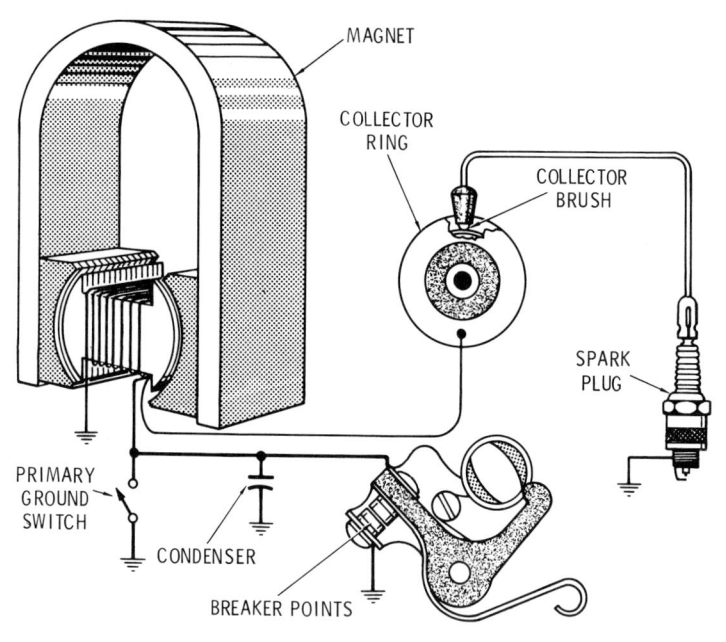

Fig. 13. Diagrammatic view of shuttle wound high tension magneto. This type of magneto is of the rotating conductor class, that is, the windings of an armature is rotated in a magnetic field. Although this type of magneto was once the predominant type it is now largely superseded by the rotating magnet type.

Rotating Magnet Magnetos

Because of improved magnetic materials the rotating conductor type is now largely superseded by the more efficient rotating magnet magneto.

IGNITION SYSTEMS

In a rotary conductor magneto the windings are subjected to considerable centrifugal stress due to high rotative speed, resulting in winding failure and added maintenance and repairs.

In the rotating magnet type, on the other hand, only the solid magnet and pole pieces rotate. The result is a magneto capable of higher speed, and producing more energy for a given size. Fig. 14 illustrates schematically the various components of a rotating magnet magneto.

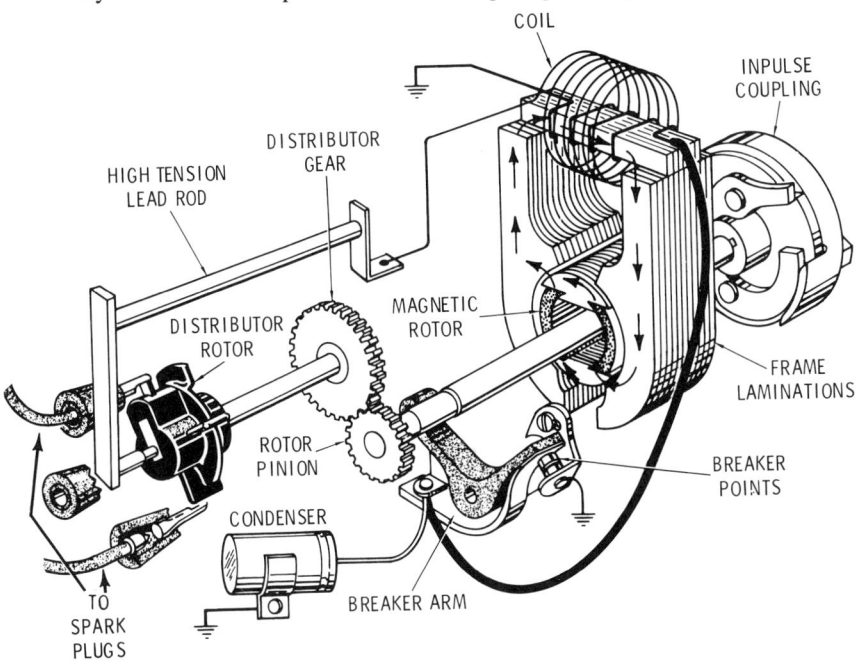

Fig. 14. Schematic diagram of a rotating magnet magneto with jump spark distribution.

Spark Action

Voltage requirements at the spark gap in the engine cylinder, as noted in the battery ignition system, is estimated at 15,000 to 20,000 volts. The problem when magnets are used is to raise the low voltage induced in the conductor to the required high voltage. This is accomplished in a way similar to that used in the battery ignition system.

The armature winding is connected to the primary winding of an ignition coil. When the current in the primary winding or conductor is at its maximum flow, the circuit is suddenly broken, collapsing the elec-

247

tromagnetic field set up in the primary circuit as the result of current flow. The lines of force in the field collapse at an extremely high rate of speed across the secondary winding, which is made up of many turns of fine wire, whereas the primary winding is composed of relatively few turns of heavier wire.

This rapid movement of lines of force across the secondary winding induces a momentary high voltage in the secondary winding, in proportion to the ratio of the number of turns of the two windings. It is in this manner that sufficient high voltage is obtained to create the spark strength necessary to ignite the charge in the engine.

Breaker Points

The interrupting device which breaks the primary circuit at the moment the high voltage spark is desired consists of a set of breaker points, Fig. 14. As noted in the illustration, one end of the primary winding is connected to ground and the other end is connected to the insulated breaker point. When the points are closed, the circuit is completed through them to ground. When they are open, the circuit is broken. Lobes on a cam actuate the breaker points, interrupting the primary circuit and timing the induction of maximum voltage in the secondary circuit. The cam is mounted on either the armature or the rotating magnet.

Condenser Function

The condenser inserted in parallel with the breaker points, has the same function as in the battery ignition circuit. That is, it absorbs the self induced voltage produced in the primary circuit by the collapsing magnetic field. In other words, when the primary circuit is interrupted, the condenser receives the surge of current, and on discharging, reverses the normal current flow. It is in this manner that the condenser hastens the collapse of the magnetic field, around the primary winding, and increases the induced voltage.

Distributor

The distributor rotor, Fig. 14, which controls the proper ignition timing, is fastened to the distributor gear. It is driven by a smaller gear (rotor pinion) located on the drive shaft of the rotating magnet or armature, depending upon the type of magnets used. The ratio between the two gears is such that the distributor cylinder is always driven at one-half the crankshaft speed. This ratio insures that each cylinder will be fired at the correct moment of the engine cycle.

Ignition Systems

With reference to Fig. 14, it will be noted that one end of the secondary winding is grounded to the primary, while the other terminates at the high tension lead rod. The high tension voltage developed in the secondary coil passes through the lead rod to a carbon brush and then internally through the distributor rotor to the surface electrodes. The rotor is timed so that these electrodes will line up with other electrodes on distributor blocks to which spark plug leads are connected.

Edge Gap

The edge gap or edge distance may be defined as the distance between the pole shoe and the edge of the magnet rotor at which point it is most desirable to interrupt the primary circuit in the high tension coil for maximum field strength.

This exact distance, Fig. 15, is determined by the magneto manufacturer using electrical measuring instruments and which after proper determination becomes a service specification. For maximum spark strength the primary contacts should just start to open when the magnetic rotor is at the specified edge gap distance.

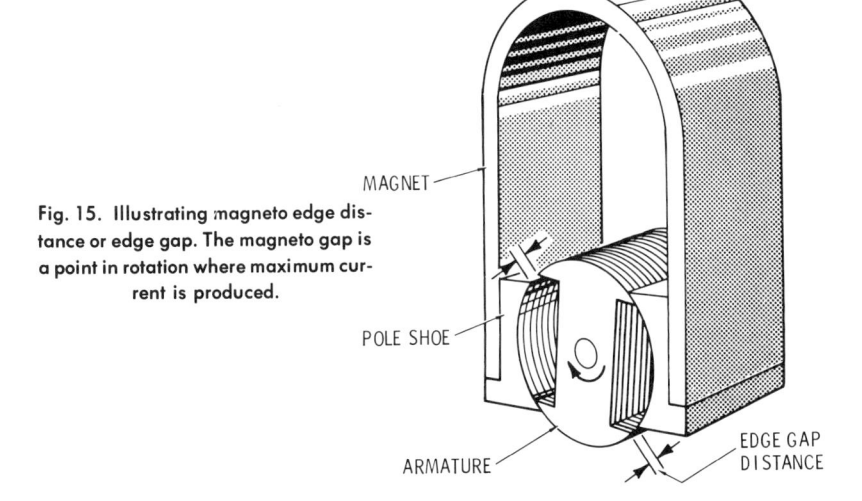

Fig. 15. Illustrating magneto edge distance or edge gap. The magneto gap is a point in rotation where maximum current is produced.

Timing

In magneto installations the timing can be varied within certain limits, by manual rotation of the breaker plate as illustrated in Fig. 16. The amount of advance or retard obtained by this method, which does not

249

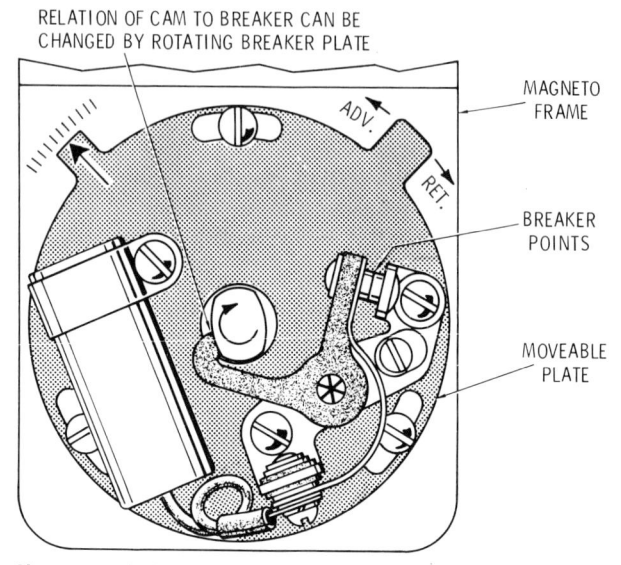

Fig. 16. Showing method of obtaining a limited amount of spark advance (change in ignition timing) on a magneto by rotating the breaker plate.

change the relationship of the cam with respect to the rotating magnet is limited.

This limitation is due to the fact that the spark strength falls very rapidly as the points open farther away from the position of maximum field strength.

Impulse Couplings

The function of the impulse coupling in magneto installations is to increase the spark strength at starting by a temporary increase in the rotative speed of the magnetic rotor at that point in the cycle where ignition should occur. This temporary assist is necessary, since the spark strength of the magneto is directly proportioned to the speed of rotation of the magnetic rotor which is weakest at cranking speed.

Flywheel Magnetos

These are of the rotating magnet type and are used exclusively for ignition on small internal combustion engines such as those used on outboard motors, lawn mowers and similar applications. Flywheel magnetos are commonly used on single cylinder engines, although they are occasionally used on two and four cylinder engines as well.

Ignition Systems

In the typical flywheel type, Fig. 17, the magneto is a self contained electrical generating unit consisting of an armature plate with ignition coil and lamination assembly, condenser and breaker mounted on it. A permanent magnet built into the flywheel completes the assembly.

In operation, as the permanent magnet poles pass over the pole shoes of the coil laminations, a magnetic field causes a current to flow through the primary winding of the coil. This current is normally grounded through closed breaker points.

When the breaker points open, actuated by a cam on the crankshaft, the flow of the primary current is broken and the magnetic field about the coil breaks down instantly. As the current tends to continue flowing, the condenser which is connected across the breaker points, momentarily absorbs this current and hastens the collapse of the magnetic field by creating a high frequency oscillation in the current. The condenser also reduces pitting of breaker points by absorbing any sparking across them.

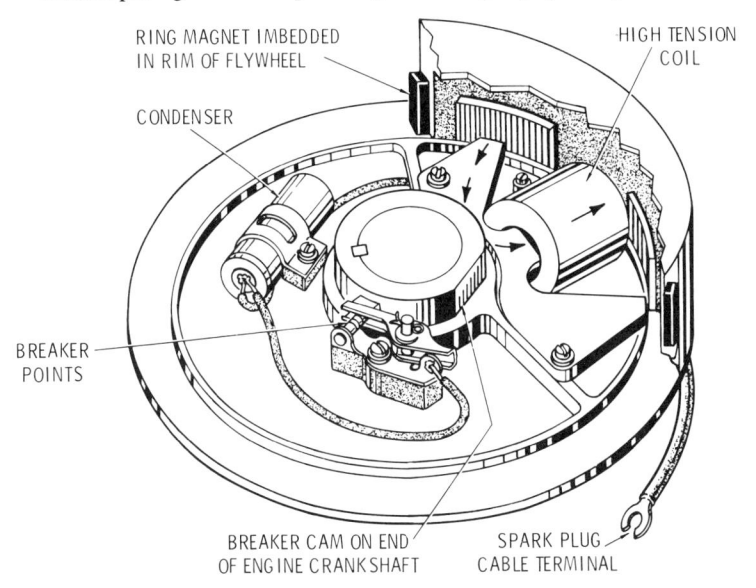

Fig. 17. Typical flywheel magneto with permanent rotating magnet imbedded in rim of engine flywheel.

CHAPTER 20

Electrical System

The electrical system as employed on modern internal-combustion engines consists of numerous electrical components connected together in various ways to form a circuit or circuits. Principally, the electrical system consists of the following:
1. The ignition system (Battery or Magneto),
2. The generating system,
3. The starting system,
4. The lighting system (when used).

NOTE: Most modern vehicles use an alternator instead of a generator. In the following text, the word "generator" is used to denote either a dc generator or an ac alternator.

STORAGE BATTERY

The storage battery may be considered as the central unit of the electrical system, because the various circuits or paths which carry electricity to the operating units, begin and terminate there. Thus, in tracing circuits and in hunting trouble, the battery is the reference point from which other observations and tests are conducted.

The battery supplies the electric current for operating the starting motor and such other units as are needed, until the generator comes into action. Also, when the generator is operating, but is not producing sufficient current for all purposes, the battery supplements its output. The battery is normally maintained in a charged condition by the electric current produced by the generator. See Fig. 1.

GAS ENGINE MANUAL

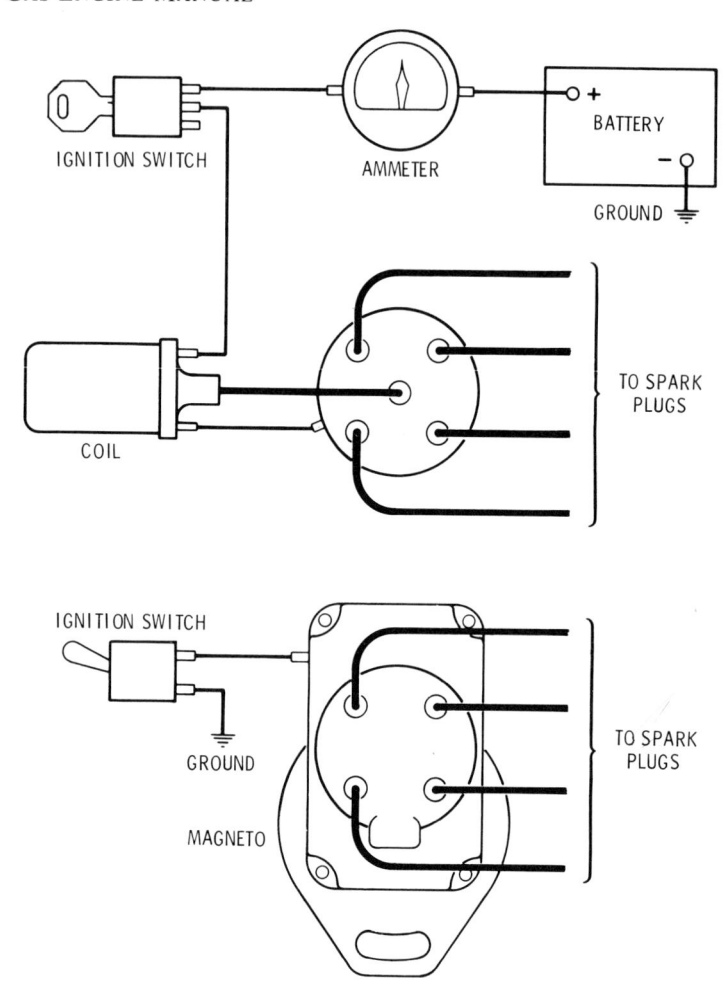

Fig. 1. Schematic wiring diagram showing essential parts of a typical battery and magneto ignition system for a four cylinder engine.

The storage battery is an electrochemical device for converting chemical energy into electrical energy. The amount of electrical power in a storage battery is determined by the amount of chemical substance in the battery and when these substances have been used up, they are restored to their original chemical condition by passing an electric current through the battery. This is known as *charging* the battery. When the generator is not producing the necessary electrical energy, the battery, through chem-

ELECTRICAL SYSTEMS

ical reaction, can supply the energy required. The battery is then said to be *discharging*.

Storage Battery Construction

The storage battery of the automotive type consists of three or more cells, depending upon the voltage desired. Thus, a battery of three cells, of two volts each, connected in series, is known as a *6 volt battery* and one of six cells connected in series, is known as a *12 volt battery*.

Each cell of the battery consists of a hard rubber compartment into which are placed two kinds of lead plates, known as *negative* and *positive*. These plates are insulated from each other by suitable separators and are submerged in a solution of sulfuric acid and water. Fig. 2 illustrates the parts of a storage battery.

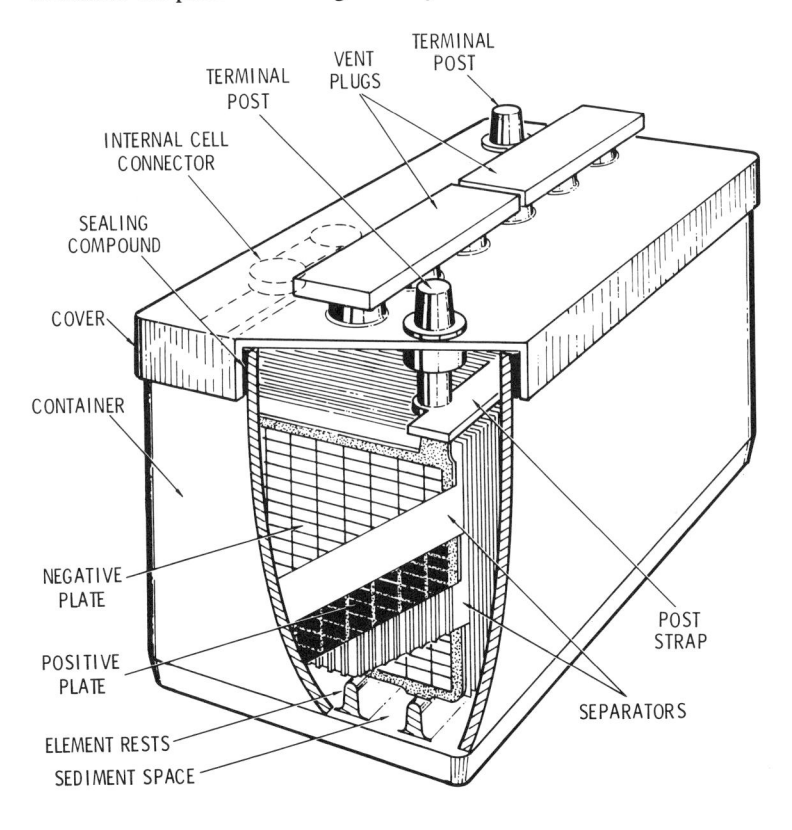

Fig. 2. Illustrating construction details of typical lead-acid storage battery.

Gas Engine Manual

After the plates have been formed, they are connected into *positive* and *negative* groups. The negative group of plates has one more plate than the positive group to provide a negative plate on both sides of all positive plates.

The assembly of a positive and negative group together with the separators, is called an element. Because storage battery plates are more or less of standard size, the number of plates in an element is roughly a measure of battery capacity.

The distance between the plates of an assembled element is approximately one eighth of an inch. To prevent the plates from touching one another and causing a short circuit, sheets of insulating material usually wood, porous rubber or spun glass, are inserted between the plates. These easily pass between the plates.

With the elements in place, the covers are pressed on and the compartments are sealed. The cells are then connected together by short heavy bars of lead, called top connectors as shown in Fig. 3. A top connector is

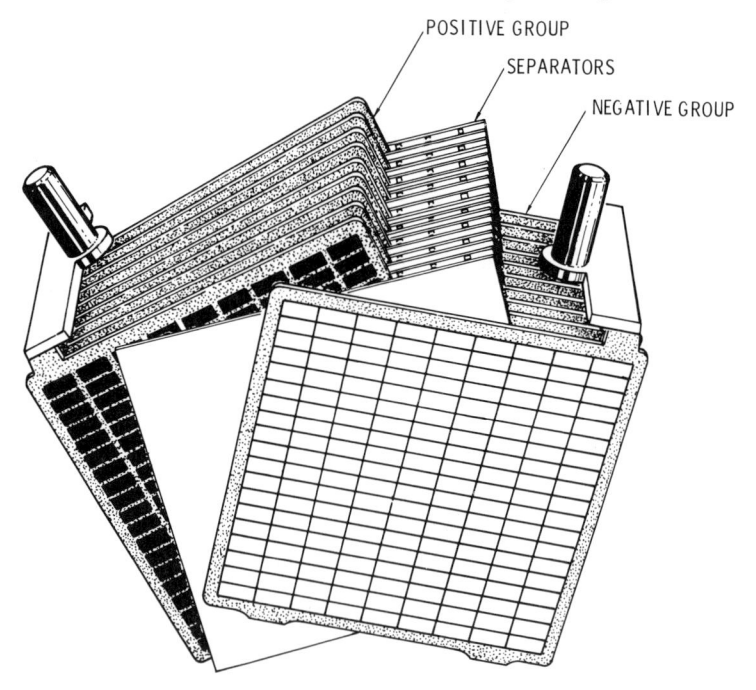

Fig. 3. Showing negative and positive group of storage battery plates and separators respectively.

fused or "burned" to the positive post of one element and the negative post of the element in the adjoining cell.

When all top connectors are in place, there will be one unconnected positive and negative cell at each end of the assembly. These are known as the terminal posts. It is to these terminal posts that the cables of the electric circuit are attached.

Electrolyte

When the assembly is complete, the electrolyte is poured into the cells to cover the plates and insulation. The electrolyte is prepared by mixing chemically pure sulfuric acid and pure water. In a fully charged battery, the proportions are approximately five parts water to two parts of sulfuric acid. After the plates have soaked in this solution for a short period of time, the battery is connected to a suitable source of electric current and charged.

Specific Gravity

By specific gravity is meant the weight of a substance compared to the weight of the same volume of chemically pure water at a temperature of 4°C. Since the specific gravity of sulfuric acid is 1.835, it simply means that sulfuric acid is 1.835 times heavier than an equal volume of water when both liquids are at the same temperature. The electrolyte of a storage battery is a mixture of water and sulfuric acid in such proportions that when the battery is fully charged it has a specific gravity of 1.280.

Because the amount of sulfuric acid in the electrolyte changes with the amount of electrical charge, the specific gravity of the electrolyte also changes with the amount of charge. This provides a convenient way of measuring the degree of charge in a battery.

Hydrometer

Specific gravity of an electrolyte can conveniently be measured by a hydrometer syringe. The hydrometer used for testing batteries is provided with a specific gravity scale graduated from 1.100 to 1.300. The heavier the liquid drawn into the tube, the greater its buoyancy, and the higher the float will extend above the surface of the liquid. Liquids having a low specific gravity are less buoyant, and the hydrometer float sinks deeper into the liquid. See Fig. 4.

A fully charged battery has a specific gravity reading of 1.280 to 1.300, while the specific gravity of a discharged battery may be as low as 1.150. A specific gravity reading of 1.200 and 1.215 indicates that the

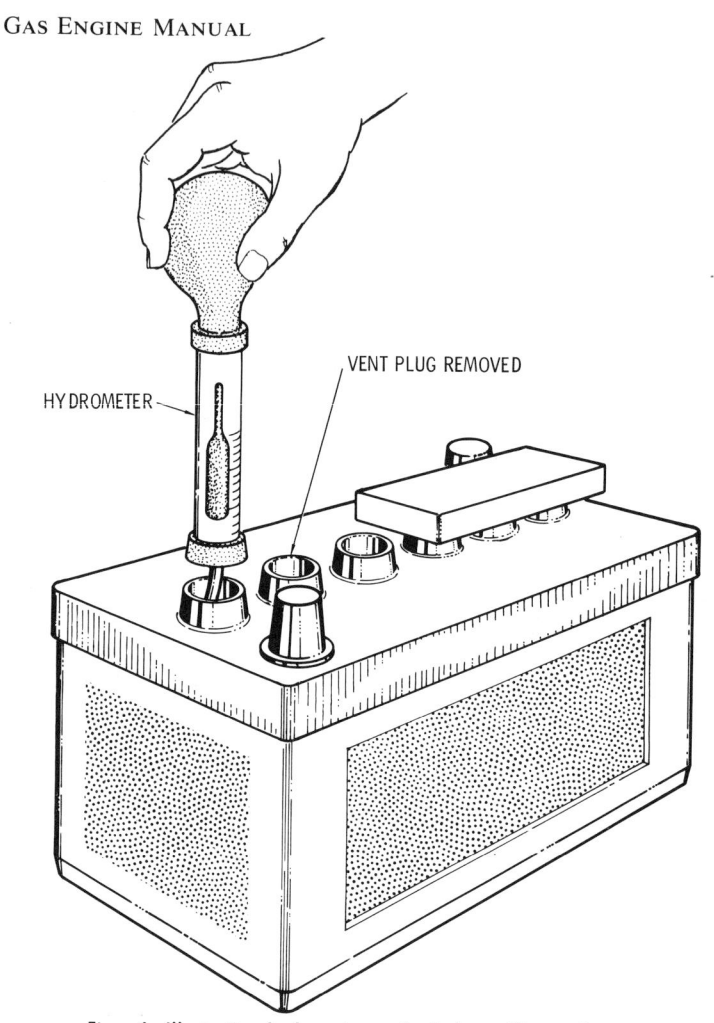

Fig. 4. Illustrating hydrometer method of specific gravity test.

battery is more than half discharged. For convenience, the reading is spoken of as being 1150, 1200, 1280 etc., instead of 1.150, 1.200, 1.280 etc., which is the true specific gravity. After measurement the electrolyte is returned to the cell by compressing the bulb, and the reading of the next cell can be taken.

Temperature Corrections

In this connection it should be noted that the specific gravity reading of a battery varies with the temperature of the electrolyte. Hydrometers are

Electrical Systems

generally calibrated so as to give accurate readings at 80°F. for the electrolyte. This refers to the temperature of the liquid itself and not the temperature of the surrounding atmosphere.

Correction can be made for temperature by adding .004, usually referred to as 4 "points of gravity," to the hydrometer reading for every 10°F. that the electrolyte is above 80°F. or subtracting .004 for every 10°F. that the electrolyte is below 80°F.

If the electrolyte temperature is not too far from the 80°F. standard, or if only an approximate idea of the specific gravity reading is required, it will not be necessary to make the temperature correction. There are hydrometers available which have built in thermometer and temperature scale correction, which hydrometer, if used, will simplify the operation of obtaining a true specific gravity reading. See Fig. 5.

READ ELECTROLYTE TEMPERATURE IN FAHRENHEIT

USE THIS READING FOR SPECIFIC GRAVITY CORRECTION

Temperature	Correction
120°	+.016
110°	+.012
100°	+.008
90°	+.004
80°	0
70°	−.004
60°	−.008
50°	−.012
40°	−.016
30°	−.020
20°	−.024
10°	−.028
0°	−.032
−10°	−.036
−20°	−.040

STEP 1: TAKE TEMPERATURE OF ELECTROLYTE WITH THERMOMETER

STEP 2: TAKE SPECIFIC GRAVITY OF ELECTROLYTE WITH HYDROMETER

STEP 3: CHANGE HYDROMETER READING BY AMOUNT SHOWN ON THIS SIDE OPPOSITE THE ELECTROLYTE TEMPERATURE

EXAMPLE A:
TEMPERATURE .... 120° F
HYDROMETER READING 1.230
CORRECTION AT 120° F .... +0.016
TRUE SPECIFIC GRAVITY 1.246

EXAMPLE B:
TEMPERATURE .... 0° F
HYDROMETER READING 1.230
CORRECTION AT 0° F .... −0.032
TRUE SPECIFIC GRAVITY 1.198

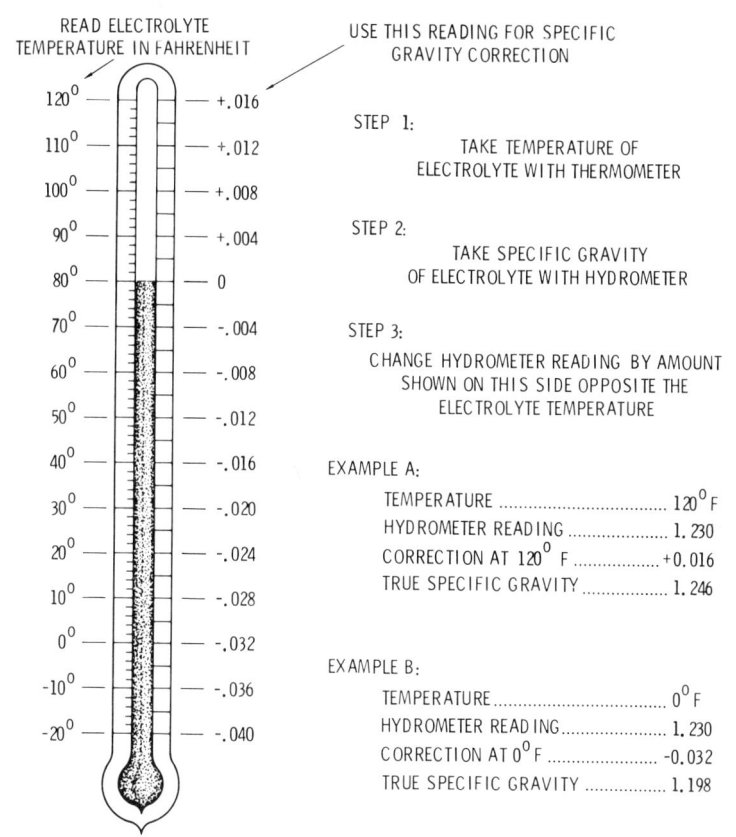

Fig. 5. Showing specific gravity temperature correction scale.

Chemical Action

When a cell is fully charged, the active material of the negative plates consists of spongy lead (Pb) and the active material on the positive plates is lead peroxide (PbO_2). The specific gravity of the electrolyte (sulfuric acid, H_2SO_4 and water H_2O) is then at its maximum and the cell is capable of delivering electricity when connected to a circuit.

Discharge

As the cell is delivering current, that is discharging, the chemical action that takes place changes both the lead (Pb) of the negative plate and the lead peroxide (PbO_2) of the positive plate to lead sulfate ($PbSO_4$) and the sulfuric acid (H_2SO_4) to water (H_2O).

The decomposition of the sulfuric acid and the formation of water dilutes the remaining acid, thus lowering the specific gravity of the electrolyte. As the discharge progresses, the negative and positive plates finally contain considerable lead sulfate. The discharge should always be stopped before the plates have become entirely changed to lead sulfate.

Charge

To charge the cell, an external source of direct current must be connected to the battery terminals. The chemical reaction is then reversed and the lead sulfate on the positive plates is converted into lead peroxide (PbO_2) again, while the lead sulfate on the negative plates is changed back to sponge lead (Pb). The SO_4 from the plates combines with hydrogen to form sulfuric acid (H_2SO_4) and the electrolyte gradually becomes stronger until no more sulfate remains on the plates. The electrolyte will then be of the same specific gravity as before the discharge.

Charging Methods

A storage battery can be charged with direct current only. If only alternating current is available, a motor-generator or a rectifier must be used to convert it into direct current.

When connecting the battery terminals to the charging equipment, the positive wire of the charging circuit must always be connected to the positive terminal of the battery, and the negative wire to the negative terminal. The electrolyte in each cell should be brought to the proper level by addition of pure water before the battery is connected for charging.

If several batteries are to be charged at the same time and connected in series, the positive terminal of one battery should be connected to the

ELECTRICAL SYSTEMS

negative terminal of the next battery. The positive terminal of the end battery of the series is then connected to the positive terminal of the charging source, and the negative terminal of the series of batteries is connected to the negative terminal of the charging source.

There are two methods of charging batteries, namely the *constant current (Fig. 6)* and the *constant potential*. The *constant current* method

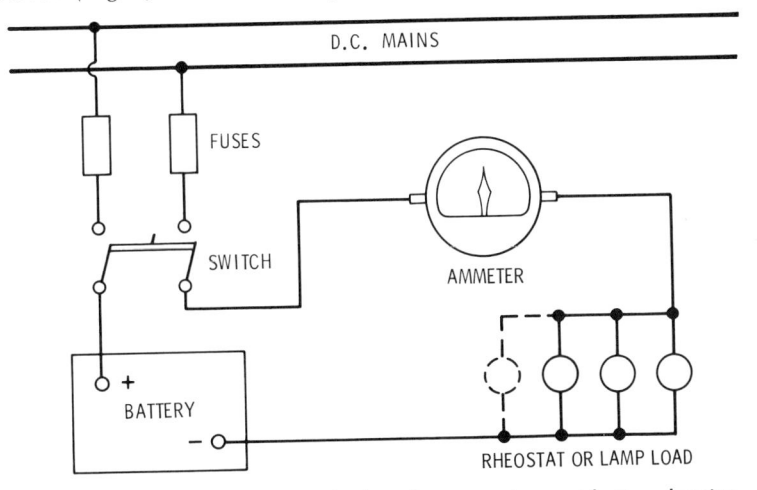

Fig. 6. Wiring diagram illustrating hook-up for constant current battery charging when direct current is available.

is used extensively, particularly where the condition of the battery is not fully known. There are no exact values as to the charging rate, but a safe rate would be equal to one half of the number of plates in the cell. Thus, for example, the charging current of a 13 plate cell would be 6.5 amperes, and for a 17 plate cell 8.5 amperes approximately. Also, where several batteries are connected in series for charging, the charging current is determined by the size of the smallest battery in the circuit. See Fig. 7.

The temperature of the battery should be watched carefully during all stages of the charging process. It should be checked frequently with a thermometer and if it rises above 110°F. either the current should be shut off until the battery is cool or the charging rate reduced. Proper ventilation should be provided when charging batteries.

As a battery approaches a charged condition, gas bubbles commence to appear at the surface of the electrolyte. This is known as "gassing." All cells should gas freely when the battery is fully charged. If a cell does not gas, either the cell is not charged or else there is some internal trouble.

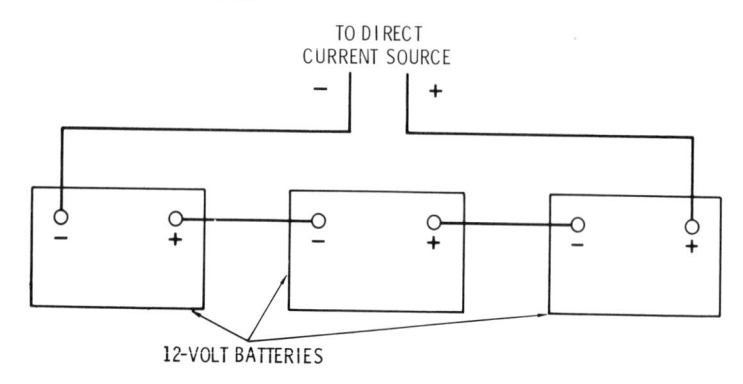

Fig. 7. Wiring diagram of a series hook-up for constant current battery charging. A series connection of batteries as noted consists in connecting positive and negative terminal posts together throughout the circuit. When charging on a series constant current line, connect each battery in series and adjust current rate to a safe value for every battery in the circuit.

Excessive charging will damage the battery, particularly the positive plates. Depending on the charging rate, most batteries can be charged in 12 to 16 hours, although batteries with sulfurated plates may require a charging period of up to 24 hours.

Constant potential charging, as the name implies, maintains the same voltage throughout the charging period, and as a result the current is automatically reduced as the battery approaches full charge. When properly done, this method has been found generally satisfactory for recharging of batteries which are in good working condition, and it has the advantage of completing the charging process in a minimum of time. A badly sulfurated battery may not, however, come up to charge when this charging method is being used.

Battery Ratings

All batteries are given a normal capacity rating according to the number of ampere-hours obtained from the battery under certain working conditions. The 20 hour rating has been accepted as standard by the *American Society of Automotive Engineers* for automotive batteries. To measure the capacity, a battery is discharged continuously at a specific rate until the voltage drops too low for efficient use. Thus, for example, a battery that will deliver 6 amperes for 20 hours is said to have a capacity of 120 ampere-hours.

This measurement is of particular interest because it indicates what may be expected of a battery in the way of satisfactory performance. The

ELECTRICAL SYSTEMS

capacity of a battery depends upon the amount of active material that can react with the electrolyte. Obviously this depends upon the thickness and design of the plates, hence the number of plates is not always an accurate index of the capacity.

Also, one of the characteristics of a storage battery is that its total ampere-hour capacity is dependent upon the rate of discharge. The lower the rate of discharge, the greater the ampere-hour capacity will be; whereas, the higher the discharge rate, the lower will be the capacity. Thus, a battery having a 120 ampere-hour capacity at a 6 ampere discharge rate will ordinarily have a capacity of over 120 ampere-hour at a lower discharge rate.

GENERATING SYSTEM

The function of the generating system is to restore to the battery the energy used during periods when the generating system is *not* in operation (as when the engine is being cranked to a start), and to take-over from the battery and supply all of the vehicle's electric energy requirements whenever the engine is running at sufficient speed to make the generating system fully operative. The engine speed at which the generating system will produce all the energy needed usually varies from an idle to somewhat more, depending upon the electrical devices (lights, radio, etc.) in use at the moment.

The generating circuit is designed so that the battery "floats on the line"; that is, when the system is delivering sufficient current, the excess above vehicle requirements serves to keep the battery in a charged condition. When the system, due to low engine speed, is *not* delivering sufficient current, the battery furnishes the excess current needed.

There are two general types of generating systems. The type first used for vehicles consists of a *dc generators,* a *generator-regulator* (which is one unit incorporating three devices; a *cutout relay, current regulator* and *voltage regulator),* and the battery and wires or cables needed for connecting these various units. Now mostly used is the type of system that consists of an *ac alternator,* a *voltage regulator,* a *rectifier,* and the battery and wires or cables needed. This latter system is preferred because the output of an alternator is directly related to the voltage applied to its field, whereas a generator produces voltage and current in relation to the speed at which it is being operated. Consequently, an alternator can be designed to produce current even at engine low-idling speed, and no engine power is wasted producing unneeded current. On the other hand, a

Gas Engine Manual

generator must be operated at an engine fast-idle speed (or better) to produce even enough current for the ignition system, builds up the output rapidly so that both voltage and current regulation are required, and wastes some engine power. Moreover, a generator requires a cutout relay (to prevent reversal of current from the battery through the generator), whereas an alternator has a rectifier circuit which accomplishes this without mechanical operation.

D C Generator System

Most automotive type generators are of the shunt-wound type with an outside means of regulating the voltage output. A typical generator, Fig. 8, consists essentially of an armature revolving within a magnetic field,

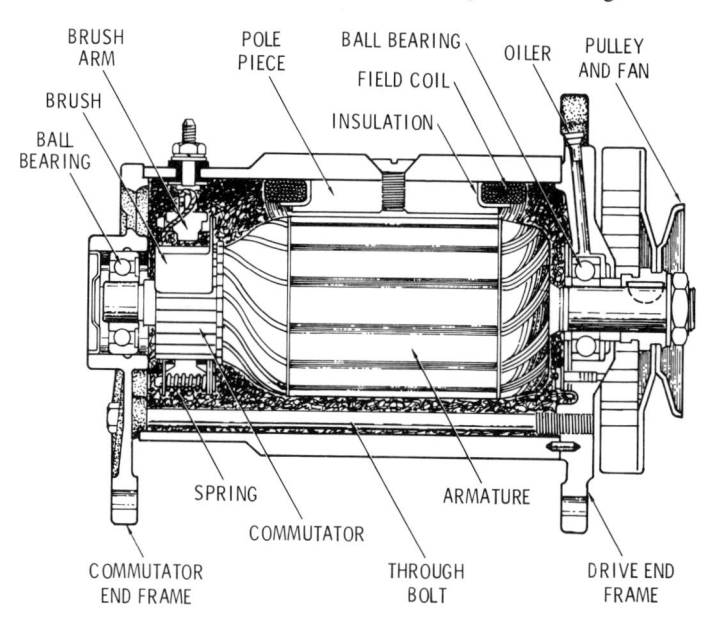

Fig. 8. Cross-section showing construction of typical direct current generator driven by a fan belt.

and its components consist of a commutator, spring-loaded brushes, shaft, bearings, etc., all mounted within a sturdy frame.

The generator is commonly mounted on a bracket on the side of the engine and is driven by the fan belt. This method of mounting permits the generator to be moved in or out to adjust tension of the fan belt.

Essential Controls

In any electric system such as that shown in Fig. 9, where there is a generator and battery, two control elements are necessary for the proper working of the system. They are:
1. Means of preventing reversal of current when generator is charging the battery,
2. Means for limiting generator voltage.

The generator cutout relay acts as an automatic switch to connect the generator to the battery when the generator voltage exceeds that of the battery. When the battery voltage exceeds that of the generator, the cutout relay points open to prevent the battery from discharging through the generator.

The voltage and current regulators controls the amount of current the generator produces, allowing the generator to produce a high current when the battery is in a discharged condition, and the lights and other

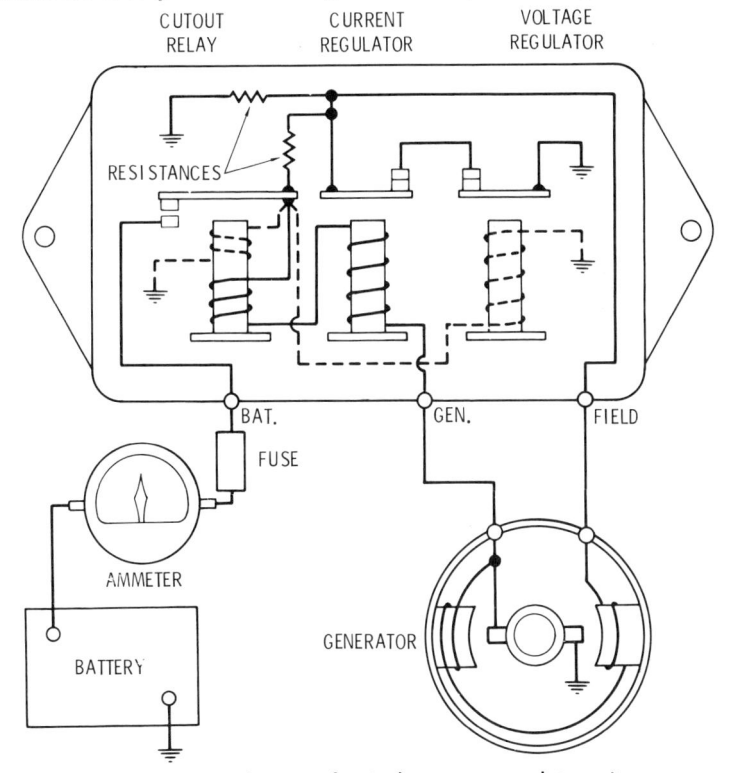

Fig. 9. Wiring diagram of typical generator regulator unit.

accessories are turned on. When the battery is charged and the electrical accessories are disconnected, the regulator reduces the current produced to the amount needed to meet the operating requirements of the system.

Cutout Relay

The cutout relay has a series or current winding of a few turns of heavy wire, and a shunt or voltage winding of many turns of fine wire, both assembled on the same core. The shunt winding is connected between generator armature and ground so that generator voltage is impressed upon it at all times. The series winding is connected so that all generator output current must pass through it. It is connected to a flat steel armature which has a pair of contact points through which current passes to the battery and other electrical units. The contact points are held open by armature spring tension when the unit is not operating.

When the generator begins to operate, voltage builds up and forces current through the shunt winding, thereby magnetizing the core. When the voltage reaches the value for which the relay is set, the magnetism is strong enough to overcome the armature spring tension and pull the armature toward the core, thereby closing the contact points. Generator current now flows through the series winding of relay in the right direction to add to the magnetism holding the points closed, and passes on to the battery and other electrical units in operation.

When the generator slows to engine idling speed, or stops, current begins to flow from the battery back through the generator, reversing the current flow through the series winding. This reduces the magnetism of the relay core to the extent that it can no longer hold the contact points closed against armature spring tension. The points are separated and the circuit broken between the generator and battery.

All regulators have a fuse in the generator charging circuit. This fuse connects to the battery terminal of the regulator and the battery lead connects to it in turn. The purpose of the fuse is to protect the generator and wiring should a stuck or welded cutout relay occur. Shorts or grounds occurring in the charging circuit or reverse polarity conditions of the generator can cause the cutout relay points to weld together. This allows the battery to discharge through the generator when the generator is not developing greater than battery voltage.

Current Regulator

The current regulator, Fig. 9 automatically controls the maximum output of the generator. When the current requirements of the electrical

ELECTRICAL SYSTEMS

system are large and the battery is low, the current regulator operates to protect the generator from overload by limiting its output to a safe value.

The current regulator has one series winding of heavy wire through which the entire generator output flows at all times. This winding connects to the series winding in the cutout relay, previously described. Above the winding core is an armature, with a pair of contact points which are held together by spring tension when the current regulator is not operating. When current regulator is not operating and the contact points are closed, the generator field circuit is directly grounded so that generator may produce maximum output, unless further controlled by the voltage regulator.

When the generator output increases to the value for which the current regulator is set, the magnetism of the current winding is sufficient to overcome the armature spring tension. The armature is pulled toward the winding core so that the points are separated. The generator field circuit must then pass through a resistance, which reduces the flow through the field coils and thereby reduces the output of the generator.

This reduces the magnetic strength of the current winding so that spring tension again closes the contact points, directly grounding the generator field circuit and increasing generator output. This cycle is repeated many times a second, and the action limits the generator output to the value for which the regulator is set.

The current regulator has a bimetal hinge on the armature for thermostatic temperature control. This automatically permits a somewhat higher generator output when the unit is cold, and causes the output to drop off as the temperature increases.

The current regulator operates only when the condition of battery and the load of current-consuming units in operation require maximum output of the generator. When current requirements are small, the voltage regulator controls generator output. Either the current regulator or the voltage regulator operates at any one time; both regulators never operate at the same time.

Voltage Regulator

The voltage regulator, Fig. 9, limits the voltage in the charging circuits to a safe value, thereby controlling the charging rate of the generator in accordance with the requirements of the battery and the current-consuming electrical units in operation. When the battery is low, the generator output is near maximum but as the battery comes up to charge and other requirements are small, the voltage regulator operates to limit

Gas Engine Manual

the voltage, thereby reducing the generator output. This protects the battery from overcharge and the electrical system from high voltage.

The voltage regulator unit has a shunt winding consisting of many turns of fine wire which is connected across the generator. The winding and core are assembled into a frame. A slat steel armature is attached to the frame by a flexible hinge so that it is just above the end of the core. When the voltage regulator unit is not operating, the tension of a spiral spring holds the armature away from the core so that a point set is in contact which allows the generator field circuit to complete the ground through them.

When the generator voltage reaches the value for which voltage regulator is set, the magnetic pull of the voltage winding is sufficient to overcome the armature spring tension, so that the armature is pulled toward the core and the contact points are separated. The instant the points separate, the field current flows only through the resistance to ground. This reduces the current flow through the field coils and decreases generator voltage and output.

The reduced voltage in the circuit causes a weakening of the magnetic field of the voltage winding in the regulator. The resulting loss of magnetism permits the spring to pull the armature away from the core and close the contact points again, thereby directly grounding the generator field so that generator voltage and output increases.

This cycle is repeated many times a second, causing a vibrating action of the armature, and holds the generator voltage, the voltage regulator continues to reduce the generator output as the battery comes up to charge. When the battery reaches a fully charged condition, the voltage regulator will have reduced the generator output to a relatively few amperes.

The voltage regulator has a bimetal armature hinge for thermostatic temperature control. This automatically permits regulation to a higher voltage when the unit is cold, and a lower voltage when hot, because a high voltage is required to charge a cold battery.

As previously stated, the current and voltage regulators do not operate at the same time. When current requirements are large, the generator voltage is too low to cause voltage regulator to operate, therefore the current regulator operates to limit maximum output of generator. When current requirements are large, the generator voltage is too low to cause voltage regulator to operate. The generator output is then reduced below. the value required to operate the current regulator, consequently all control is then dependent on the operation of voltage regulator.

ELECTRICAL SYSTEMS

Resistance

The current or voltage regulator circuit each uses its own resistance which is inserted in the field circuit when either regulator operates. A third resistance is connected between the cutout relay base plate and the voltage regulator to operate. The generator output is then reduced below winding.

The sudden reduction in field current occurring when either the current or voltage regulator contact points open, is accomplished by a surge of induced voltage in the field coils as the strength of the magnetic fields change. These surges are partially dissipated by the two resistances, thus preventing excessive arcing at the contact points.

A C ALTERNATOR SYSTEM

Though similar in construction to a generator (and mounted on an engine in the same way), an alternator differs somewhat in its electrical design. The rotating part, called a *rotor*, acts as the alternator's field, and the generating component (comparable in function to a generator's armature) is composed of stationary coils surrounding the rotor and called the *stator*. Since the generated "output" current passes through the stationary stator, a generator-type commutator and brushes are *not* needed. Current to activate the rotor is passed to it through one brush and a (circular) slip ring on the rotor, then from the rotor to ground through a second brush and slip ring.

To control output voltage maximum it is necessary only to control the voltage applied to the rotor, and the alternator then tends to limit its output current to the "demands" of the circuits it serves. Voltage control may be accomplished either by a remote electromechanical regulator similar to the voltage regulator used with a generator, or by a (usually integral) transistor regulator. See Fig. 10.

Automotive alternators generally have a 3-phase, delta-connected (fig. 13) stator which, because current is generated within one or the next of the three windings during each successive 120° of rotation, tends to "level out" the peaks and valleys of alternating current generation, and to produce a more consistently uniform output voltage. The output is ac, *not* suitable for charging a battery or the ignition system, and must, therefore be converted to dc. Conversion is accomplished by a *rectifier bridge,* which accomplishes "full" rectification (that is, *all* of the ac is channeled through the circuit so that it comes out "flowing" in the same

GAS ENGINE MANUAL

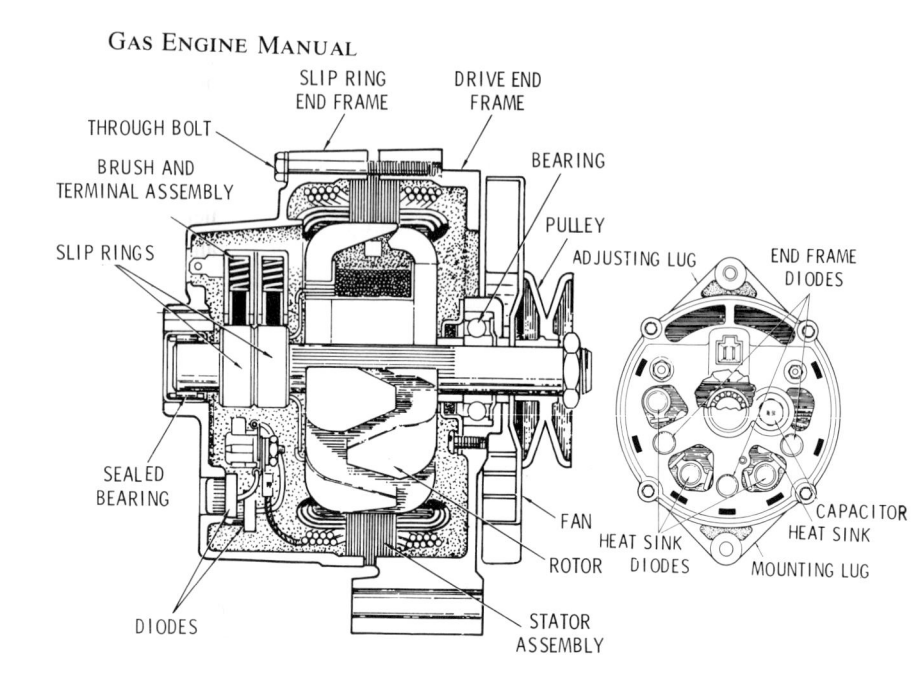

Fig. 10. A typical alternator without an integral regulator.

direction (instead of in first one then the other direction) to become an uninterrupted (constant "flowing") dc. See Fig. 11.

A full-rectification bridge requires two rectifiers for each of the three 3-phase windings—a total of six minimum. (In some cases, where a separate starting-amps current is desired) an additional pair—for a total of eight—is added into the common connecting line of the delta connection.) The rectifiers used are called *diodes*. These are compact units made of silicon and have the property of freely conducting a current flow in one direction while opposing its flow in the opposite direction. The electrical circuitry of a bridge circuit is such that, while flowing in one direction ac will "flow out through the front door" then, during the part of its cycle that causes it to flow in the opposite direction, will "flow out through the back door" into the same external circuit.

Diodes are part of an alternator assembly. They usually press-fitted into one of the alternator-housing end shields (or an insulated attachment called a *heat sink),* and wiring to them is internally accomplished. Half of the total number are on the grounded circuit side and are referred to as *positive* diodes; the other half are *negative* diodes.

ELECTRICAL SYSTEMS

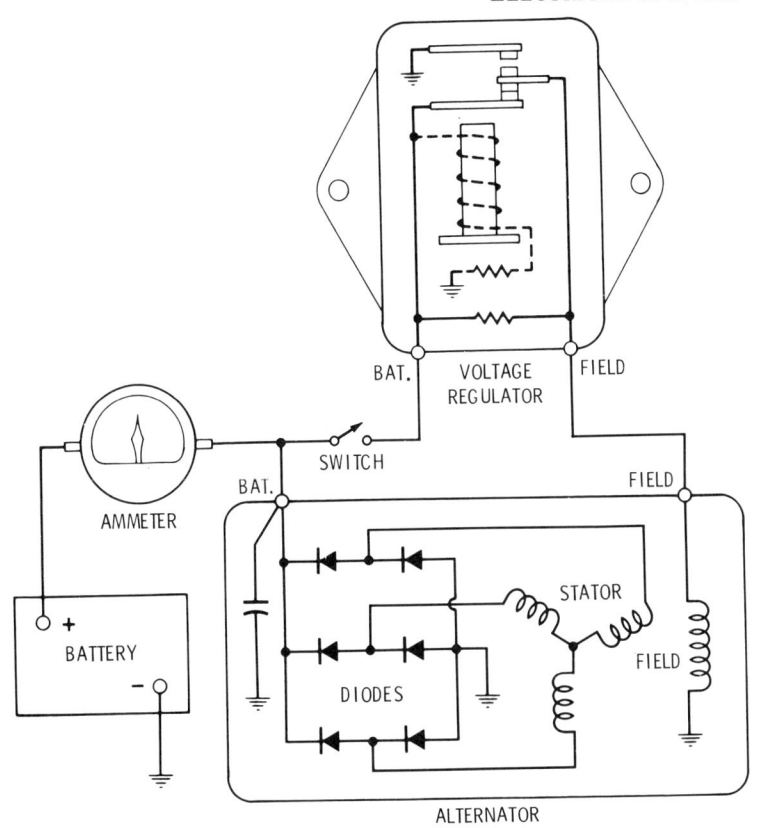

Fig. 11. Alternator charging circuit with external regulator.

NOTE: Because each diode, depending upon its size, will "pass" only so much current, in some cases two diodes in parallel may be used in place of one.

A *transistor regulator,* as illustrated, in Fig. 12, consists of an electrical assembly composed of transistors, diodes, resistors, a capacitor and a thermistor. The transistors are devices that limit voltage by controlling current, and the other components assist the transistors. The thermistor adjusts for temperature. Compact in size, this type of regulator generally is mounted within the alternator housing, and wired internally into the rotor circuit. Such a unit has provisions for manual adjustment.

When an electromechanical regulator is used, one of the two rotor brushes is grounded as previously explained. When a transistor regulator is used, either an *insulated brush* or an *isolated field* circuitry may be

271

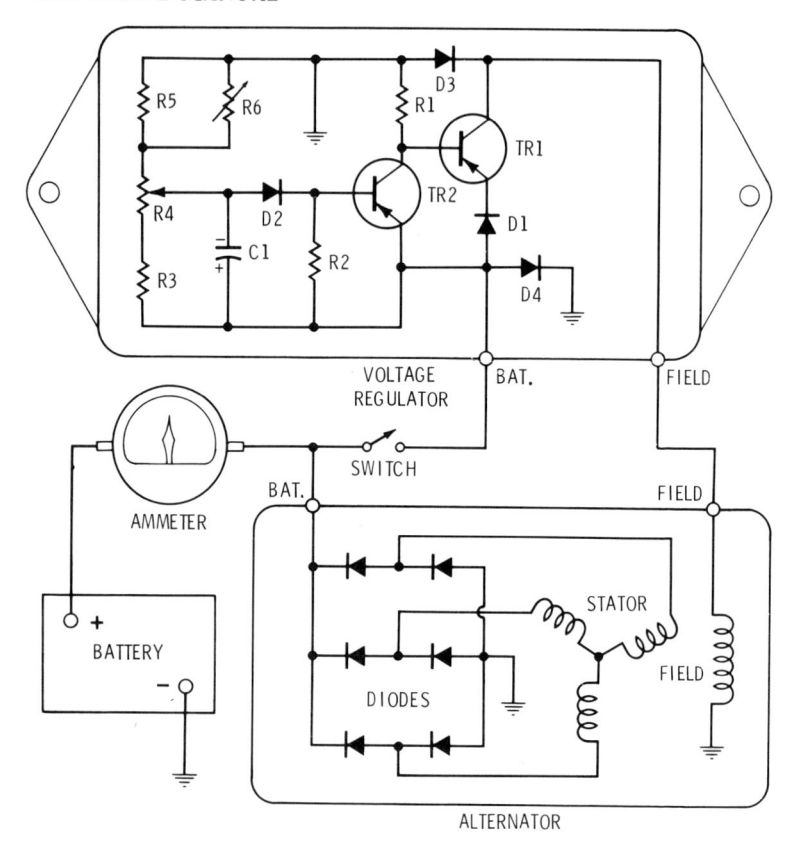

Fig. 12. Alternator charging circuit with a transistor regulator.

used. These differ only in the manner in which grounding of the rotor circuit is accomplished.

STARTING SYSTEM

Although starting motors resemble generators, they are different in construction and operation. Many general parts, like field coils, armature and brushes are common to both but the design of these are different. Also, in a generator mechanical energy is converted into electrical energy, whereas in a motor electrical energy is converted into mechanical energy.

ELECTRICAL SYSTEMS

The starting motor is a low voltage series wound direct current motor, that converts electrical energy from the battery into mechanical energy for cranking the engine when the circuit between the battery and the motor is closed.

The means for coupling the motor to the engine when starting is known as the *cranking system*. The cranking system is generally composed of the following units:
1. Battery and battery cables,
2. Cranking motor including the drive assembly which engages the flywheel ring gear during cranking operation,
3. Cranking motor solenoid switch mounted on the cranking motor, for shifting drive assembly and closing the motor circuit,
4. Solenoid switch relay.

Overrunning Clutch Drive

In the overrunning clutch drive, the starter motor may drive the engine through a pinion or by a dog clutch attached to the motor armature shaft, which is brought into mesh with teeth cut on the rim of the flywheel or into mesh with the mating half of the dog clutch. See Fig. 13.

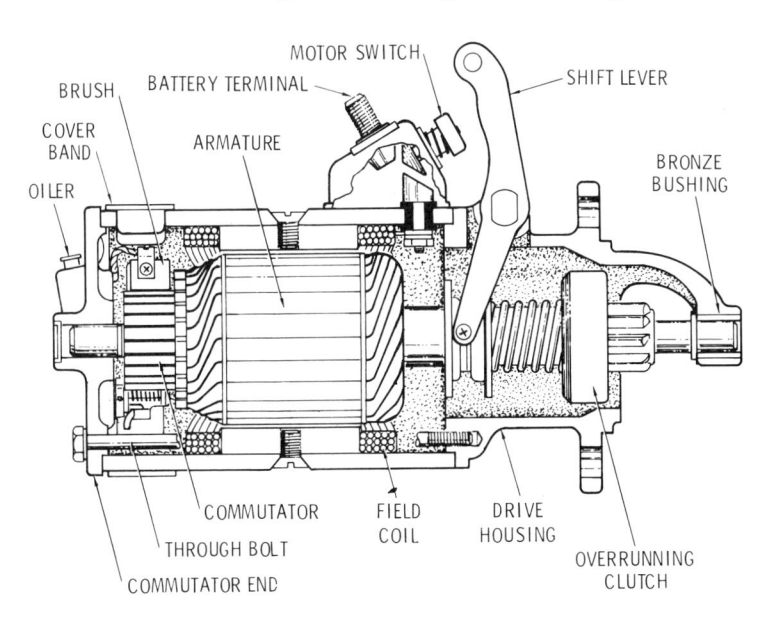

Fig. 13. Sectional view of starting motor with over-running clutch drive.

GAS ENGINE MANUAL

Shifting the pinion gear into mesh with the flywheel gear is commonly made automatically by the use of a solenoid. For operation of the solenoid shift a remotely operated control switch is necessary. The ignition is connected to the control circuit so that the starting motor will not operate until the ignition is on. The solenoid shift unit is rigidly mounted on the starting motor field frame. Inside the solenoid coil is a heavy plunger connected to the shift lever as noted in Fig. 14.

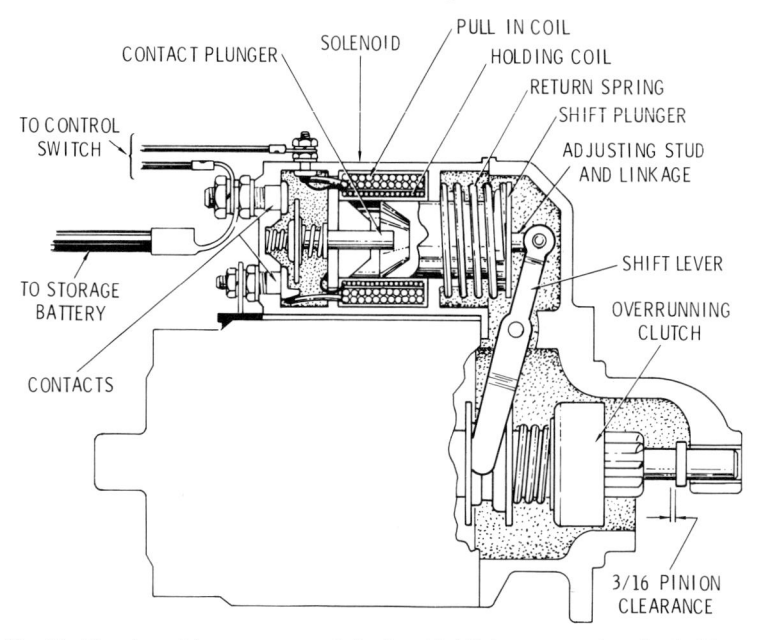

Fig. 14. Showing wiring arrangement of solenoid shift in over-running clutch drive.

When the remote control circuit is closed, the solenoid exerts a pull on the shift plunger which shifts the pinion into mesh with the flywheel teeth. After the pinion shift lever has moved the required distance for meshing the pinion gear, the pointed end of the shift plunger presses against the end of a contact plunger and pushes a contact disk on the contact plunger across the switch contacts to operate the motor. An overrunning clutch is required with this system to prevent damage to the starter at the time the engine fires. See Fig. 15.

Push Button Control

One method of controlling the solenoid shift is by means of a push button on the instrument panel. Pushing the button closes the control circuit so that the current can be applied to the solenoid coil.

ELECTRICAL SYSTEMS

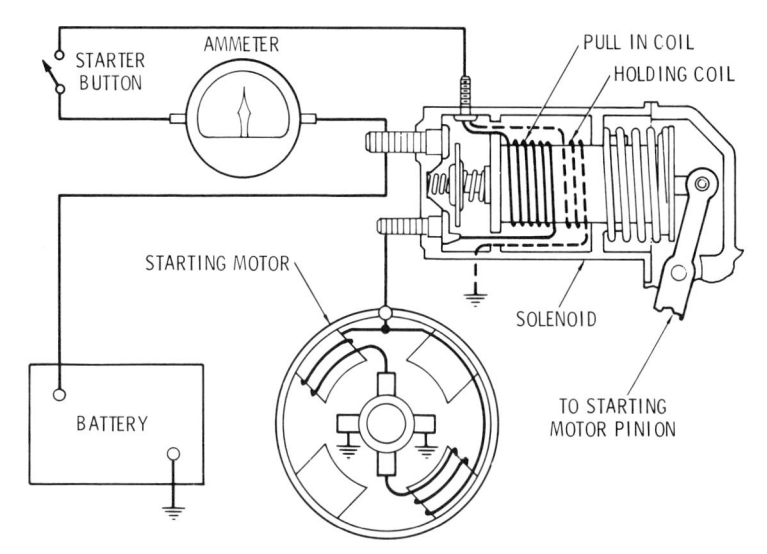

Fig. 15. Schematic diagram showing solenoid wiring method.

A relay is frequently used in the control circuit to supply current to the solenoid coil. Only a low current control circuit to the instrument panel push button is then necessary. The relay will close the circuit through the solenoid coil which carries a larger current.

Bendix Drive

The Bendix is a popular starting mechanism. This automatic screw pinion shift mechanism is designed in two types; one is known as the *inboard type,* in which the pinion shifts towards the motor to disengage the flywheel; the other is an *outboard type* in which the pinion shifts away from the motor. A typical assembly view of the outboard drive is shown in Fig. 16.

The same general construction, however, is used in both types. A sleeve having screw threads, with stops at each end to limit the lengthwise travel of the pinion, is mounted on the extended armature shaft. The pinion gear which is unbalanced by the weight on one side, has corresponding internal threads for mounting upon this sleeve. The sleeve is connected to the motor armature shaft through a special drive spring attached to a collar pinned to the armature shaft.

In operation, when the starting motor is not running, the pinion is out of mesh and entirely away from the flywheel gear. When the starting switch is closed and the battery voltage impressed on the motor, the

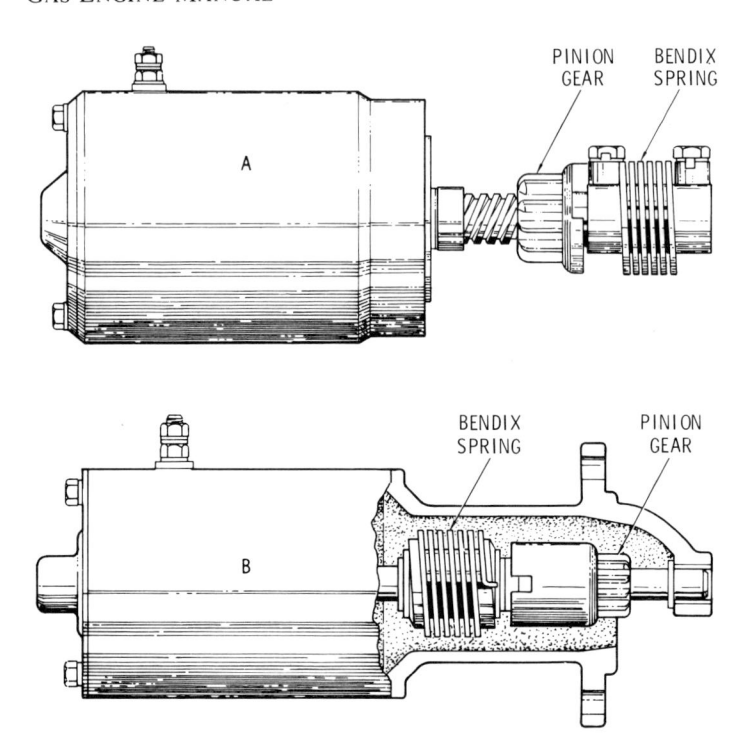

Fig. 16. Bendix drive starter showing gear arrangement. (A) Inboard Type. (B) Outboard Type.

armature starts to rotate at high speed. The pinion, being weighted on one side and having internal screw threads, does not rotate immediately with the shaft, but because of its inertia, runs forward on the revolving threaded sleeve until it meets and meshes with the flywheel gear.

If the teeth of the pinion and the flywheel meet instead of meshing, the drive spring allows the pinion to revolve and forces it into mesh with the flywheel. When the pinion gear is fully meshed with the flywheel gear, the pinion is then driven by the motor through the compressed drive spring and cranks the engine.

The drive spring acts as a cushion while the engine is being cranked against compression. It also breaks the severity of the shock on the teeth when the gears mesh and when the engine kicks back due to early ignition.

When the engine fires and runs on its own power the flywheel drives the pinion at higher speed than does the starting motor, causing the pinion to turn in the opposite direction on the threaded sleeve and automatically demesh from the flywheel. This prevents the engine from driving the starting motor. When the pinion is automatically demeshed from the flywheel, it is held in a demeshed position by a latch until the starting switch is again closed.

CHAPTER 21

Spark Plugs

By definition, a spark plug is a device whose duty is to create a spark between its electrodes to ignite the fuel charge in the combustion chamber.

With reference to Fig. 1, the spark plug consists of a center electrode which is connected to the ignition coil secondary through the distributor as shown in Fig. 2.

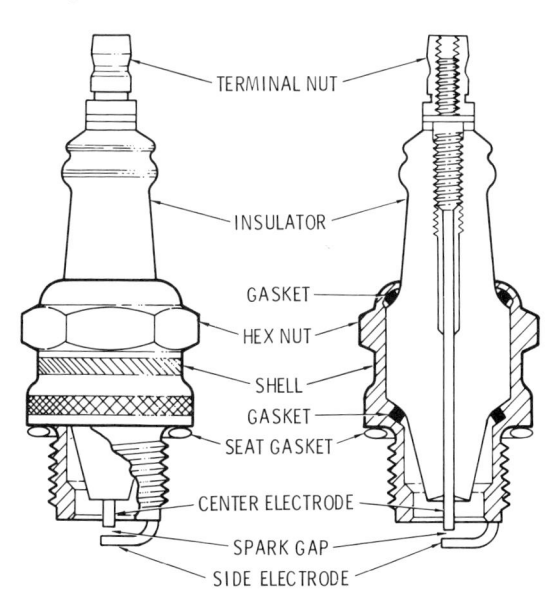

Fig. 1. Illustrating construction of typical spark plug with names of parts.

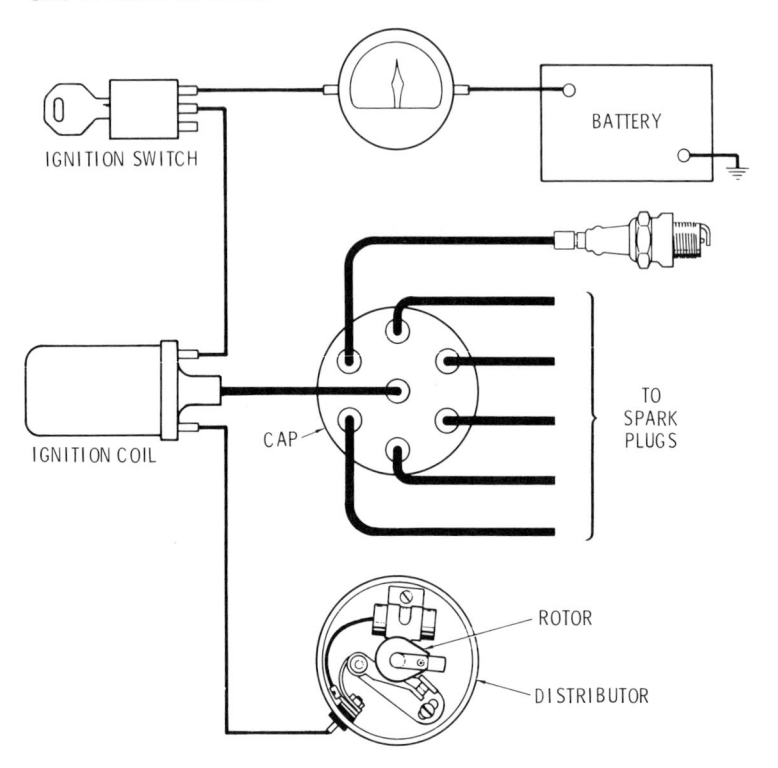

Fig. 2. Typical battery ignition circuit for six cylinder engine showing spark plug connections.

The center electrode is insulated from the spark plug shell by means of a molded insulator. The side electrode projects from the bottom edge of the spark plug shell and is positioned in such a manner that there is a gap between it and the center electrode. *This gap* is known as the *spark plug gap,* the size of which is determined by the manufacturer to suit the various types of engines. The spark plug recommended by the manufacturer is generally one that will give the best service under normal operating conditions.

Spark Plug Gaps

Spark plug gaps are set accordance with manfacturer's specifications and range from approximately 0.020 in. to 0.040 inch. The size of the gap depends upon the engine compression ratio, the ignition system and the characteristics of the combustion chamber. In modern high voltage igni-

Spark Plugs

tion systems, spark plug gaps of 0.035 in. are common. Advantages of wider gaps are obvious, since they permit additional gas-air mixture with better chance for ignition.

Thread Sizes

The shell of the spark plug is threaded to facilitate the plugs removal for inspection or replacement. Spark plugs are generally made in four thread sizes as: 10mm, 14mm, 18mm and ⅞ inch, to fit the corresponding sizes in the cylinder head. Ten millimeter plugs carry AC numbers beginning with "10," such as 104, 106, etc. Fourteen mm. plugs carry numbers beginning with "4" such as 44, 46, 47, etc. Eighteen mm. carry numbers beginning with "8" such as 82, 83, 85. Seven-eights inch plugs carry numbers beginning with "7," 73, 74, 75.

Heat Range

Modern spark plugs are designed for perfect functioning at certain predetermined temperatures. They are designed this way because different engines run at different temperatures. The temperature of an engine varies with its revolutions per minute (rpm) and load.

Thus, engines carrying heavy loads at high speed will run hotter than similar engines carrying light loads. Engines subjected to hard work for an extended period of time will ordinarily operate better and give longer spark plug life if operated with a colder plug.

Spark plug manufacturers provide for these conditions of cold or hot engines by making plugs with longer insulators for use in cold engines and shorter insulators for use in hot engines.

Fig. 3 illustrates the path the heat must take from the tip of the spark plug to reach the water jacket of the engine. The longer the path the hotter the plug.

As a guide for proper plug selection, manufacturers employ certain code numbers. Thus, the last numeral in each plug number indicates the heat range. The spark plug with low numbers are those made for engines which use cooler plugs. Plugs with higher numbers are those for engines which need hotter plugs.

Also, the installation of a *cold* plug in a low speed engine will result in consistent fouling, while the installation of a *hot* plug in a high speed engine will result in pre-ignition, causing the engine to "surge" that is, to run at full rpm for short periods, then at a noticeable reduction in speed. In extreme cases, the plug may become hot enough to result in the engine stopping entirely or even cause damage to the piston head.

GAS ENGINE MANUAL

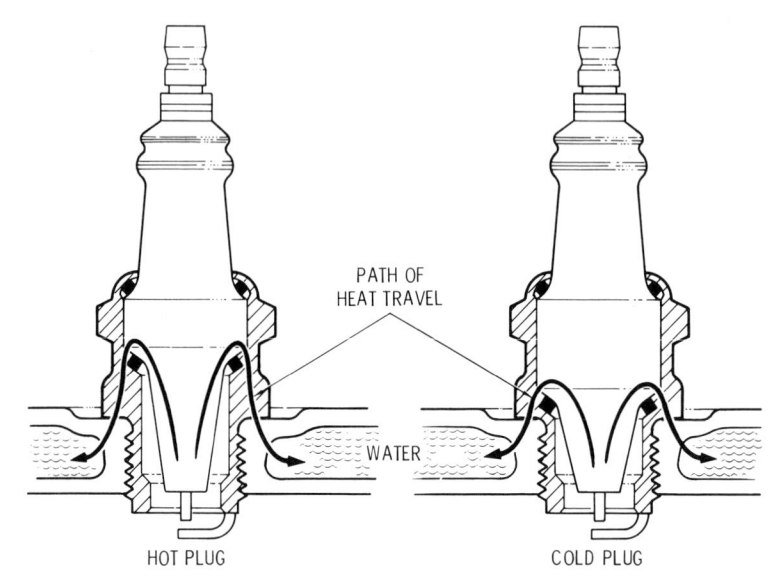

PATH OF
HEAT TRAVEL

WATER

HOT PLUG   COLD PLUG

Fig. 3. Illustrating hot and cold spark plugs. Heat ranges of spark plugs are determined by the distance heat must travel from the electrode to the cylinder head coolant.

Cleaning and Gap Adjustment

Plugs which are in good condition except for carbon or oxide deposits should be thoroughly cleaned and adjusted. To clean plugs soak them in a carburetor cleaning solvent from 15 to 30 minutes. Thoroughly dry the interior of plugs with compressed air, then scrape out all carbon and oxide deposits from the shells and insulators with a pointed steel scraper. Blow out all residual matter with sand blasting equipment.

When adjusting spark gaps, use round wire feeler gauges to check the gap between spark plug electrodes. The feeler gauge should be of the same diameter as recommended by the engine manufacturer. Flat feeler

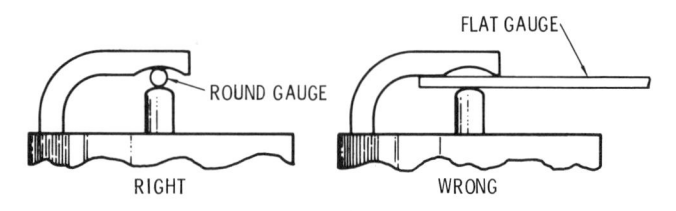

ROUND GAUGE   FLAT GAUGE
RIGHT   WRONG

Fig. 4. Showing correct and incorrect method of measuring spark plug gap.

Spark Plugs

gauges will not give a correct measurement if the electrodes are worn. Adjust gap by bending the side electrodes only; bending the center electrode will crack the insulator. See Fig. 4.

Before installing a spark plug, make sure that the spark plug seat in the cylinder head is clean and free from obstructions. It is strongly recommended that a new seat gasket be used each time a plug is installed. The plug should be screwed into the cylinder head to fully compress the gasket.

CHAPTER 22

Troubleshooting

Because an internal combustion engine must perform the work of drawing in and compressing its charge before energy is developed in its cylinder, some special device is required to start it. To start an engine, therefore, requires the use of power from an external source. It is also necessary to disengage the load during the starting period.

Small two stroke cycle engines as used on lawn mowers, outboard motors, etc., are usually started by starter cord, by turning the flywheel or by a special hand gear. The latter must have a ratchet or clutch which will release or throw it out of gear as soon as power is developed.

The electrical, or self-starting system is usually employed on large engines, particularly those used in automotive service. This starting system consists of a small electric motor which is mechanically connected to the engine shaft through a set of gears, and electrically connected to a storage battery which furnishes the current required for starting.

The battery is kept charged by a direct current generator which is commonly driven by means of a set of pulleys and belts from the engine. The generator also assists the battery in supplying the current requirements for lights and ignition under normal operating conditions.

When an engine fails to start, the trouble in the majority of cases is to be found in the ignition system, a run down battery, an open electrical circuit and other miscellaneous causes, but only rarely in the fuel or cooling systems.

All internal combustion engines whether two, or four stroke cycle, depend for their proper performance upon a supply of correct fuel mixture, good compression, and an adequate spark to ignite the mixture at the proper time.

Troubles and remedies in two-stroke cycle engines are similar to those for the four-stroke cycle automotive type gasoline engines, the difference being mainly due to the method of getting the fuel and air mixture into the combustion chamber. In most small two-stroke cycle engines, the fuel mixture passes through the crankcase and enters the combustion chamber through ports uncovered by the piston.

It should be distinctly understood that this chapter in the "Remedies" sections tells you *what to do*—not *how to do*. To give service technique here would be a useless repetition of instructions given in other sections of the book.

SERVICE DIAGNOSIS
Conditions—Possible Causes—Remedies

1. ENGINE WILL NOT START

Possible Causes:
- *a.* Weak Battery.
- *b.* Corroded or loose battery terminal connections.
- *c.* Dirty or corroded distributor contact points.
- *d.* Weak coil.
- *e.* Broken or loose ignition wires.
- *f.* Moisture on ignition wires, cap or plugs.
- *g.* Fouled spark plugs.
- *h.* Improper spark plug gap.
- *i.* Improper timing (ignition).
- *j.* Dirt or water in gas line or carburetor.
- *k.* Carburetor flooded.
- *l.* Fuel level in carburetor bowl not correct.
- *m.* Supply of fuel insufficient.
- *n.* Defective fuel pump.
- *o.* Vapor lock.
- *p.* Defective starting motor.
- *q.* Open ignition switch circuit.
- *r.* Inoperative breaker points in magneto (when used).

Remedies:
- *a.* Recharge and test battery, as outlined in Electrical Chapters, using open circuit battery tester. If necessary, replace battery.

b. Clean, inspect and tighten battery terminals and clamps. Replace battery cables and clamps if badly eroded.

c. Clean and inspect contact points; if badly burned or pitted, replace points and condenser. Adjust gap to manufacturer's specifications and check timing.

d. Replace weak coil with a new one. Then, check condition of contact points and replace if necessary. See *c* above.

e. Replace broken ignition wires or those with cracked insulation. Tighten all connections at distributor, coil, ammeter and ignition switch. Be sure the spark plug wires are secure in distributor cap and coil tower.

f. Dry the wet ignition system with compressed air or a clean dry cloth. Remove the individual spark plug wire from cap; dry cavity and wire ends thoroughly. Inspect inside of cap and remove all traces of moisture and dirt.

g. Clean and tighten spark plugs. Adjust gaps to manufacturer's specifications.

h. See *g* above.

i. Check ignition timing. Replace parts as necessary to correct this condition.

j. Disconnect lines and clear with compressed air. Remove and clean carburetor. Drain tank and refill.

k. Check carburetor float level, and needle seat assembly. Check float for leaks and replace parts as necessary to correct this condition.

l. Fuel level should be ⅝ inch from top of body on *Stromberg* carburetor, and from top float $^5/_{64}$ inch (1.98 mm) plus or minus $^1/_{64}$ inch (.397 mm) below top surface of *Carter* carburetor body casting.

m. Refill fuel tank and then check gauge.

n. Replace defective fuel pump with a new one to correct this condition. Fuel pump pressure should be from 3½ to 5½ pounds.

o. Check for air and fuel restrictions around fuel pump and check for a misplacement of heat shield. Repair as necessary to correct this condition.

p. Repair or replace defective starting motor. Replace worn or damaged parts as required.

q. Turn on ignition switch and if ammeter shows a slight discharge it indicates that current is flowing. A glance at the fuel guage will indicate whether or not there is fuel in the tank. If no indication is obtained when turning the ignition switch the circuit if faulty and should be repaired.

r. See *c* above.

GAS ENGINE MANUAL

2. ENGINE STALLS

Possible Causes:
- *a.* Idling speed too low.
- *b.* Idle mixture too lean or too rich.
- *c.* Dirt or water in gas line or carburetor.
- *d.* Incorrect carburetor float level.
- *e.* Leak in intake manifold.
- *f.* Defective accelerator pump. (Stall occurs on acceleration.)
- *g.* Improper choke adjustment.
- *h.* Carburetor icing (cold wet weather).
- *i.* Dash pot ineffective.
- *j.* Corroded battery terminals.
- *k.* Weak battery.
- *l.* Spark plugs dirty, damp, or gaps incorrectly set.
- *m.* Coil or condenser defective.
- *n.* Distributor points dirty, burned or incorrectly spaced.
- *o.* Trailing edge of rotor worn.
- *p.* Leaks in ignition wiring.
- *q.* Burned or pitted valves.
- *r.* Engine overheating.

Remedies:

a. Reset throttle adjustment screw until engine idles at approximately 450 to 500 *rpm*.

b. Reset idle adjustment screw ½ to 1½ turns open for correct idle mixture. For richer mixture, turn screw out.

c. Disconnect lines and clear with compressed air. Remove and clean carburetor.

d. Fuel level should be ⅝ inch from top of body on *Stromberg* carburetor, and from top of float, $^5/_{64}$ inch (1.98 mm) plus or minus $^1/_{64}$ inch (.397 mm) below top surface of *Carter* carburetor body casting.

e. Check intake manifold, gasket and heat riser gasket. Replace parts as required to correct this condition. Tighten manifold stud nuts to a torque of 15 to 20 foot pounds.

f. Replace or repair defective accelerator pump. Replace parts as required.

g. Readjust automatic choke.

h. Open throttle as engine starts to stall. Keep motor at fast idle until condition clears.

TROUBLESHOOTING

    i. Check carburetor dash pot piston, leather and valve. Replace parts as required. Check travel; should be not less than $5/16$ inch or more than $11/32$ inch.

    j. Clean, inspect and tighten battery terminal posts and clamps. Replace battery cables and clamps if badly eroded.

    k. Recharge and test battery. If necessary, replace battery with a new one of the same type and capacity.

    l. Clean and tighten spark plugs. Adjust plug gaps to manufacturer's specifications.

    m. Replace defective coil and condenser with new ones, then check condition of distributor points and replace if necessary. Adjust gap to manufacturer's specifications.

    n. Replace distributor points and condenser. Set gap to manufacturer's specifications.

    o. Replace worn rotor with a new one. Check contacts in cap for burning or pitting. If necessary, replace cap.

    p. Replace broken ignition wires or those with cracked insulation. Tighten all connections at coil, distributor, ammeter and ignition switch. Be sure the spark plug wires are secure in distributor cap and coil tower.

    q. Replace or reface and grind valves.

    r. Refer to Cooling Chapter, for various causes of engine overheating. Correct as indicated.

3. ENGINE HAS NO POWER
Possible Causes:
- *a.* Incorrect ignition timing.
- *b.* Coil or condenser defective.
- *c.* Trailing edge of rotor worn.
- *d.* Defective mechanical or vacuum advance (distributor).
- *e.* Excessive play in distributor shaft.
- *f.* Distributor cam worn.
- *g.* Spark plugs dirty or gap incorrectly set.
- *h.* Low grade of fuel.
- *i.* Carburetor in poor condition.
- *j.* Dirt or water in gas line or carburetor.
- *k.* Improper carburetor float level.
- *l.* Defective fuel pump.
- *m.* Valve timing incorrect.
- *n.* Blown cylinder head gasket.
- *o.* Low compression.

p. Plugged or restricted muffler or tail pipe.
q. Clutch slipping.
r. Engine overheating.

Remedies:
  a. Check and reset ignition timing. Replace parts as necessary to correct this condition.
  b. Replace defective coil and condenser with new ones, then check condition of contact points and replace if necessary. Adjust gap to manufacturer's specifications.
  c. Replace worn rotor with a new one. Check contacts in gap for burning or pitting. If necessary, replace distributor cap.
  d. Check vacuum advance mechanism. Make adjustments or replace parts as necessary to correct this condition.
  e. Check distributor shaft play. Replace parts as required to correct this condition.
  f. Replace worn distributor cam.
  g. Clean and tighten spark plugs. Adjust plug gaps to .035 inch.
  h. Drain fuel tank and refill with a fuel that will give a more complete combustion.
  i. Remove and recondition carburetor. Replace parts as required to correct this condition.
  j. Disconnect lines and clear with compressed air. Remove and clean carburetor. Drain tank and refill.
  k. Fuel level should be ⅝ inch from top of body on *Stromberg* Carburetor, and from top of float, $5/64$ inch (1.98 mm) plus or minus $1/64$ inch (.397 mm) below top surface of *Carter* carburetor body casting.
  l. Replace defective fuel pump with a new one to correct this condition. Fuel pump pressure should be from 3½ to 5½ pounds.
  m. Reset valve timing.
  n. Replace blown cylinder head gasket with a new one. Tighten cylinder head stud nuts to a torque of 65 to 70 foot pounds, in proper sequence.
  o. Replace, or reface and grind valves.
  p. Remove plugged or restricted muffler or tail pipe and replace with a new one. Check for excessive carbon in combustion chamber.
  q. Refer to Clutch, for possible causes and remedies. Correct as indicated.
  r. Refer to Cooling, for possible causes and remedies. Correct as indicated.

4. ENGINE "SKIPS" OR MISSES
Possible Causes:
- *a* Incorrect carburetor idle adjustment.
- *b.* Dirt or water in gas line or carburetor.
- *c.* Dirty jets or plugged passages in carburetor.
- *d.* Burned, warped or pitted valves.
- *e.* Incorrect ignition timing.
- *f.* Leaks in ignition wiring.
- *g.* Moisture on ignition wires, cap or plugs.
- *h.* Defective spark advance mechanism.
- *i.* Excessive play in distributor shaft.
- *j.* Distributor cam worn.
- *k.* Spark plugs dirty, damp, or gaps set too close.
- *l.* Weak battery.
- *m.* Low grade of fuel.

Remedies:

a. Reset idle adjustment screw ½ to 1½ turns open, for correct idle mixture. For richer mixture, turn screw out.

b. Disconnect lines and clear with compressed air. Remove and clean carburetor. Drain tank and refill.

c. Remove carburetor and recondition. Replace parts as necessary to correct this condition.

d. Replace, or reface and grind valves.

e. Check and reset ignition timing.

f. Replace broken ignition wires or those with cracked insulation. Tighten all connections at distributor, ignition coil, ammeter and ignition switch. Be sure the spark plug wires are secure in distributor cap and coil tower.

g. Dry the wet ignition system with compressed air or a clean dry cloth. Remove the individual spark plug wires from cap; dry cavity and wire ends thoroughly. Inspect inside of cap and remove all traces of moisture and dirt.

h. Check vacuum advance mechanism. Make adjustments or replace parts as required to correct this condition.

i. Check distributor shaft play. Replace parts as required to correct this condition.

j. Replace worn distributor cam. Replace parts as required to correct this condition.

k. Clean and tighten spark plugs. Adjust plug gaps to .035 inch.

GAS ENGINE MANUAL

 l. Recharge and test battery. If necessary, replace battery with a new one of the same type and capacity.
 m. Drain fuel tank and refill with a fuel that will give a more complete combustion.

5. ENGINE MISSES WHILE IDLING
Possible Causes:
 a. Spark plugs dirty, damp, or gap incorrectly set.
 b. Broken or loose ignition wires.
 c. Burned or pitted contact points, or set with insufficient gap.
 d. Coil or condenser defective.
 e. Weak battery.
 f. Distributor cap cracked.
 g. Trailing edge of rotor worn.
 h. Moisture on ignition wires, cap or plugs.
 i. Excessive play in distributor shaft.
 j. Distributor shaft cam worn.
 k. Burned, warped or pitted valves.
 l. Incorrect valve tappet clearance.
 m. Incorrect carburetor idle adjustment.
 n. Improper carburetor float level.
 o. Low compression.

Remedies:
 a. Clean and tighten spark plugs. Adjust plug gaps to manufacturer's specifications.
 b. Replace broken ignition wires or those with cracked insulation. Tighten all connections at distributor, coil, ammeter and ignition switch. Be sure the spark plug wires are secure in distributor cap and coil tower.
 c. Clean and inspect contact points; if badly burned or pitted, replace points and condenser. Adjust gap to manufacturer's specifications and check timing.
 d. Replace defective coil and condenser with new ones, then check the condition of contact points and replace if necessary. Adjust gap to manufacturer's specifications.
 e. Recharge and test battery. If necessary, replace battery with a new one of the same type and capacity.
 f. Replace cracked distributor cap, then check rotor for burned conductor. Replace if necessary.
 g. Replace worn rotor with a new one. Check contacts in cap for burning or pitting. If necessary, replace distributor cap.

TROUBLESHOOTING

  h. Dry the wet ignition system with compressed air or a clean dry cloth. Remove the individual spark plug wires from cap; dry cavity and wire ends thoroughly. Inspect inside of cap and remove all traces of moisture and dirt.
  i. Check distributor cam play. Replace parts as required to correct this condition.
  j. Replace worn distributor shaft. Replace parts as required to correct this condition.
  k. Replace, or reface and grind valves.
  l. Adjust valve tappet clearance; intake .008 and exhaust .010, with hot engine.
  m. Reset idle adjustment screw ½ to 1½ turns open, for correct idle mixture. For richer mixture, turn screw out.
  n. Fuel level should be ⅝ inch from top of body.
  o. Refer to other sections, for possible causes and remedies. Correct as indicated.

6. ENGINE MISSES ON ACCELERATION
Possible Causes:
  a. Distributor points dirty or incorrectly spaced.
  b. Coil or condenser defective.
  c. Incorrect ignition timing.
  d. Spark plugs dirty, damp, or gap set too wide.
  e. Abnormal resistance in spark plugs.
  f. Dirty jets in carburetor, especially economizer jet or accelerator pump operating improperly.
  g. Burned or pitted valves.
  h. Low grade of fuel.

Remedies:
  a. Clean and inspect contact points; if badly burned or pitted, replace points and condenser. Adjust point gap to manufacturer's specifications and then check timing.
  b. Replace defective coil and condenser with new ones, then check the condition of contact points and replace if necessary. Adjust gap to manufacturer's specifications.
  c. Check and reset ignition timing.
  d. Clean and tighten spark plugs. Adjust plug gaps to manufacturer's specifications.
  e. Replace faulty plugs with new resistor plugs. Before installing, check gaps and if necessary, adjust to .035 inch. Clean seats, install new gaskets, and tighten with a torque wrench.

GAS ENGINE MANUAL

    f. Remove carburetor and recondition. Replace parts as required to correct this condition.
    g. Replace, or reface and grind valves.
    h. Drain fuel tank and refill with a better combustion.

7. ENGINE MISSES AT HIGH SPEED
Possible Causes:
    a. Dirt or water in gas line or carburetor.
    b. Dirty jets in carburetor, especially the economizer jet.
    c. Coil or condenser defective.
    d. Incorrect ignition timing.
    e. Distributor points dirty or incorrectly spaced.
    f. Trailing edge of rotor worn.
    g. Loose ignition wiring.
    h. Excessive play in distributor shaft.
    i. Spark plugs dirty, damp, or gaps set too wide.
    j. Abnormal resistance in spark plugs.
    k. Distributor shaft cam worn.
    l. Engine overheating.
    m. Low grade of fuel.

Remedies:
    a. Disconnect lines and clear with compressed air. Remove and clean carburetor. Drain tank and refill.
    b. Remove carburetor and recondition. Replace parts as necessary to correct this condition.
    c. Replace defective coil and condenser with new ones, then check condition of contact points and replace if necessary. Adjust gap to manufacturer's specifications.
    d. Check and reset ignition timing.
    e. Clean and inspect contact points; if badly burned or pitted, replace points and condenser. Adjust point gap to manufacturer's specifications and then check timing.
    f. Replace worn rotor with a new one. Check contacts in cap for burning or pitting. If necessary, replace distributor cap.
    g. Replace broken ignition wires or those with cracked insulation. Tighten all connections and be sure the spark plug wires are secure in distributor cap.
    h. Check distributor shaft play. Replace as required to correct this condition.

i. Clean and tighten spark plugs. Adjust plug gaps to manufacturer's specifications.

j. Replace faulty plugs with resistor plugs as required. Before installing, check gaps and if necessary, adjust to .035 inch. Clean seats, install new gaskets and tighten with a torque wrench.

k. Replace worn distributor shaft, and/or bushings. Replace parts as required to correct this condition.

l. Refer to Cooling Chapter, for possible causes and remedies. Correct as indicated.

m. Drain fuel tank and refill with a fuel that will give a better combustion.

HIGH OIL CONSUMPTION

8. EXTERNAL OIL LEAKAGE
Possible Causes:
- *a.* Outside oil lines.
- *b.* Timing gear case cover oil seal.
- *c.* Rear main bearing oil seal.
- *d.* Oil pan gaskets.
- *e.* Oil pan drain plug.
- *f.* Oil filter gasket.
- *g.* Clogged rear camshaft bearing drain hole.
- *h.* Tappet cover gaskets.
- *i.* Fuel pump or gasket.
- *j.* Timing chain cover gasket.

Remedies:

a. Check for oil leaks at filter tubes and oil gauge lines. Replace tubing or fittings to correct this condition. Be sure filter mounting bracket is fastened directly next to the cylinder head.

b. Replace chain case cover oil seal. Be sure and use a new cover gasket.

c. Replace rear main bearing oil seal. Be sure seal and gaskets are in correct location in the cap before installation.

d. Replace faulty oil pan gaskets to correct this condition.

e. Replace worn oil pan plug, using a new gasket.

f. Clean filter and cover, removing all traces of old gasket. Install new gasket, using care to be sure gasket is centered. Tighten, then run engine for five minutes. Then, inspect for leakage.

GAS ENGINE MANUAL

  g. Remove rear tappet chamber cover and open drain hole. Check gasket on cover and replace if necessary.
  h. Remove gaskets from tappet chamber covers and replace with new ones. Before installing, be sure all traces of old gasket have been removed from the machined faces of block. Wipe surfaces dry and install covers.
  i. Replace fuel pump to block gasket. Check fuel pump for oil leaks. Correct as necessary to relieve this condition.
  j. Replace timing chain cover gasket. Inspect oil seal and if necessary, replace.

9. OIL PUMPING AT RINGS
Possible Causes:
  a. Worn, scuffed or broken rings.
  b. Incorrect size rings.
  c. Out-of-round rings.
  d. Rings fitted too tight in grooves.
  e. Carbon in oil ring slots.
  f. Insufficient tension in rings.
  g. Stuck rings.
  h. Compression rings installed incorrectly.

Remedies:
  a. Replace worn, scuffed rings after a careful inspection of cylinder walls. Worn, wavy or scored walls are a contributing factor to high oil comsumption. Recondition walls.
  b. Replace incorrect size rings with new piston rings of the proper type.
  c. Replace out-of-round rings, after checking cylinder bore.
  d. Replace improperly fitted rings with new piston rings. Oil ring clearance in groove should be from .001 to .0025 inches, with a .007 to .015 inch gap.
  e. Remove rings and clean piston ring slots with a suitable cleaning tool. Check cylinder bore and if necessary, recondition before installing new rings.
  f. Replace weak rings with new rings after checking condition of cylinder walls.
  g. Free up stuck rings either by disassembly or a suitable "upper lubricant." If necessary, replace rings.
  h. Remove and replace incorrectly installed rings.

10. OIL PUMPING AT VALVES
Possible Causes:
- *a.* Worn valve stems or guides.
- *b.* Too much oil spray in tappet chamber.
- *c.* Intake valve stem guide in inverted position.

Remedies:
   a. Replace worn valves and guides as necessary to correct this condition.
   b. Reduce main and connecting rod clearances.
   c. Replace valve guide whenever this condition is apparent.

11. HIGH OIL CONSUMPTION DUE TO LUBRICATING OIL
Possible Causes:
- *a.* Oil level too high.
- *b.* Contaminated oil.
- *c.* Poor grade of oil.
- *d.* Thin, diluted oil.
- *e.* Oil pressure too high.
- *f.* Sludge in engine.

Remedies:
   a. Add oil only when level reaches add oil mark. If oil level is over "full", drain sufficient oil to obtain correct level.
   b. Drain and refill crankcase with a good quality oil of the proper type and grade. Replace filter cartridge or filter after thoroughly cleaning filter can.
   c. Drain and refill crankcase with a good quality oil. Replace filter cartridge or filter after thoroughly cleaning filter can.
   d. Drain and refill crankcase with a good quality oil. Replace filter cartridge or filter after thoroughly cleaning filter can. Check operation of automatic choke.
   e. Replace oil pressure relief valve spring, as listed below: Springs are available in light (painted red) standard (not painted) and heavy (painted green). Make selection to suit need.
   f. Drain and refill crankcase with a good quality oil. Replace filter cartridge or filter after thoroughly cleaning filter can. Check thermostat. A thermostat that remains in the open position allows the engine to operate below normal temperatures, thus allowing sludge formation.

Gas Engine Manual

12. HIGH OIL CONSUMPTION—MISCELLANEOUS
Possible Causes:
 a. Overheated engine.
 b. Sustained high speeds.
 c. Misadjusted breather cap, causing excessive crankcase ventilation.

Remedies:
 a. Refer to the Cooling Chapter for correction of this condition.
 b. Avoid sustained high speeds at wide open throttle whenever possible.
 c. Check breather for correct position and readjust if necessary. Inspect crankcase ventilator outlet tube and oil drain passage in block for restrictions. Clean or repair as required to correct this condition.

ENGINE NOISES

13. PISTON RING NOISE
Possible Causes:
 a. Broken ring.
 b. Top ring striking cylinder ridge.
 c. Broken ring lands.

Remedies:
 a. Replace broken ring as required. Check to determine cause of breakage and correct as necessary.
 b. Remove ridge at top of cylinders as required using suitable ridge reamer. Check rings and piston for possible damage and replace parts as necessary.
 c. Replace pistons as needed. Check for ridge at top of cylinder wall and remove using suitable ridge reamer. Replace parts as required.

14. PISTON NOISE
Possible Causes:
 a. Piston pin fit too tight.
 b. Excessive piston to bore carbon accumulations in head.
 d. Collapsed piston skirt.
 e. Insufficient clearance at top ring land.
 f. Broken piston, skirt or ring land.

Remedies:

 a. Refit piston pins as required. Fit pins at 70 degree normal room temperature thumb press fit.

 b. Replace pistons as required. Check cylinder walls for excessive wear, if necessary, recondition cylinder walls and install new pistons and rings.

 c. Remove cylinder head and clean carbon from chamber, pistons, and valves. Drain and refill crankcase with a good grade of lubricating oil.

 d. Replace pistons as required. Check cylinder walls for possible scoring, recondition as necessary to correct.

 e. Check piston clearance. If necessary, refit pistons to correct this condition.

 f. Remove pistons as required. Check cylinder walls for possible scoring or damage. Recondition walls if necessary and install new pistons.

15. BURNED VALVES
Possible Causes:
- *a.* Insufficient valve clearance.
- *b.* Weak valve springs.
- *c.* Gum formation on stem causing valves to stick.
- *d.* Deposits on valve seats.
- *e.* Warped valves.

Remedies:

 a. Check and adjust valves with engine at normal operating temperature.

 b. Valve springs can be checked with testing gauge. Discard springs that do not test 107 to 115 lbs. when compressed to 1⅜ ins.

 c. Remove gum from valve stems, reinstall and adjust. Replace valves, if necessary.

 d. To remove deposits on valve seats, a valve grinding operation is necessary.

 e. A visual inspection will determine if valves be warped. Such valves should be replaced.

16. NOISY VALVES
Possible Causes:
- *a.* Incorrect tappet clearance.
- *b.* Worn tappets or adjusting screws.

c. Wear in cam lobes.
   d. Worn valve guides.
   e. Excessive runout of valve seat or valve face.

Remedies:
   Refer to Valve and Valve Gear Service for Noisy Valves.

17. CONNECTING ROD NOISE
Possible Causes:
   a. Low oil pressure.
   b. Insufficient oil supply.
   c. Thin or diluted oil.
   d. Misaligned rods.
   e. Excessive bearing clearance.
   f. Eccentric or out-of-round crank pin journal.

Remedies:
   a. Refer to lubrication chapter for possible causes of low oil pressure. Correct as indicated.
   b. Check oil level in crankcase, if necessary add oil to obtain correct level, or drain and refill. Test for possible loose or damaged rod bearings.
   c. Drain and refill crankcase and then test for possible loose or damaged rod bearings.
   d. Check rods for alignment. If necessary, straighten rod or install new one to correct this condition. Check bearing and journal for excessive wear. Replace parts as required.
   e. Replace worn bearings as required. Fit connecting rod bearings to the desired clearance of .0005 to .0015 and from .006 to .011 inches end play.
   f. Replace or regrind crankshaft as necessary. Replace with new undersize bearings after grinding operation is completed.

18. MAIN BEARING NOISE
Possible causes:
   a. Low oil pressure.
   b. Insufficient oil supply.
   c. Thin or diluted oil.
   d. Loose flywheel or fluid coupling.
   e. Excessive bearing clearance.
   f. Excessive end play.

g. Eccentric or out-of-round journals.
h. Sprung crankshaft.

Remedies:
 a. Refer to lubrication chapter for possible causes of low oil pressure. Correct as indicated.
 b. Check oil level in crankcase; if necessary, add oil to obtain correct level, or drain and refill. Test for possible loose or damaged main bearings.
 c. Drain and refill crankcase; then test for possible loose or damaged main bearings.
 d. Tighten flywheel or fluid coupling bolts to a torque of 55 to 60 foot pounds, then check engine for noise.
 e. Replace worn bearings as required. Fit main bearings to the desired clearance of .0005 to .0015 and from .003 to .007 inches end play.
 f. Refer to remedy *(e)* above for correction of this condition.
 g. Replace crankshaft or regrind journals as necessary. Replace with new undersize bearings when grinding operation is completed.
 h. Replace or straighten crankshaft as necessary, then check condition of bearings; replace as required.

19. BROKEN PISTON RINGS
Possible Causes:
 a. Wrong type or size.
 b. Undersize pistons.
 c. Ring striking top ridge.
 d. Worn ring grooves.
 e. Broken ring lands.
 f. Insufficient gap clearance.
 g. Excessive side clearance in groove.
 h. Uneven cylinder walls (particularly due to a previous ring breakage in same cylinder).

Remedies:
 a. Replace rings as required, after checking cylinder walls for possible scoring or grooving. When replacing rings, use only those that are factory engineered and inspected and the correct type and size for the engine being worked on.
 b. Fit new pistons and rings. Check cylinder walls for possible scoring or grooving. Recondition walls as required.

    c. Replace rings as required, after checking cylinder walls for possible scoring or grooving. Remove ridge and recondition walls, if necessary.
    d. Fit new pistons and rings, after checking cylinder walls for possible scoring or grooving. Recondition cylinder walls as required.
    e. Fit new pistons and rings as required, after checking cylinder walls for possible scoring or grooving. Recondition cylinder walls if necessary.
    f. Replace rings as required. Check walls for damage and recondition if necessary. Correct ring gap should be .007 to .015 inch.
    g. Replace broken rings as required. Inspect cylinder walls for damage and recondition if necessary. Correct side clearance in groove should be: oil rings .001 to .0025. Compression upper—.0025 to .004. Intermediate .002 to .0035 inches.
    h. Fit new pistons and rings, after reconditioning cylinder walls.

20. BROKEN PISTONS
Possible Causes:
    a. Undersize pistons.
    b. Eccentric or tapered cylinders.
    c. Misaligned connecting rods.
    d. Engine overheating.
    e. Cracks at expansion slots (excessive engine speed with no load).
    f. Water or fuel leakage into combustion chamber.
    g. Detonation.
    h. Resizing of pistons.

Remedies:
    a. Recondition cylinder walls if necessary; then "mike" walls and fit new pistons and rings.
    b. Recondition cylinder walls and fit new pistons and rings.
    c. Recondition cylinder walls if necessary. Then fit new pistons and rings. Realign connecting rods.
    d. Recondition cylinder walls if necessary; then, fit new pistons and rings. Refer to cooling chapter for possible causes and remedies of engine overheating.
    e. Recondition cylinder walls if necessary; then, fit new pistons and rings. Avoid the practice of "racing" the engine during the warm-up period or thereafter.
    f. Recondition cylinder walls if necessary, then fit new pistons and rings. Check cylinder head, gasket and cylinder block for leaks. Repair as necessary to correct this condition.

TROUBLESHOOTING

    g. Recondition cylinder walls if necessary, then fit new pistons and rings. Check for excessive spark knock and preignition or detonation, then adjust as required to correct this condition.

    h. Recondition cylinder walls if necessary, then fit new pistons and rings. Avoid the use of resized pistons whenever possible.

21. LOW OIL PRESSURE

Possible Causes:

    a. Thin or diluted oil.
    b. Oil relief valve spring broken or weak.
    c. Restricted oil pump screen.
    d. Excessive clearance in main or connecting rod bearings.
    e. Excessive clearance in camshaft bearings.
    f. Low oil level.
    g. Loose connections or restricted oil lines.

Remedies:

    a. Drain, flush and refill crankcase.

    b. Replace broken or weak relief spring with a new one of identical color.

    c. Remove strainer from car and wash in gasoline. Dry with an air hose and reinstall.

    d. and *e.* Check clearance in main, connecting rod and camshaft bearings, using bearing oil leak detector. Correct as necessary.

    f. Add oil to bring capacity up to the proper level.

    g. Check for restricted lines and clean out or replace where necessary.

CHAPTER 23

Engine Tune-Up

The term "tune-up" as applied to a gas engine is defined as: *The testing and servicing of the engine's various mechanisms upon whose proper functioning, satisfactory and efficient operation of the engine depends.*

These various mechanisms are the *starting, ignition, carburetor,* and *cooling* systems in addition to the *valves* and *valve gears.*

There are two kinds of tune up termed *minor* and *major.* A minor tune-up is confined principally to the ignition system, whereas a major tune-up comprises a complete engine diagnosis or overall check and servicing where necessary.

MINOR ENGINE TUNE-UP

A minor engine tune-up is intended as a preventive measure for engines which are in fairly normal condition. This tune up should be performed frequently in order to maintain the standard performance originally built into the engine. If the engine does not perform satisfactorily after a minor tune-up, a major tune-up including a compression test may be necessary.

A minor tune-up includes tests and servicing of:
1. Battery,
2. Spark plugs,
3. Distributor,
4. Magneto (when used),
5. Wiring circuits,
6. Ignition timing,

Gas Engine Manual

    7. Carburetion,
    8. Fuel filter.

Battery

Inspect the battery cable and ground strap for broken insulation, corroded or broken strands and loose or corroded terminals. Repair broken or chafed insulation with loom or tape. If cable strands are broken, corroded, or loose in the terminals the cables should be replaced with new cable of adequate current carrying capacity.

Clean and tighten all connections. Test for weak or discharged battery. Make a voltage test of the battery cells. Add water if necessary. Tighten all primary and high tension wire connections, particularly at the ignition starter switch, ammeter and fuel gauge behind the instruments.

Spark Plugs

Probably more fuel is wasted by faulty spark plugs than any other cause. Such alleged tests as the screw driver test or by laying the plug on top of the cylinder indicate nothing except that the plug is absolutely dead; and while a faulty plug may spark on very low compression as with engine idling, it will cease sparking with increasing compression as when load is put on the engine.

Most gas and service stations have spark testers to indicate conditions of plugs. In testing, the pressure applied should equal the maximum compression pressure as rated by the manufacturer. Before testing, however, the points should be adjusted to the proper gap as specified by the engine manufacturer.

REASONS WHY SPARK PLUGS FAIL

Effect Noticed	Probable Cause of Trouble	Suggested Remedy
Plugs fouled with oily carbon deposit — piston slap — poor compression — crankcase dilution — excessive oil passing pistons.	Loose pistons or leaking rings.	Use oil plugs until repairs are made — install new pistons, refinish cylinder walls if necessary and install new rings.
Fouled or carbonized plugs — excessive oil consumption — carbon in combustion chamber — sticking valve stems.	Unsuitable oil — perhaps low grade.	Change to suitable grade oil — Some engines require heavier oil than others — remove carbon and clean sticking valve stems.
Apparent misfiring — loss of power — popping back at carburetor — engine overheating — irregular running.	Sticking valves.	Clean valve guides and stems — set tappets at correct clearance — install new valves where stems are bent, or heads are warped.
Sharp metallic knock in engine, sometimes resulting in cracked insulators.	Low test gas.	Use better grade of gas.
Spark plugs fouled — covered with soot — engine runs irregularly — black smoke with disagreeable odor from exhaust.	Carburetor mixture too rich.	Readjust carburetor by cutting down on gasoline adjustment.

ENGINE TUNE-UP

REASONS WHY SPARK PLUGS FAIL (Continued)

Effect Noticed	Probable Cause of Trouble	Suggested Remedy
Plug insulator clean — popping back through carburetor — hard starting — engine heats up — possible cracking of insulators.	Carburetor mixture too lean.	Readjust carburetor to give richer mixture by opening needle valve slightly — look for air leaks in intake manifold.
Engine fails to start — carburetor floods — engine runs irregularly — spark plugs foul.	Excessive choking.	Do not use choke unnecessarily.
Engine misses at all speeds — inadequate spark at high tension cable terminals — spark plugs fouled.	Faulty ignition cables.	Replace leaky cables — shorten cables if possible — unnecessary length causes losses.
Weak spark at plugs — starter fails to turn engine, or turns it very slowly — dim headlights — engine refuses to start, or starts hard.	Weak battery.	Recharge battery — if it does not charge in a reasonable time, inspect cells and replace if necessary.
Engine misfires — weak or irregular spark at plugs.	Worn contact breaker points.	Redress with file or oil stone until points meet squarely — if tungsten tips on points are gone, replace with new ones.
Engine knocks — detonation — may result in cracked insulators.	Excessive spark advance.	Reset spark advance, if full automatic — retard spark, if manually operated.
Engine sluggish — lacks power and overheats.	Excessive spark retardation.	Change spark setting to proper position.
Misfiring, or weak and irregular spark.	Film of dirt, or moisture on distributor, coil, insulators, or cables.	Clean and dry thoroughly.
Misfiring — cutting out at high speeds or loads.	Excessive spark gap.	Adjust gap.
Preignition — back firing through carburetor — electrodes wear rapidly — blistered insulators — possible breaks at insulator tips.	Too hot a plug.	Replace with colder plug.
Engine misfires — starts hard — lacks power — plugs may foul badly.	Too cold a plug.	Replace with hotter plug.
Misfiring or fouling.	Possible losses in radio suppressors.	Try performance with suppressors removed from plugs.
Engines miss at low speeds — Plugs fouled.	Corroded high tension terminal or defective distributor cap.	Clean terminal — Replace defective cap.
Arcing from terminal to gland nut or ground.	Wide gap.	Use smaller spark gap — switch primary ignition wires on coil.
Engine misses or runs irregularly with radio plugs.	Open circuit in radio resistor or too much resistance.	Replace plugs having defective resistors.
Engine fails to start or plugs are wet or carbonized.	Defective condenser.	Replace with a new condenser.
Engine misses at high speeds.	Weak or loose breaker point spring.	Replace with new points or tighten spring end.
Engine misses at high speeds.	Weak valve springs — worn valve guides — fuel pump diaphragm leaking.	Replace spring and guides if defective — replace pump diaphragm.

Carefully inspect the insulators and electrodes of all spark plugs. Replace any plug which has a cracked or broken insulator or with loose electrodes. If the insulator is worn away around the center electrode, or if the electrodes is burned or worn so they cannot be adjusted to the proper gap, the plug is worn out and should be replaced.

Distributor

Adjust breaker points. Inspect distributor cap and rotor for cracks and corrosion. Inspect small lead wires for breaks and damaged insulation. Inspect distributor advance plate bearing for excessive play.

Every tune-up job must include a complete distributor test. The testing should be done in accordance with the testing equipment manufacturer's instructions and recommendations giving necessary attention to all parts of the distributor, including the breaker points.

Condenser Test. An open circuited condenser will cause ignition trouble. Trouble of this nature is usually caused from a broken lead, bad connections inside the condenser or a poor ground connection. A condenser with an open circuit can also be checked with test points and when charge has been placed on condenser touch condenser lead to the condenser shell to see if spark occur. See Figs. 1 and 2.

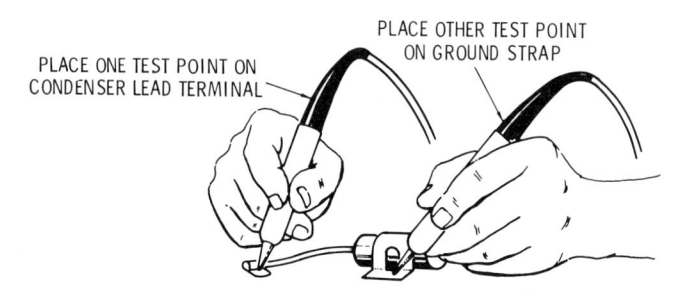

Fig. 1. Testing condenser for short circuit. The test lamp will light up if the condenser is short circuited. When using test points to check condenser, use 220 volt a-c. A short circuited condenser is regarded as a broken down condenser and should be replaced.

With *ac* it is sometimes necessary to apply the voltage more than once. Sometimes the points are removed from the condenser terminal just at the moment the alternating current is changing direction; therefore, spark will not be obtained upon discharge.

Sometimes a condenser will not be completely open, but will have a high resistance connection which may be due to a poorly soldered lead. Condenser testers are being used that will detect a high resistance connection inside the condenser. This check or condition is referred to as "dampening."

Engine Tune-Up

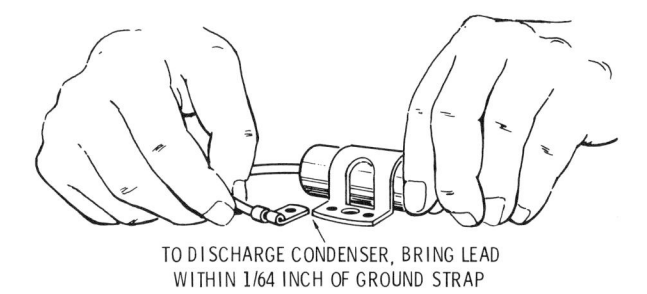

TO DISCHARGE CONDENSER, BRING LEAD
WITHIN 1/64 INCH OF GROUND STRAP

Fig. 2. Testing a condenser for open circuit. Trouble of this nature is usually caused from a broken lead, bad connections inside the condenser or poor ground connection.

Capacity Requirements. It has been found that a lower capacity condenser is suitable for continuous high speed operation while a higher capacity condenser is suitable for continuous low speed operation.

The capacity of a condenser used to satisfy average operating conditions is between .20-.25 microfarads. Condensers in automotive use range between .15-.45 microfarads capacity. Fig. 3 shows pitted contact points caused by using condenser with incorrect capacity for normal operating conditions.

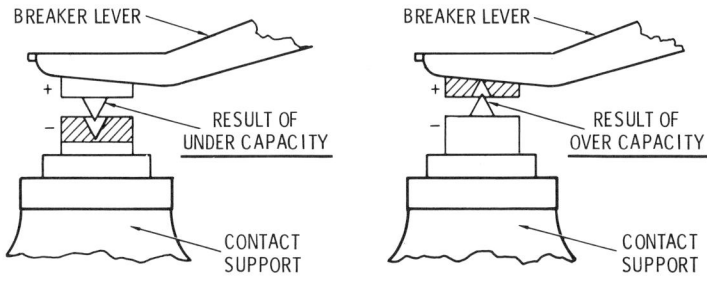

BREAKER LEVER — RESULT OF UNDER CAPACITY — CONTACT SUPPORT

BREAKER LEVER — RESULT OF OVER CAPACITY — CONTACT SUPPORT

Fig. 3. Illustrating pitting of breaker contacts due to condenser of wrong capacity. Condenser capacity is determined largely by the type of operation and the rapidity of contact point pitting. A condenser with a capacity that is suitable for one operation may not be suitable for another operation.

High Tension Distributor Switch. This comprises a distributor cap and rotor forming a rotary switch, which ordinarily has as many contacts as there are engine cylinders.

The rotor connects a central contact to each of these contacts in turn, and from there the current follows a secondary cable to the spark plug. These secondary cables are connected to the various spark plugs according to the firing order of the engine.

GAS ENGINE MANUAL

Distributor Cap and Rotor Troubles. Distributor caps and rotors are made of bakelite and very often a dirty tract is formed inside the cap where the rotor contact travels. This should be wiped clean, and if the insulation of the cap be burned to any extent, the cap should be replaced with a new one.

When replacing either the distributor cap or the rotor, care must be taken to use the proper replacement, otherwise the distributor may not make contact at the same time the spark occurs, which would result in rapid burning of the new parts and poor performance of the engine.

Distributor caps that have no protective cover soon become covered with dust. A good way to clean out the distributor cap is to use a string to work the dirt out from between its terminals. By renewing the string frequently, it is possible to get the distributor cap quite clean and dry. Absorbent twine of a rather large diameter is suitable for this purpose.

Testing Breaker Arm Spring. The breaker arm spring should hold the contacts firmly closed, except when opened by the cam, and without sufficient tension to cause rapid wear.

The tension can be tested with a small spring scale and when this method is used the reading should be taken with the points just separated, the scale hooked over the contact arm at the breaker point, and held so that it is perpendicular to the current arm.

Magnetos

In most two stroke cycle engines, the magneto is seldom the cause of trouble. Although magnetos have so-called permanent magnets, sometimes due to abuse or carelessness such as dropping the magnet rotor, most of the magnetism may be lost, thus making remagnetizing necessary.

To remagnetize a magnet rotor proceed as follows:
    a. Remove magnet motor from magneto.
    b. Determine polarity of both magnet rotor and magnetizer by means of compass.
    c. Place magnet rotor between jaws on tester as noted in Fig. 4. Note that unlike poles of both the magnet rotor and magnetizer must be placed together, that is, the jaw at the North pole end of the magnet rotor must rest on top of the South pole of the magnetizer.
    d. With magnetizer properly connected to switch, allow current from batteries to flow through magnet rotor for about five seconds, disconnect current for about three seconds. Repeat the foregoing procedure three or four times.

ENGINE TUNE-UP

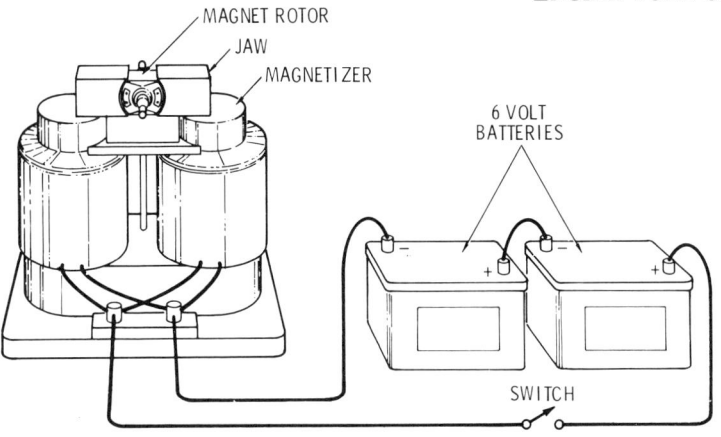

Fig. 4. Typical magnet rotor remagnetizer showing wiring of essential components.

After disassembly of magneto all metal parts should be thoroughly cleaned in gasoline and dried with compressed air. All parts should be inspected for damage or wear.

The breaker contacts should be adjusted for an opening as specified by the manufacturer (usually 0.020 in.).

Wiring Circuits

The circuits and parts that should be tested and replaced if necessary are:
1. Starting circuit,
2. Ignition circuit,
3. Lighting circuit.

Starting Circuit Test. When testing the starting circuit, a voltmeter should be used to determine its condition under actual operating conditions.

Attach the negative voltmeter test lead to the engine for the ground connection, and the positive lead to the starting motor switch, where the cable from the battery fastens for the positive connection. (This connection is used when the negative post of the battery is grounded; reverse the connection if the opposite is true).

By cranking the engine with the starter, a discharge load will be put on the starter circuit. If starter turns the engine at a good rate of speed, the average voltage reading should be between 4½ to 5 volts and double for 12 volts circuits.

311

Gas Engine Manual

For satisfactory circuits the starter should crank engine for 15 seconds without any perceptible voltage drop because of the drain on the battery. Such performance indicates a satisfactory circuit.

Battery Cable Test. Connect positive voltmeter test lead to the positive battery post, and the negative test lead to the battery cable terminal on the starter switch. Crank the engine for 15 seconds while observing the voltmeter reading. If the voltmeter shows more than 0.2 volts drop, recheck loose and dirty terminals. If the terminals are tight and clean, replace the cable.

The battery ground cable test is shown in Fig. 5. This test is made in the same manner as the battery cable test except that the negative voltmeter lead should be connected to engine frame, and the positive voltmeter lead connected to the negative battery post.

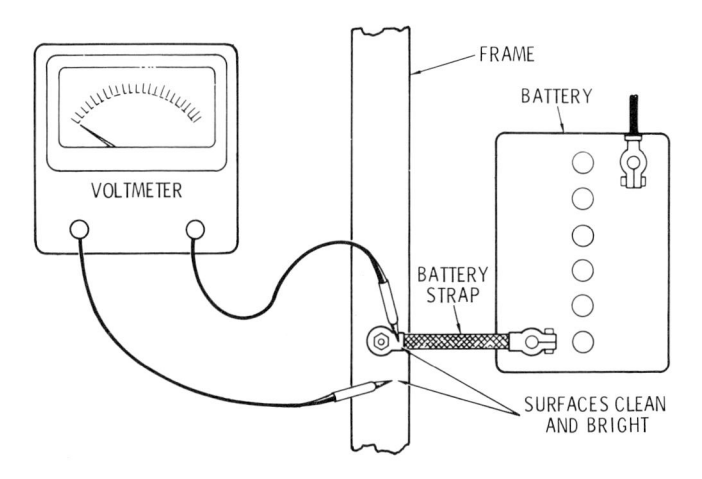

Fig. 5. Testing efficiency of battery ground connections by means of voltmeter.

Generator and Starter Circuit Tests. Whenever the starter, generator or voltage regulator requires servicing, the wiring circuit should be checked for loose or defective connections and frayed or damaged wires.

High resistance is frequently the underlying cause of many electrical difficulties that cannot be permanently repaired until the cause is located and corrected.

To check for resistance (voltage drop) in the starter and generator circuits, the following equipment is required:

An accurately calibrated voltmeter with 10 volt scale graduated in 0.1 volt divisions or a millivolt meter of 500mV range.

ENGINE TUNE-UP

An ammeter with 0 to 50 ampere scale graduated in 1 ampere divisions.

A battery hydrometer.

The battery should be checked with a hydrometer to establish its specific gravity.

If the battery is not fully charged, replace it temporarily with one that is.

The ammeter should be connected with heavy short leads between the B, terminal of the voltage regulator and its base ground screw.

All checks should be made with the engine running and the generator charging at 10 amperes. The voltmeter reading should remain stable.

If no readings can be obtained on the voltmeter, clean and tighten all ground connections, especially the generator frame bracket where it is bolted to the engine, and the battery ground strap.

MAJOR ENGINE TUNE-UP

A major tune-up comprises an overall check and service as required. In addition to the tests and servicing included in a minor tune-up, a major tune-up includes such items as:

1. Battery,
2. Tappet adjustment,
3. Valve tuning,
4. Compression test,
5. Vacuum gauge test,
6. Carburetor adjustment,
7. Cooling system.

Battery

Clean and tighten connections. Tighten all primary and high tension wire connections, particularly at the ignition starter switch, ammeter and fuel gauge behind the instrument panel.

Specific gravity reading of the electrolyte must be taken before adding water, as water will not mix with the electrolyte immediately and a true reading will not be obtained.

A battery in good condition should have a specific gravity reading of not less than 1.250. A battery with a specific gravity of less than 1.235 must be recharged. Add pure distilled water to bring level of electrolyte to one quarter inch above the plates in each cell.

Gas Engine Manual

Tappet Adjustment

Values for tappet clearances are usually given with the engine hot, that is, running temperature. Follow manufacturer's recommendations for the particular engine to be serviced. Typical method used to measure tappet clearance is shown in Fig. 6. Here the clearance and thickness of feeler stock is greatly exaggerated so that they may be clearly visible.

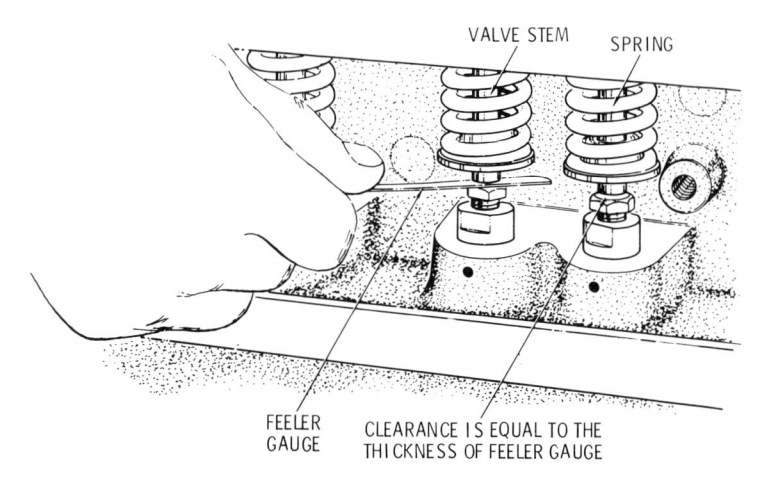

Fig. 6. Testing adjustment of tappets by feeler stock.

Compression Test

Satisfactory engine operation depends upon adequate and uniform compression in all cylinders. Loss of compression results in loss of power and nonuniform compression in cylinders causes unsatisfactory or jerky operation. The compression test is therefore important.

When making the test it is essential that the engine be at operating temperature and that the engine oil be of the proper grade. To make the compression test proceed as follows:

Remove spark plug of each cylinder to be tested and with engine warmed to working temperature, throttle open and ignition switch off, apply compression test gauge to spark plug hole and crank engine by hand or with starter. A check valve in the tester holds the compression in the gauge until released by the operator, permitting an accurate reading to be made. See Fig. 7.

Test each cylinder and record each reading. A variation in pressure in some makes of engines is small and large in others. The variation is

ENGINE TUNE-UP

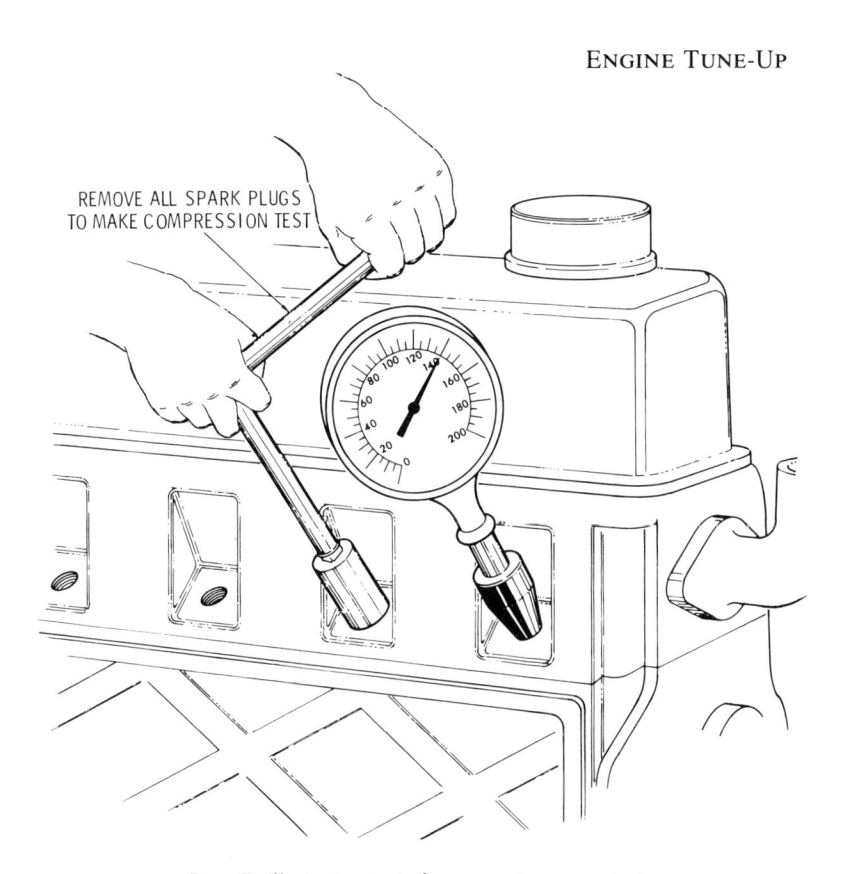

Fig. 7. Illustrating typical compression gauge test.

principally due to lack of uniformity in combustion chambers in a cylinder exactly the same size.

Summing up, it may be said that a pressure variation of two to four pounds in a high compression engine is permissible, whereas five to seven pounds are objectionable, and if more the cause for the low compression should be found and remedied.

Starting Motor

Inspect brushes, commutator and switch. Check tension of brush spring by comparing old springs and new ones. Test voltage and current flow at cranking speed.

Distributor

In a testing fixture, check distributor performance at various speeds. Check the automatic governor advance and the vacuum advance. Test condenser in suitable testing equipment if available.

Gas Engine Manual

Ignition Coil

Test with coil tester for output at high and low speeds, and for shorts or open circuits. Test coil at normal operating temperature since a cold coil may appear to be satisfactory under test and yet may not be operating properly when warmed up to its normal working temperature.

Generator

Test generator and voltage regulator with voltmeter and ammeter.

Fuel Pump

Check the fuel pump pressure with a low reading pressure gauge. Replace diaphragm, check valves or entire pump assembly as necessary.

Muffler and Tail Pipe

Inspect for clogged or choked muffler, damaged baffles, kinks in tail pipe or other conditions which may affect engine performance.

Cylinder Head and Manifold

Tighten cylinder head cap screws and manifold nuts to specified torque while engine is at its normal operating temperature.

Carburetor Adjustment

When adjusting the carburetor, be sure the engine is at normal operating temperature. Use either a vacuum gauge or combustion analyzer for accurate adjustment. To insure normal engine performance the following adjustments should be made or checked:

1. Idle air mixture,
2. Idle speed control,
3. Accelerator pump,
4. Dash pot pump.

Idle Air Mixture Adjustment. The idle needle valve controls the fuel mixture and when turned clockwise gives a leaner mixture, while counterclockwise a richer mixture.

Idle Speed Control Adjustment. If a tachometer is available, connect to the engine, then adjust idle speed control screw either in or out until a speed of 400 to 500 *r.p.m.* has been established.

Accelerator Pump. In order to provide the additional fuel required for rapid acceleration, the carburetor is equipped with a pump which supplies an extra charge of fuel momentarily as the throttle is opened.

ENGINE TUNE-UP

There are usually three positions provided on the accelerator pump lever to give a greater or lesser discharge of fuel depending upon climatic conditions. Adjust accelerator pump lever to long, medium or short stroke as conditions require.

Dash Pot Pump Adjustment. To adjust, turn the adjusting screw in or out until a maximum travel of $5/16$ to $11/32$ ins. is obtained on the mechanical dash pot carburetor. Unless the adjustment is known to be improper, check with a rule before making adjustment.

Cooling System

In order to get the maximum efficiency from the cooling system, it must be kept clean. There is a tendency toward corrosion of parts due to electrolytic action of water containing minerals and also deposits of minerals when water is heated.

Both the corrosive scale and the mineral deposits tend to coat the cooling surfaces, reducing radiation and in time will clog the radiator surfaces, unless special steps are taken to prevent these deposits.

The cooling system should be cleaned at least twice a year. This cleaning is most effective when reverse flushing is used to remove deposits after they have been loosened by the use of a good cleaning solution.

Cooling System Protectors. The regular use of a cleaning and inhibiting fluid in the cooling system and periodic reverse flushing will greatly reduce the formation of rust, scale and corrosion.

Fan Drive Clutch

When engine is cold and not operating, the fan should be easily rotated by hand (be sure fan belt is tight). Start engine and allow it to warm-up to normal operating temperature. After it has warmed-up, allow engine to run for *at least* five minutes. Turn off engine and *immediately* check to see if fan can be rotated (use gloves or a cloth to protect your hands). If it takes a great amount of effort to rotate the fan, the *fan drive clutch* is operating properly. If little effort is required, the clutch should be replaced.

CHAPTER 24

Cylinder Block Service

In operation, the cylinder of a gas engine wears out of trueness because of the angularity, or various angular positions passed through by the connecting rods during the compression and power strokes. The causes of wear are the lateral thrusts of the piston against the cylinder walls, due to compression and power impulses.

The compression thrust acts on one side of the cylinder and the power or impulse thrust on the other as indicated in Fig. 1. The result is to change the cross sectional shape of the cylinder from circular to elliptical. This results first in a loss of compression which together with the greater leakage during the power stroke is one of the causes of power loss in old engines.

Fig. 1 is drawn under the assumption that the thrust is constant during the entire stroke in order to show more clearly that the compression thrust wear on the right side less than the power thrust on the left side. In actual operation, however, the thrust is anything but constant, varying at all points of the compression and power strokes.

When the scoring is too deep to be removed by honing, it will usually become necessary to regrind the cylinder bore slightly oversize, and refit the cylinders with oversize pistons and rings. In some cases of cylinder scoring it may be possible to bore the cylinder to a considerable oversize and press a prepared sleeve of the exact size required into the cylinder.

RECONDITIONING CYLINDER BORES

There are various methods of reconditioning cylinders, as by:
1. Boring,
2. Grinding,
3. Reaming,
4. Lapping,
5. Honing.

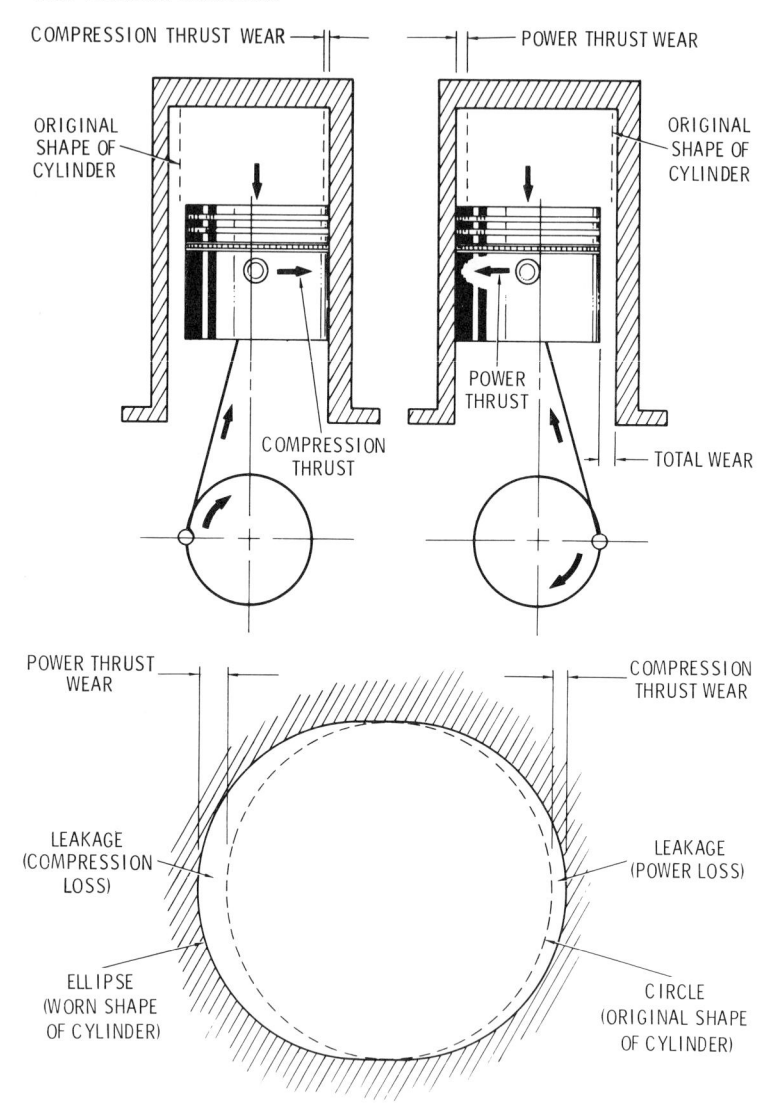

Fig. 1. Greatly exaggerated view of compression and power thrusts of piston which cause wear.

Boring

Before using the cylinder boring bar, the top of the cylinder block should be filed off to remove any dirt and burrs. The holding tool is then

Cylinder Block Service

placed lengthwise into the cylinder bore next to the one that is going to be worked on. It is expanded with a special wrench so that it is tight enough to hold its own weight.

Wipe off the top of the block and face of the boring machine. Set the boring machine on the block and adjust the height of the hold-down screw until it will just enter the saddle.

Center the boring bar over the cylinder to be bored; raise the feed screw lever to the off position; crank down the boring bar until the centering tools on the boring bar are below the ring travel in the bore; expand the centering tools to center the boring bar in the cylinder bore. After these preliminaries, set the cutter to the oversize required, say .020 in.

Example.—

	3.3125 in. Standard size of cylinder bore
plus	.0200 in Size of oversize piston
	3.3325 in Total
less	.0020 in. Allowance for honing
	3.3305 in Size of hole to be bored

Having determined the size of the hole to be bored, set the boring bar cutter micrometer to 3.3305 in. Bring the lock over against scale and tighten screw to set micrometer. Loosen the plunger lock in the cutter holder; install cutter and holder in micrometer; turn micrometer handle up until it locks; hold in place and tighten plunger lock. Cutter is now set to bore a 3.3305 in. hole.

Install cutter in boring bar and lock in place. Run the boring bar down until the cutter is ready to enter the cylinder and place feed lever in the on position. Turn on the machine and bore through cylinder.

When finished boring, place the feed lever in the off position and crank boring bar up. Continue as previously described until all holes are bored and then finish cylinder bores with a hone.

Grinding

The operation of truing up a cylinder by grinding is performed on a special machine which is equipped with an abrasive wheel geared to revolve at a high rate of speed on the end of a rigid spindle. The spindle is arranged so that it moves at the same time in a circular path and in so doing, the revolving wheel travels around the cylinder bore to be ground. The path of the spindle is adjustable to the cylinder diameter. Grinding gives a true smooth surface. A lathe cut is not as smooth as is obtained by grinding.

Reaming

A reamer consists of a number of cutting blades mounted radially in a substantial body, which can be adjusted in or out radially to cut the required diameter. By means of a micrometer they may be adjusted to .001 inch.

The reamer is self-aligning, self-centering and self-piloting. In use, the cylinder reamer is inserted into the full length of the bore and expanded until it will hold its own weight. The reamer blades are then backed off one full turn and the reamer removed. Next the reamer is expanded one full turn, to the exact graduation on the scale started from, and .005 in. more. The reamer is then passed through the cylinder bore and backed off two full turns and removed.

Wipe out the bore and check its size with a dial indicator. This will show the exact size of the bore, and from this it can be determined just how much stock has yet to be removed for the size of piston being fitted. **Example.**—*Assume that the dial indicator shows a .006 in. oversize bore. There remains .003 in. stock to be removed to bring the bore to .009 in. oversize for a .010 in. oversize piston, which leaves .001 in. for finishing. The reamer is expanded the two full turns, which was backed off at the first cut for removing from the bore, and an additional .003 in. The reamer is then passed through the bore and the cylinder is ready for finishing. The same procedure is followed for the other cylinders.*

Lapping

With this method only a very small amount of metal can be removed from the cylinder walls as the operation is very tedious. It is a form of grinding in which a metal surface coated with a grinding compound is rubbed against the cylinder walls with both up and down rotary motions.

Honing

This operation is performed by a tool which consists of cylindrical frame having mounted grinding stones which are pressed against the cylinder walls by springs or expanders.

In operation, the hone is placed into the cylinder bore and expanded until it can just be turned by hand. It is then revolved and moved up and down in the bore until it begins to run freely. Then the expanding nut on the top of the hone is tightened and the operation repeated until the desired amount of metal is removed, that is, until the piston being fitted can be pushed through the cylinder on a .002 feeler and locked on a .003 feeler. The feeler gauge must be at right angles to the wrist pin.

Cylinder Block Service

REMOVING CARBON

When cylinder walls and valve chambers during operation have become coated with carbon, it must be removed to avoid pre-ignition and resulting unsatisfactory operation.

To remove carbon, it is scraped off with a hard, sharp edged tool. For cleaning out the ring grooves, a special tool should be used made to fit so closely as to leave no deposit under the end or by the edges. Keeping the deposits moist with kerosene will facilitate their removal; soaking with kerosene for hours or even days will still be better. For surfaces that can be reached in this manner, and that will not be injured by the wear it will cause, finishing may be done with coarse emery cloth, held in the hand or around a stick.

Scraping is the best method of removing carbon. It can also be removed by burning but this method is not recommended unless properly done by an expert, otherwise damage may result.

Importance of Cleaning

The serious trouble experienced by not throughly cleaning a reconditioned job cannot be too overemphasized. Recondition tool manufacturers have united in their efforts to overcome this bad practice of carelessly finishing and cleaning reconditioned jobs.

A cylinder that is rough or has not been throughly cleaned acts as a lap and causes excessive wear to the entire piston assembly and results in dissatisfaction to all parties concerned.

During the first half of the stroke (approximately) where the mean effective pressure is high and the angularity of the connecting rod is increasing, the wear is greatest and this is called the *zone of maximum wear*.

During the second half of the stroke where the mean effective pressure is low and the angularity of the connecting rod is decreasing, the wear is less and this is called the *zone of minimum wear*. This action tends to wear the walls partly cone shaped or tapered, being made more pronounced on account of the lower cylinder walls receiving better lubrication than the upper walls. From the foregoing it follows that when cylinders are worn both out of round and tapered, refinishing to true surface becomes necessary.

SCORING OF CYLINDERS

Seizing and sticking of pistons in the cylinders is commonly due to overheating or lack of lubrication or both. In almost every case of this sort piston rings will be damaged and cylinders scored.

Such scratches or scores will run lengthwise in the cylinder walls and if not too deep may be honed out with a suitable cylinder hone, prior to substitution of new piston rings.

CHAPTER 25

Piston and Piston Rings Service

When cylinders and pistons are properly machined and there is proper clearance and lubrication, there is no reason for excessive wear. When any one of the foregoing requirements are not obtained, abnormal wear will result.

There may be ample lubrication, but if the oil is dirty or mixed with water, its efficiency is greatly reduced. Under good lubrication conditions with clean oil of the proper viscosity, no appreciable amount of wear will take place.

EXPANSION OF PISTONS

Each piston type has its own expansion characteristic under variable temperatures in operation. This governs the method of finishing to be used for best results. Operating temperatures to which a piston is subjected depend upon several conditions:

1. Working loads,
2. Fuel and lubricant,
3. Climatic conditions,
4. Cooling.

The expansion of a piston depends upon the rise in temperature, hence this must be taken into account in determining the clearance to be given in fitting.

The temperature range will depend upon the service conditions. Thus engines of light duty cars will not be subjected to so great a temperature range as heavy duty engines, and it follows that more clearance must be given pistons for heavy duty than for light duty.

REMOVING PISTONS FROM CYLINDERS

On some engines, pistons may be removed from the top; on others they must be removed from the bottom.

For a top removal, take off cylinder head and oil pan and turn crankshaft until the connecting rod is on lower dead center. Remove connecting rod cap and push up rod clear of crank pin; place cap in position and bolt together loosely so that it will not get mixed with the other caps.

With any tool such as a screw driver, rod, or hammer handle placed inside the piston, push up piston until the first ring is above cylinder and has sprung out so it will hold piston suspended. By grasping the end of piston, the assembly of piston and rod may be lifted out of cylinder.

On engines where the piston must be removed from the bottom of the block, rotate the crankshaft until the counterweights are crosswise of the cylinder block and opposite the camshaft.

Pistons may be ordered from manufacturers in three different stages of completion known as:
 1. Rough,
 2. Semi-finished,
 3. Finished.

Semi-finished are completely machined with exception of the outside diameter which is left larger so that it may be ground down to proper size to fit the cylinder. The wrist pin holes are rough reamed and must be further reamed to size. A good adjustable hand reamer may be used to finish.

FITTING PISTONS

First, check piston and pins removed from engine with new pistons and pins before installing, to avoid possible errors in ordering. Make certain that cylinders are round, the same diameter top and bottom, and at right angles to (square with) the base.

As recommended for *Arrowhead* iron pistons they must have extra clearance on top ring lands, a maximum of .004 in. (four thousandths of an inch) for each inch of diameter on top land and .003 inch (three thousandths of an inch) for each inch of diameter on the next to the top land.

PISTON AND PISTON RINGS SERVICE

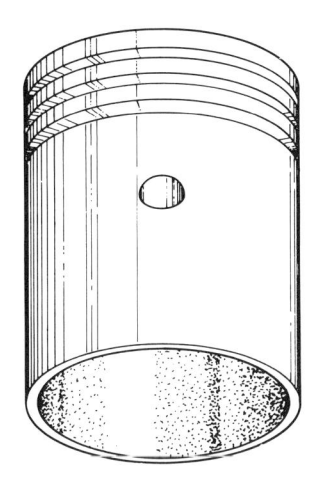

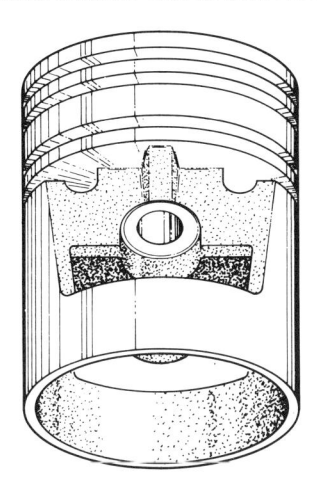

Fig. 1. Illustrating oval or cam ground piston and spring type piston with supporting band respectively. In the spring type when the piston heats up, the band expands enough to bear on the cylinder walls. In this way a bearing located at the maximum distance from the wrist pin is obtained which gives minimum bearing loads at this point.

The skirt of a cast-iron piston should be fitted to the cylinder with a clearance of at least three-quarters of a thousandth (.00075 per inch of diameter) and not more than one thousandth per inch of diameter clearance is necessary with pistons of larger diameters. Failure to provide proper clearances may cause either scored cylinders or "freezing" which seizure might pull off the head of a piston or cause the breakage or distortion of other parts such as pins, connecting rods, etc.

Particular care must be used to give the rings sufficient end and side clearance, for failure to do this might also cause "freezing." In reaming the pin holes, the better method is to clamp the reamer and turn the piston around the reamer. If the piston is clamped, it may be distorted causing pin holes to be out of alignment when the pressure is released. Furthermore, if a piston is put in a vise, it is liable to be cracked or forced out of round. It is best to buy fitted assemblies.

In the installation of a piston, never use force on any part of it or in any way. Should it be necessary to place pistons in a lathe or grinder, it is difficult to secure the greatest accuracy with any device that exerts either inside expansion or outside pressure.

While bell centers or chucks may be used, for best results a universal grinding and turning arbor or any similar device that will positively

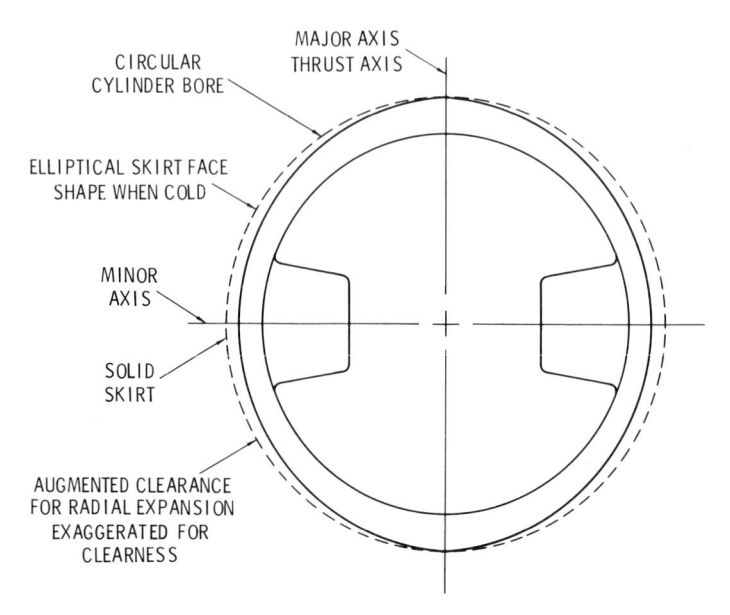

Fig. 2. Bottom view of piston illustrating elliptical or cam grinding, exaggerated for clearness.

maintain uniform roundness while the pressure of tool or grinding wheel is being applied, should be used. This permits an experienced workman to remove accurately even as little as a thousandth of an inch of material from the outer surface.

When reassembling the engine, cleanliness is most important. Dirt and chips must be eliminated or all the care used in fitting various parts may be nullified. Just before installing, dip piston in oil and be sure that the entire engine is properly lubricated.

PISTON RING SERVICE

In ring installations, the selection of the proper type piston rings is of great importance in assuring a satisfactory job. Great advances have been made in the manufacture of rings and there is a great number of types from which to select.

The adaptation of the leading types as here briefly given will be found helpful to the serviceman in selecting the proper rings for best results.

Compression Rings. In comparatively new engines where cylinders and pistons show little or no wear, or when the clearances between

Piston and Piston Rings Service

pistons and cylinders are in accordance with manufacturers' recommendations, use standard compression rings.

Wide Channeled Oil Rings. In modern high speed high compression engines, either new or slightly worn, the wide channeled oil ring should be used in the oil ring grooves.

In four ring groove pistons, it is always advisable to use two oil rings. Make sure that all oil ring grooves are drilled or slotted for oil passage.

Single Ventilated Oil Ring. This ring is frequently substituted for the wide channeled ring because of its lower cost. It should be used in the oil ring grooves as recommended for wide channeled oil ring installation.

Flexible Rings. In cylinders worn out of round, flexible compression and flexible oil rings should be used. If the wear and taper is sufficient to cause piston slap, use flexible compression rings in all compression ring grooves. If slap is not apparent, a plain compression ring may be used in the top groove. Flexible oil rings should be used with flexible compression rings in all oil ring grooves. In four ring groove pistons install two flexible oil rings.

Compression Inner Ring. This type ring is designed for use with conventional type compression rings. Conditions governing its use vary the same as those for flexible compression rings.

Ventilated Inner Ring. This ring can be used with all types of slotted oil rings and should be used in all oil ring grooves. Their use will reduce oil consumption if used in engines that consume excessive quantities of oil.

Fitting Rings

In fitting rings great care should be used to determine that rings have proper end clearance. Care should also be exercised in following the ring manufacturer's specifications as to the proper ring set up for various types of pistons. After removing pistons, carefully examine the cylinders.

Wash the rings thoroughly with gasoline and dry with a clean cloth. Measure the cylinders to determine the amount of wear, taper or out of roundness. Fig. 3. In badly out of round, tapered or bellied cylinders, they should be re-honed or re-bored and new pistons fitted.

Next examine the pistons (if new pistons are not being installed) using micrometers to determine if they are worn or warped. (Fig. 4.) Clean the ring grooves thoroughly, removing all carbon. Use a heavy piece of cotton twine soaked in gasoline in the corners of the gooves.—(Fig. 5.)

Fit the rings to the cylinders first, starting from the top of the bore and forcing them downward to be sure that they have enough clearance at the gap all the way through the cylinders. See Fig. 7.

Gas Engine Manual

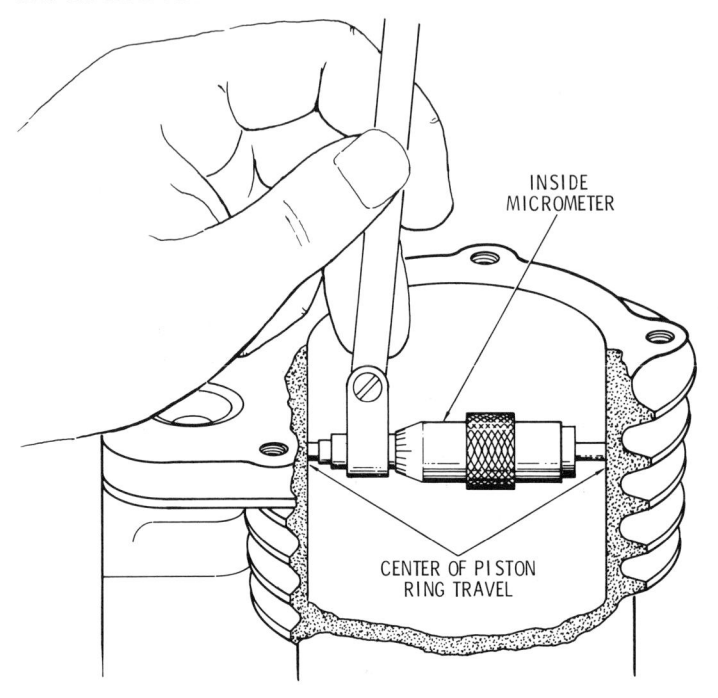

Fig. 3. Checking cylinder bore for roundness with inside micrometer.

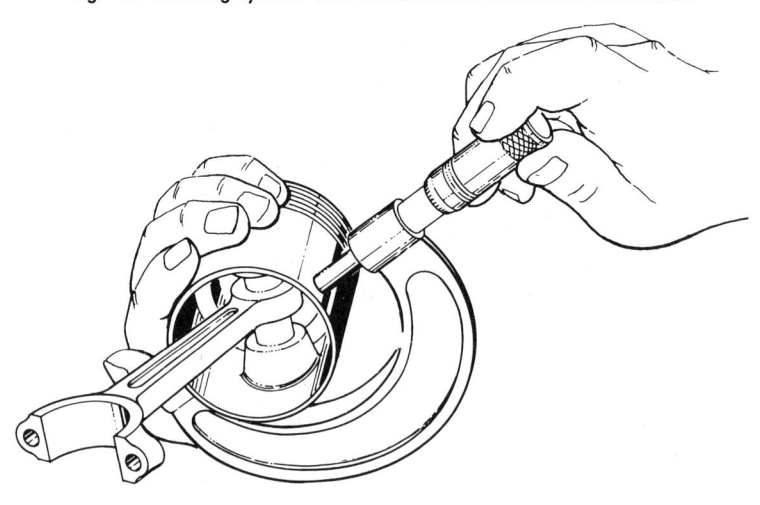

Fig. 4. Checking piston for roundness with micrometer.

PISTON AND PISTON RINGS SERVICE

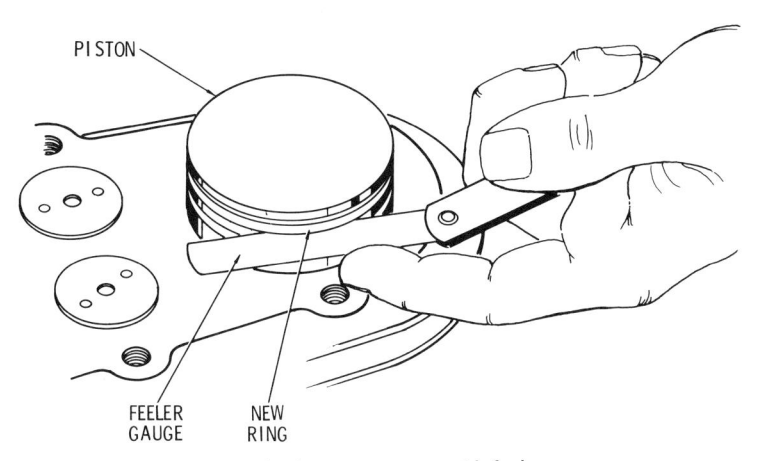

Fig. 5. Checking ring groove with feeler gauge.

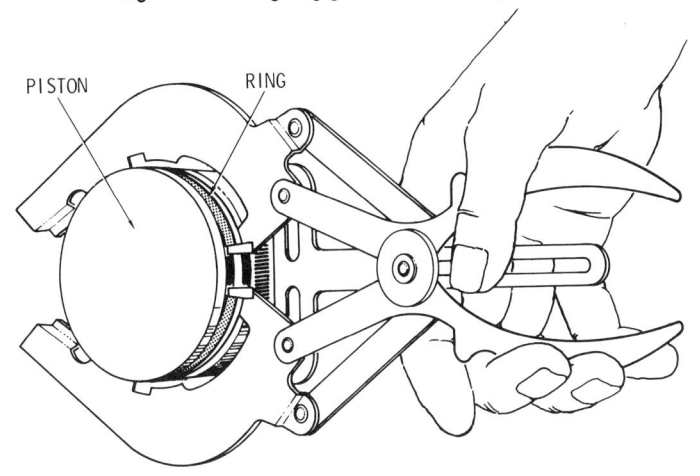

Fig. 6. Showing tool for removal of piston rings.

If the cylinders are not to be rebored measure top and bottom of the cylinder, because the cylinders always wear tapered and are larger at top than they are farther down.

Do not fit piston rings too tight. The proper end clearance should be .005 (five thousandths) to .015 (fifteen thousandths) at the joint. Cylinders should be carefully measured and rings fitted with proper clearance at the smallest point of the bore, if cylinders taper.

Check cylinders for shoulders at the top and bottom of ring travel caused by cylinder wear.

GAS ENGINE MANUAL

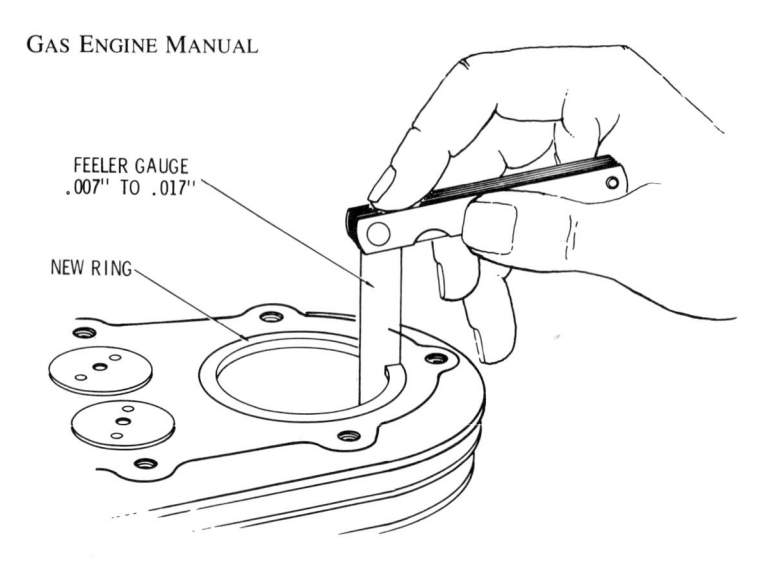

Fig. 7. Checking ring gap with feeler gauge.

After taking up, bearings at the bottom or top ring may strike such a shoulder, causing a clicking noise. Ring click is also caused by fitting the rings too tightly in the piston grooves.

Next try ring in grooves of the piston rolling it around in the grooves. Rings should be loose enough in the piston grooves to fall to the bottom of the grooves when piston is held horizontally. This means a side clearance between the ring and the piston groove of about .0015. If more than .0015 (one and one-half thousandths) of an inch, have grooves turned out to the next standard width and order rings to fit. See Fig. 8.

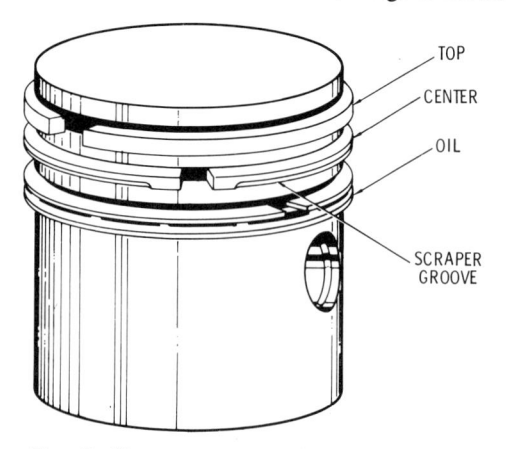

Fig. 8. Illustrating position of piston rings.

PISTON AND PISTON RINGS SERVICE

Worn piston grooves greatly reduce ring efficiency and for good engine performance the rings must fit the grooves of the piston properly.

When a test proves that the rings fit the grooves, equip the piston by installing the lowest ring first, using metal skids to slide over top grooves. After the rings have been installed in the grooves, turn them around several times to be certain that they will not stick and rotate freely.

Before returning the fully equipped pistons to the cylinders, be sure they are free from all grit and dirt and well oiled.

Old piston rings should never be refitted to the piston, as it takes twice as long for old rings to seat in again as it would for new rings to seat in.

Old rings can never be mounted in the same position as they were while in the engine. It is better to always install new rings.

Use care in ordering correct size and type rings for each job. Whenever the new piston rings are installed, drain crankcase and fill with the best grade of oil.

In fitting piston rings, note the following "don'ts" given by *Superior Piston Ring Co.*:

1. Don't fit new rings into worn piston grooves.
2. Don't file piston rings excessively.
3. Don't fit piston rings too tight.
4. Don't guess at the size required — measure the cylinders.
5. Don't fit an oversize ring into a tapered cylinder unless it has sufficient clearance at the smallest diameter.

Oil pumping and compression leakage can be traced to the following causes:

Worn out of round cylinders.	Sprung connecting rods.
Scored cylinders or pistons.	Crankshaft warped.
Worn ring grooves.	Worn bearings.
Poor piston rings.	Warped or sticking valves.
Poorly fitted pistons and rings.	Improper timing.
Piston pin hole out of round.	Leaky cylinder head gasket.
Poorly aligned rod and piston assembly.	Defective valve seats.

In automotive and small gasoline engines, the end clearance should be at least .004 in. per inch of cylinder diameter. An exception to this rule may be noted in regard to motorcycle engines, particularly those used in races. In this case, it is recommended that the end clearance be increased to at least .008 in. per inch of diameter.

CHAPTER 26

Connecting Rods and Crankshaft Service

It is important to locate engine knocks in order to avoid disassembling any more of the engine than necessary. Engine knocks due to a loose or malfunctioning connecting rod may be located by the aid of a sound rod or preferably a stethoscope.

Some mechanics locate a knock by cutting out or shorting one cylinder. This is done by short circuiting one or more spark plugs. This brings less load on the piston and connecting rod of the short circuited cylinder, thus reducing the noise.

In this manner, the shorting out of one or more spark plugs will assist in locating the cylinder or rod which is at fault in cases where the noise is not due to looseness in all cylinders or rods.

REMOVING PISTON AND CONNECTING ROD

Some engines have cylinders large enough to permit removal from the top, while in others the unit must be removed from the bottom. If the crank pin end of the rod is too large to pass through the cylinder, it must be removed from the bottom.

FITTING CONNECTING ROD BEARINGS

A connecting rod bearing consists of two halves or shells which are alike and usually are interchangeable in rod and cap. When the shells are placed in the rod and cap the ends extend slightly beyond the parting surfaces so that when the rod bolts are tightened, the shells will be clamped tightly in place to insure positive seating and to prevent turning. The ends of shells must never be filed flush with parting surface of rod or cap. See Fig. 1.

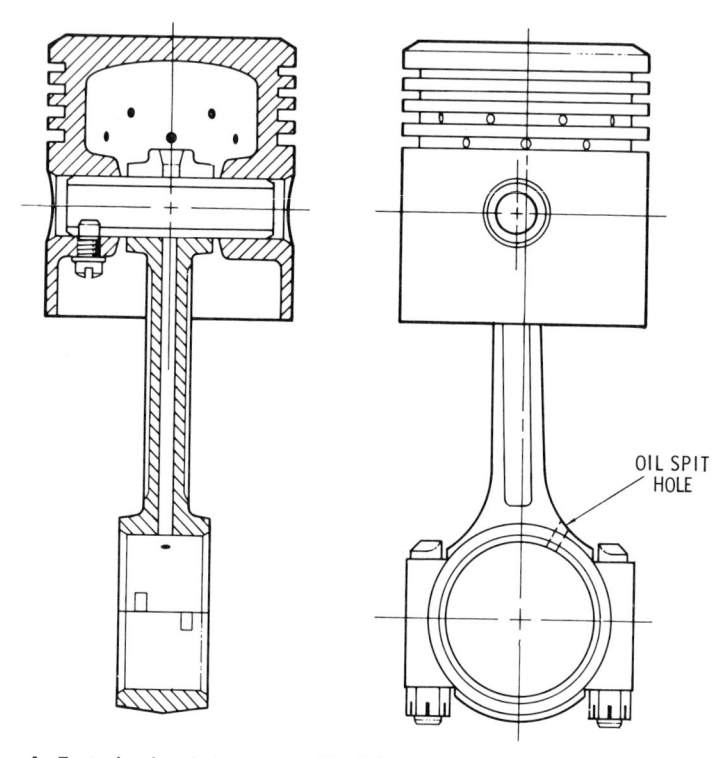

Fig. 1. Typical rod and piston assembly. When installing connecting rods, oil spit hole in upper half of connecting rod bearing must be toward the valve side of the engine.

Since these bearings are renewable, there will be no need for shims to be inserted to facilitate proper fitting. Renewable bearings at the crank end serve to reduce time and expense should the bearing material wear sufficiently to require replacement.

It also eliminates the necessity of changing connecting rod assemblies, the only operation necessary being the removal of the old bearing and installation of new ones which may be done without removing the rods from the engine. There are, however, some notable exceptions.

Connecting Rods and Crankshaft Service

CRANKSHAFT SERVICE

Modern crankshafts are a highly developed piece of mechanism made of the best material and machined with precision. As an example of construction, a typical automotive engine crankshaft is made from a drop forging of high carbon alloy steel carefully heat treated.

As previously noted, the crankshaft is one of the most important parts of an engine, as it ties together the reaction of the pistons, transforms the reciprocating motion of the pistons and connecting rods into rotary motion, and transmits the resulting torque to the flywheel and load.

Due to the foregoing the crankshaft is subject to heavy vibration and stress and as such may develop tiny cracks particularly at or near ends of the connecting rod throws or at the ends of the main bearing journals. See Fig. 2 for crankshaft and related parts.

Reconditioning of Crankshafts

When the crankshaft is free from flaws or defects and the journals are worn slightly taper or out of round, the shaft may be reground and fitted with undersize bearings. To recondition the crankshaft and to install undersized bearings reduce the journal diameter by the amount which the bearings are undersized.

For example, if .010 in. (0.254 mm) undersized main bearings are to be installed, the original journal diameter of 2.4370 in. to 2.4375 in. (61.899 mm to 61.912 mm) should be reduced to 2.4270 in. to 2.4275 in. (61.645 mm to 61.658 mm).

A crankshaft connecting rod journal which is worn, tapered, out of round, or scored in excess of .0015 in. (0,0381 mm) can be reconditioned without the removal of the crankshaft from the engine. This should, however, never be attempted unless it is known that the main bearing journals are not damaged and that the oil passages are clean.

CHECKING BEARING CLEARANCE

Limits on the taper or out-of-round of any crankshaft journal should be held to .001 inch. Undersize bearings should be installed if the crankshaft journals are worn enough to increase the bearing clearance above specifications. Never install an undersize bearing that will reduce the clearance below specifications.

The desired connecting rod clearance is .0005 to .0015 inch, with an end play of .006 to .011 inches. Whether the clearance is within these limits, it can be checked with the plastigage method or as follows:

Gas Engine Manual

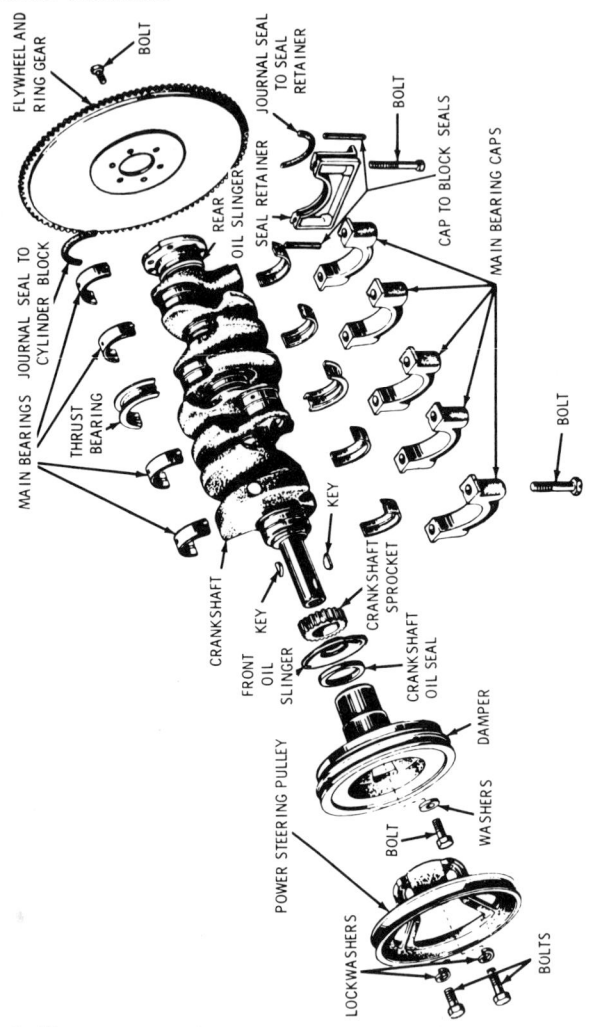

Fig. 2. Illustrating typical automotive type crankshaft and related parts.

Take each connecting rod bearing, one at a time, and remove cap. Use a piece of .002 inch feeler stock, ½ inch wide and 1 inch long. Coat feeler with oil and place it between the bearing and crankshaft journal. Tighten rod cap nuts securely. If a noticeable drag is present when the crankshaft is turned a full revolution by hand, with feeler stock in the rod bearing, clearance is less than .0015 inch. If no drag is felt, the bearing is too loose and should be fitted with an undersize bearing.

CONNECTING RODS AND CRANKSHAFT SERVICE

If the connecting rod bearing clearance is excessive, crankshaft journals should be checked for out-of-round and taper, then if necessary, the crankshaft should be removed and the journals reground. Replacement bearings are available in standard and the following under sizes: .001, .002, .010 and .012 inches. Tighten rod cap nuts to the required torque.

CLEARANCE MEASUREMENT

The accepted method of clearance measurement is to measure the diameter of the journal with a micrometer caliper. The diameter of the shaft is measured at several points around the circumference to determine the size and check for roundness.

An alternate method is to use the "plastigage" which consists of a wax-like material which will compress evenly between the bearing surfaces and journals without in any way causing damage to them. (Fig. 3.). To obtain most accurate results, certain precautions should always be observed.

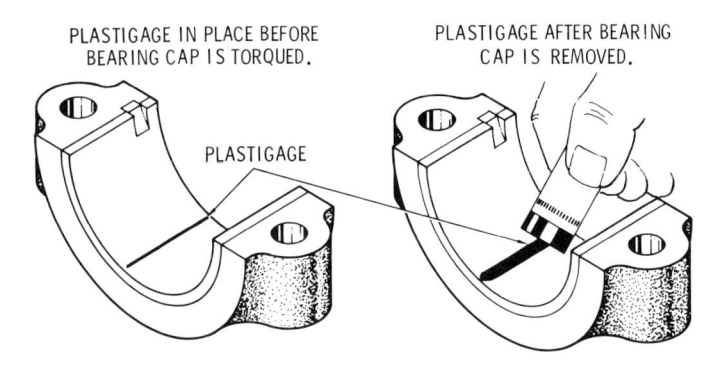

Fig. 3. Checking bearing clearance with plastigage.

When adjusting the main bearing clearance by the plastigage method, the surface of the crankshaft journal and bearing shell should be wiped clean of oil before the plastic material is placed in the cap.

Place a piece of plastigage, approximately one inch long on the bearing shell. Install and tighten the cap to recommended torque. The crankshaft should not be turned while making this check. After the correct torque is obtained the bearing cap is removed. The flattened plastigage will now be found adhering to either the bearing shell or the crankshaft.

One edge of the envelope in which the plastigage is packed is marked in graduated scales which are correlated to measure the width of the compressed wax thread in thousandths of an inch. Compare the width of the flattened plastigage at its widest point with the graduations on the envelope. The clearance in thousandths of an inch will be indicated by the number of graduations which most nearly fits the width of the flattened plastigage.

CHAPTER 27

Valves and Valve Gear Service

The prevailing cause of compression losses particularly in older engines is commonly due to valve leakage, and accordingly it is of the utmost importance that the valves be maintained in good condition so that they will seat properly. There is probably no other item which affects the performance of an engine as much as the proper seating of valves.

RECONDITIONING VALVES AND SEATS

When an engine has received proper attention at proper intervals, grinding the valves is all that is necessary. However, when in a neglected condition additional servicing may be required such as reseating and refacing.

Preliminary Operation

First the cylinder head and valve cover plates must be removed to gain access to the valves. After the cylinder head has been removed from the engine, and all of the parts disassembled from it, all carbon should be thoroughly cleaned from the combustion chamber, valve ports and guides, and the head thoroughly washed.

Removing the Valves

There are various types of lifters used depending upon the design of the valve gear. (Fig. 1.) A lifter is a tool designed to lift the valve spring cup

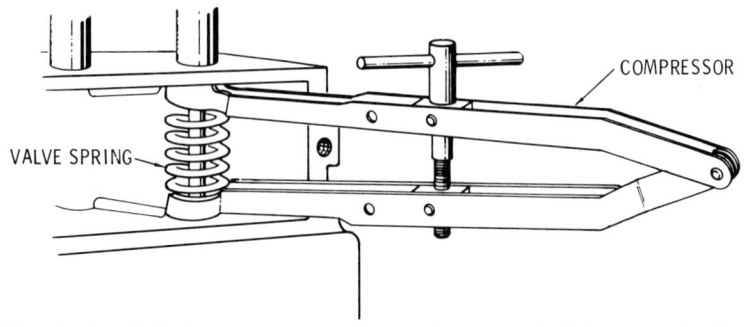

Fig. 1. A typical valve spring compressor tool showing method of compressing the valve spring to permit removal of retainer pin.

clear of the retainer pin so that the latter may be withdrawn and the valve released.

When the valves are removed they should be inserted in a board provided with 12 holes for a six cylinder engine or 16 holes for an eight cylinder engine. The holes should be numbered from 1 to 12 or 16, inclusive, so as to permit the reassembly of the valves in their original position. An inspection of the valve seats and valve faces will determine whether reseating and refacing are necessary before grinding.

VALVE GUIDES

The clearance between the valve guides and the valve stems is very important. Lack of power and noisy valves, in many instances, can be traced to worn valve guides. The admission valve guides should be checked with a new admission valve and the exhaust valve guides should be checked with a new exhaust valve, because the diameters of the stems are different on these parts.

REFACING VALVES

Valve refacing should be closely coordinated with the valve seat refacing operation so that the finished angle of the valve face will match the valve seat. This is important in order that the valve and seat will have a good compression tight fit. Be sure that the refacer grinding wheels are properly dressed.

If the valve face runout is excessive, and to remove pits and grooves, reface as necessary. Remove only enough stock to correct the runout and to clean up the pits and grooves. Note the different face angles on the exhaust and intake valves.

VALVES AND VALVE GEAR SERVICE

After refacing the valves, it is good practice to lightly lap in the valves with a medium grade lapping compound to match the seats. Be sure to remove all the compound from the valve and seat after the lapping operation. See Fig. 2.

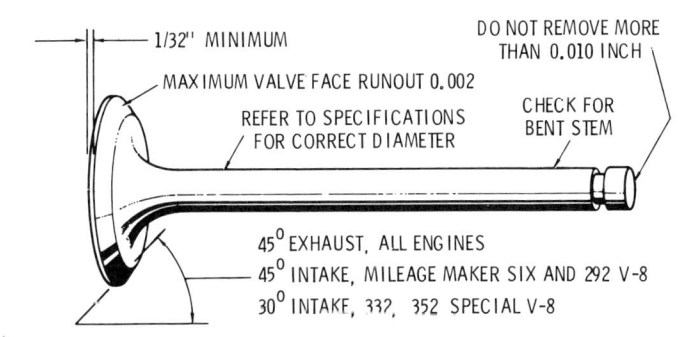

Fig. 2. Showing critical valve tolerances in typical automotive engine. Note angular difference in valve faces between intake and exhaust.

REFACING VALVE SEATS

Inspect the valve seats for cracks, burns, pitting, ridges or improper angle and reface. During any general engine overhaul it is advisable to reface the valve seats regardless of their condition. If valve guides are to be replaced, this must be done before refacing the valve seats. See Fig. 3.

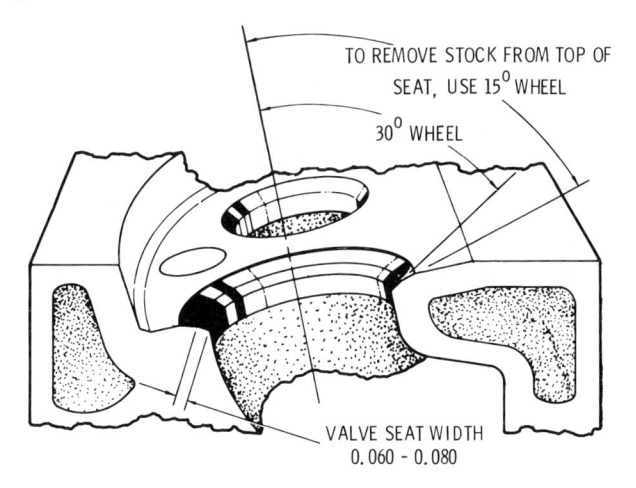

Fig. 3. Intake valve seat refacing in typical automotive engine.

The finished valve seat should contact the approximate center of the valve face. To determine where the valve seat contacts the face, coat the seat with Prussian blue, then set the valve in place. Rotate the valve with light pressure. If the blue is transferred to the top edge of the valve face, the contact is satisfactory. If the blue is transferred to the bottom edge of the valve face, raise the valve seat.

It is good practice to lightly lap in the valve with a medium grade lapping compound after refacing. Remove all the compound from the valve and seat after the lapping operation.

REAMING VALVE GUIDES

If it becomes necessary to ream a valve guide in order to install a valve with oversize stem, always use the reamers in proper sequence. Always reface the valve seat after the guide has been reamed. See Fig. 4.

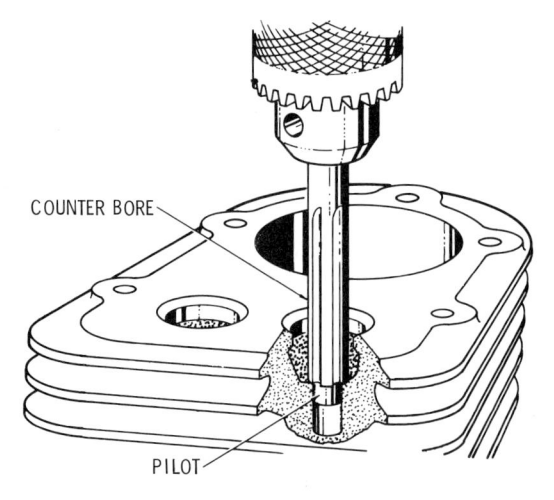

Fig. 4. Method of counterboring cylinder to permit installation of replacement valve guide. Note: Counterbore reamer should turn at a speed of approximately 250 rpm since lower speeds may result in breakage or damage.

VALVE SPRING

Weak valve springs affect the economy and power of the engine, therefore, each time the valves of an engine are ground, the valve springs should be checked to be sure they have not been weakened by the heat of the engine.

VALVES AND VALVE GEAR SERVICE

The springs are checked by comparing the springs removed, by the tension of a new spring having the identical size and characteristics as the original. Any spring that does not match up with the new spring should be replaced. In replacement use only spring specified by the engine manufacturer.

One method of checking valve springs for squareness by means of a steel square and a surface plate as shown in Fig. 5. Stand the spring and square on end on the surface plate. Slide the spring up to the square. Revolve the spring slowly and observe the space between the top coil of the spring and the square. If the spring is out of square more than the amount noted, replace it.

Broken valve springs are easily noticeable when engine is idling, while weak springs show up more during high speed performance by limiting top speed since they allow the valves to "float" or not follow the cams on the camshaft. Weak or broken valve springs must be replaced with the correct spring.

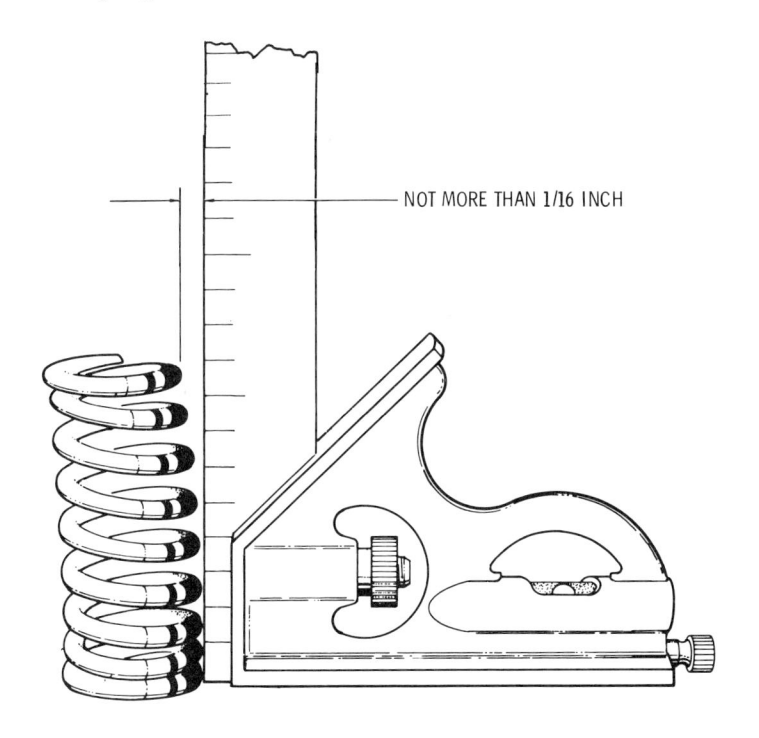

NOT MORE THAN 1/16 INCH

Fig. 5. Method of checking valve spring for squareness.

GAS ENGINE MANUAL

CAMSHAFT SERVICE

In order to facilitate removal and replacement of camshafts the usual practice is to make the bearings slightly "tapered", that is, the largest bearing is located at or near the front end of the engine, usually just back of the timing gear. See Fig. 6.

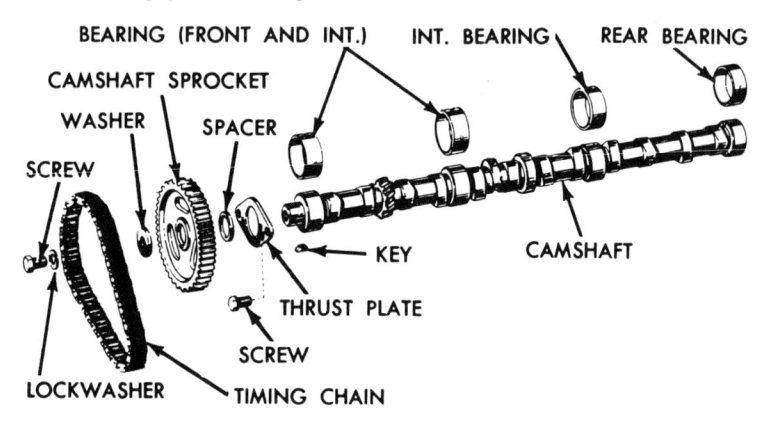

Fig. 6. Illustrating camshaft and related parts.

When the camshaft is removed, check with micrometer the bearing dimension for out of round. If the journals exceed .001 inch out of round, the camshaft should be replaced. Another important inspection operation, when the camshaft is removed from the engine, is to check it for alignment. When straightened, the camshaft should be supported on the front and rear bearing journals in V-blocks in an arbor press.

If it becomes necessary to replace the camshaft gear, a sleeve to properly support the gear on its hub is necessary. In replacing the gear on the camshaft, the back of the front journal of the camshaft must be firmly supported in an arbor press and the camshaft thrust place assembled to the camshaft after which the gear is pressed on the shaft far enough so that the camshaft thrust plate has no clearance, yet is free to revolve.

If there be an excessive amount of end play in the camshaft, it is necessary to remove the gear and shaft assembly and press the gear further on the shaft so that the thrust plate is tight yet free to revolve.

When the camshaft and gear are assembled to the engine, it is important that the punch marks on both the camshaft and the crankshaft gear be opposite each other as shown in Fig. 7.

The camshaft will then be in proper position so that the valve will open and close in proper relation to the piston. After the camshaft and the

VALVES AND VALVE GEAR SERVICE

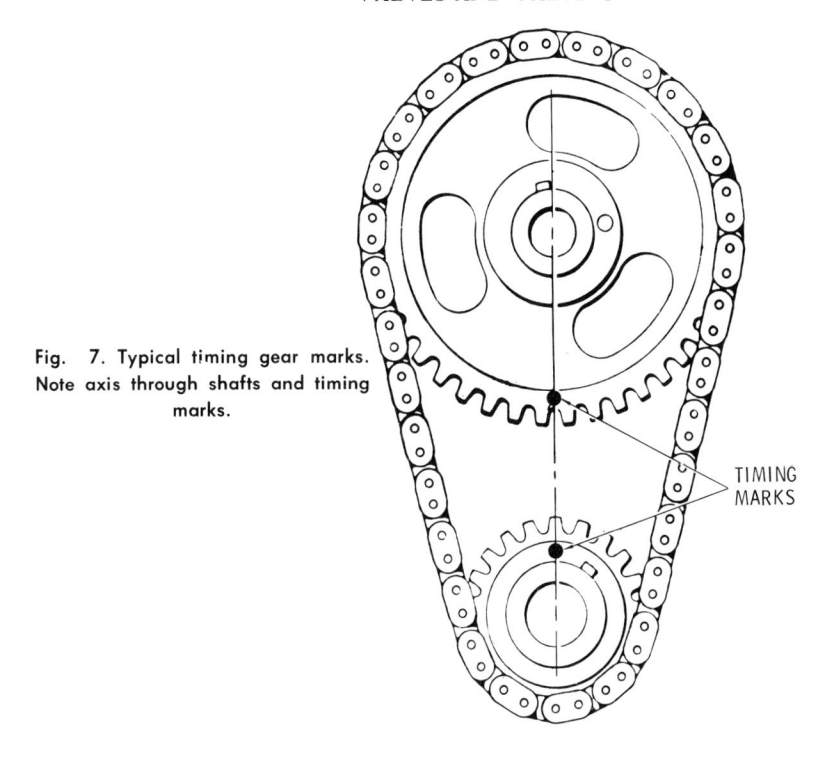

Fig. 7. Typical timing gear marks. Note axis through shafts and timing marks.

TIMING MARKS

crankshaft gears are in their proper places, check the crankshaft timing gear for run out with a dial indicator.

After the foregoing mentioned checks have been made, and if out, remove gears to be sure that burrs on the shaft and gears are not causing run-out. If necessary replace with new gears. The back lash should also be checked. This check is made with a feeler gauge placed between the teeth.

CHAPTER 28

Carburetor and Fuel-Injection

The major portion of carburetor servicing consists of cleaning, inspection and adjustment. After considerable usage, it may become necessary to overhaul the carburetor and renew worn parts to restore it to its original operating efficiency.

Before adjusting a carburetor, the ignition system should receive attention. There should be a good spark and the plugs should have the correct gap. The appearance of spark plug is a good guide in making carburetor adjustments. For precision, install a new set of plugs of the correct heat range and see that the ignition system is functioning properly. Start the engine and run until normal running temperature is obtained. Shut off the engine and remove plugs.

To test for proper carburetor adjustment proceed as follows: Draw a clean white cloth across the ends of the plug metal shell. A black mark on the cloth indicates a mixture too rich. If barely perceptible with only a faint gray cloudiness, the mixture is too lean. Correct adjustment is indicated by a distinct gray mark on the cloth.

In the absence of carburetor analyzer equipment, the general procedure to be followed will be as outlined varying with the type of carburetor. There are four adjustments to the typical air valve carburetor as follows:
1. Float feed level,
2. Gasoline valve,
3. Air valve spring,
4. Throttle idling.

Float Level Adjustment

The adjustment should be such that the liquid level will be as near the top of the nozzle as possible without overflowing. This reduces to a minimum the initial or lift vacuum necessary to cause the gasoline to issue from the nozzle. If the gasoline persists in overflowing while adjusting, a leaky float valve is indicated. Regrind or replace.

Gasoline Valve Adjustment

Run engine at an average or normal speed, and with late spark if spark is manually controlled. Reduce gasoline supply until the mixture becomes so weak that there is popping in the carburetor and note position of valve.

Next increase mixture until it becomes so rich that engine begins to choke. Set the adjustment of the gasoline valve halfway between the two points which will be the approximate setting.

Vary the adjustment slightly for rich or lean mixture until the speed picks up to maximum and engine runs smoothly. That will be the correct setting.

Air Valve Adjustment

Vary the opening of the air valve by moving it by the spindle. If engine speeds up with the additional air, loosen spring adjustment. If it slows down, the mixture was correct, or a little too lean according to the degree to which the speed is affected.

The spring should be adjusted until the engine speeds up to maximum. The spring has considerable influence on the operation of the carburetor and good adjustment cannot be obtained without the proper spring. Check adjustment by letting engine idle and then suddenly open the throttle.

If adjustment is correct, the engine should accelerate quickly and smoothly. If it back fires when the throttle is suddenly opened, the adjustment is faulty.

Throttle Idling

First, set the throttle lever adjusting screw at an engine speed corresponding to low idle. Then set the adjustment screw so that the engine fires evenly, without stalling.

SERVICING THE CARBURETOR

When overhauling a carburetor, several items of importance should be obtained to assure a good job.
 1. The carburetor must be disassembled,
 2. The various jet plugs removed,

CARBURETOR AND FUEL-INJECTION

3. Clean all parts carefully in a suitable solvent, then inspect for damage or wear,
4. Use air pressure only, to clear the various orifices and channels,
5. Replace questionable parts with new ones,
6. Use new gaskets at reassembly.

When checking parts removed from the carburetor, it is at times rather difficult to be sure they are satisfactory for further service. It is therefore recommended that *new* parts be used.

Disassembly and Assembly. A general pictorial idea of the operations involved in the removal and installation of the various carburetor parts is shown in Figs. 1, 2, 3, and 4, which will serve as a guide as to how to do it.

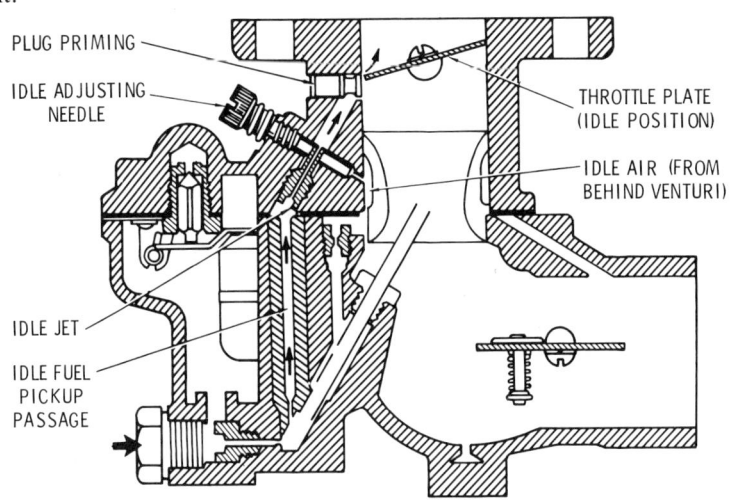

Fig. 1. Zenith series 167-7 updraft type carburetor with only an idle mixture adjustment. In some installations a main power adjustment is provided by substituting a needle valve for the plug shown by a heavy arrow.

CLEANING CARBURETOR PARTS

The recommended solvent for gum deposits is denatured alcohol which is easily obtainable. However, there are other commercial solvents which may be used with satisfactory results.

Note: Never clean jets with a wire or other mechanical means because the orifices may become enlarged, making the mixture too rich for proper performance.

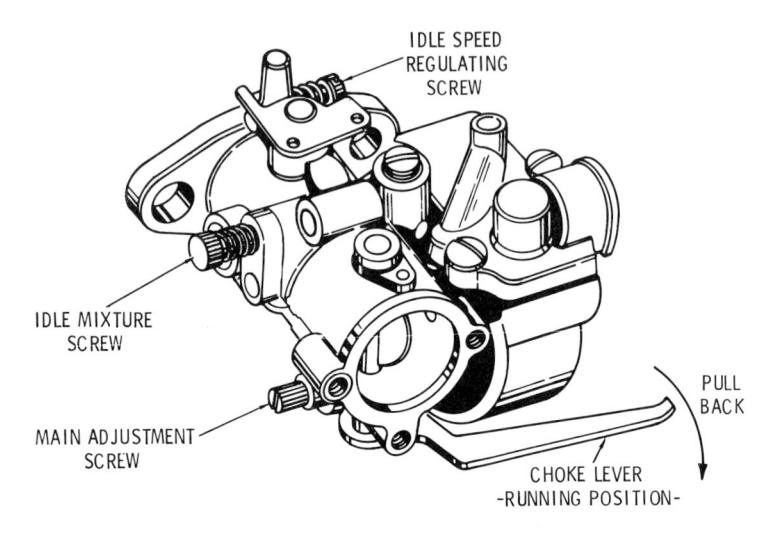

Fig. 2. View showing location of adjustment points on a typical *Tillotson* carburetor.

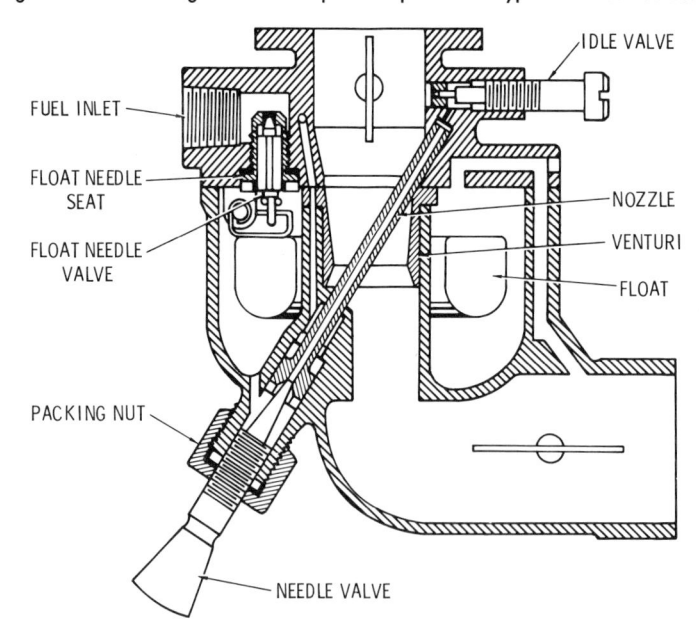

Fig. 3. Typical *Briggs & Stratton* gravity feed carburetor. Float type carburetors are provided with adjustments for both idle and power mixtures. On these, clockwise rotation of the adjusting needle leans the mixture.

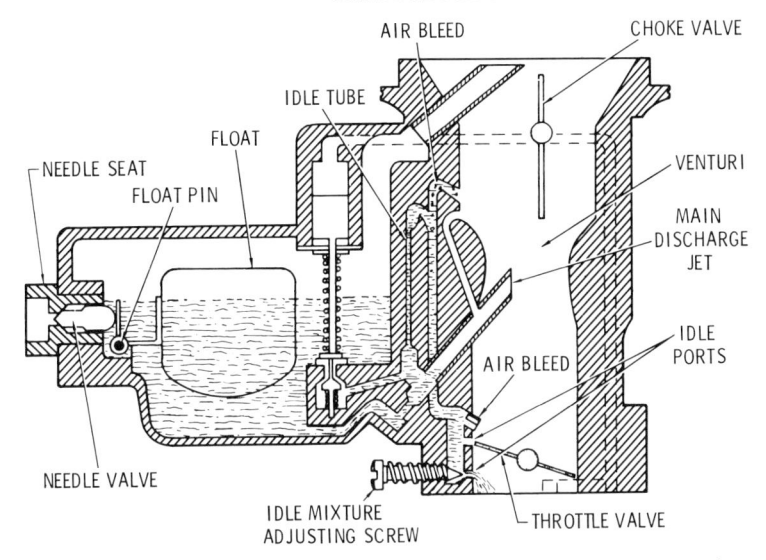

Fig. 4. Typical *Stromberg* carburetor illustrating the air bleed principle. Here the main jet controls the fuel mixture for various throttle openings. The mixture is controlled by a small amount of air which is admitted through the bleeder.

Caution: If the commercial solvent or cleaner recommends the use of water as a rinse, it should be hot. After rinsing, all trace of water must be blown from passages with air pressure. It is further advisable to rinse parts in clean kerosene or gasoline to be certain no trace of moisture remains.

To remove gum and carbon deposits, a soft brush should be used while the parts are soaking in the solvent. After cleaning, parts should be rinsed in clean solvent and then all passages thoroughly blown out with compressed air.

Inspection and reassembly.—Proceed as follows:

1. Check the throttle shaft for excessive wear in the throttle body.

If wear is extreme, it is recommended that the throttle body assembly be replaced rather than installing a new throttle shaft in the old body.

During manufacture, the location of the idle transfer port and spark advance control ports to the valve is carefully established for one particular assembly.

If a new shaft or valve should be installed in an old worn throttle body, it would be very unlikely that the original relationship of these parts to the valve would be obtained. Changing the port relationship would adversely affect normal operation.

2. Inspect the throttle lever for looseness on the throttle shaft. If the lever is loose, it will be impossible to secure proper idle speed adjustment. Silver solder or braze the joint to correct this condition, or use a new throttle body assembly.

3. Install the idle mixture needle valve and spring in the body. The tapered portion must be straight and smooth. If the tapered portion is grooved or ridged, a new idle needle valve should be used to insure having correct idle mixture control.

In making the idle adjustment, do not use a screw driver. Make the adjustment with the fingers. Turn the idle mixture needle valve against seat, then back off one full turn for approximate location.

CARBURETION

This process may be defined as *the mixing of gasoline in the form of a mist or spray with air in proper proportions to form the fuel charge for a gas engine.* The character of the fuel charge delivered by a carburetor depends on numerous conditions.

Carburetion is affected by:

1. *Fuel pump.*—Controlling the pressure of the fuel fed to the carburetor.
2. *Air cleaner.*—If not clean, will restrict the volume of air taken in by the carburetor.
3. *The carburetor.*—Including float level control, many variations of jets, Venturi air valves, throttle valves, choking devices, etc.
4. *The hot spot.*—Any device for heating the gas after it leaves the carburetor.
5. *Choke.*—The choke affects the volume of air taken in by the carburetor.
6. *Thermostats.*—Thermostats controlling choke valve, air valve, or other parts of the carburetion system affect the mixture of air and fuel.
7. *Intake manifold.*—Air leaks in the manifold or gaskets change carburetion.
8. *Muffler.*—If the muffler or exhaust pipe is restricted, a portion of the burned gases will remain in the cylinder and only a partial charge of fresh gas will be drawn in.
9. *Super charger.*—Changes in carburetion due to the use of a super charger must be taken into consideration when tests are made.

CARBURETOR AND FUEL-INJECTION
SERVICE DIAGNOSIS
Conditions—Possible Causes—Remedies

POOR PERFORMANCE—MIXTURE TOO LEAN

Possible Causes:
- *a.* Damaged, worn, incorrect type or size of main metering jet.
- *b.* Damaged tip or bad top shoulder seat or main discharge jet.
- *c.* Vacuum piston worn or stuck.
- *d.* Corroded or bad seating power jet.
- *e.* Incorrect fuel level.
- *f.* Automatic choke not operating properly.
- *g.* Worn or corroded needle valve and seat.

Remedies:
- *a.* Disassemble carburetor, then replace main metering jet if in questionable condition.
- *b.* Remove and disassemble carburetor, clean and inspect main discharge jet and replace if necessary.
- *c.* Disassemble carburetor, and free up stuck piston. If piston is badly worn, replace air horn assembly.
- *d.* Disassemble carburetor. Clean power jet and channels. If close inspection reveals a faulty seating jet, replacement is recommended.
- *e.* Check fuel level in carburetor. Adjust vertical lip of float to obtain correct level of ⅝ inch from top of fuel to top edge of fuel chamber.
- *f.* Check adjustment and operation of automatic choke. If necessary, replace choke to correct this condition.
- *g.* Clean and inspect needle valve and seat. If found to be in questionable condition, replace assembly and then check fuel pump pressure. Pressure should be from 3½ to 5½ pounds.

POOR IDLING

Possible Causes:
- *a.* Carbonized idle tube or poor seating shoulder.
- *b.* Idle air bleed carbonized or of incorrect size.
- *c.* Idle discharge holes plugged or gummed.
- *d.* Throttle body carbonized or worn throttle shaft.
- *e.* Damaged or worn idle needle.
- *f.* Incorrect fuel level.
- *g.* Loose main body to throttle body screws.

Remedies:
- *a.* Disassemble carburetor then clean idle tube and check seating shoulder. Replace idle tube if in a questionable condition.

GAS ENGINE MANUAL

    b. Disassemble carburetor then use compressed air to clear idle air bleed after soaking in a suitable solvent.

    c. Disassemble carburetor. Use compressed air to clear idle discharge holes after soaking main and throttle bodies in a suitable solvent.

    d. Disassemble carburetor. Check throttle valve shaft for wear. If excessive wear is apparent, replace throttle body assembly with a new one.

    e. Replace worn or damaged idle needle with a new unit. Adjust air mixture.

    f. Check the fuel level in carburetor. Adjust vertical lip of float lever to obtain correct level of ⅝ inch from top of fuel to top edge of fuel chamber.

    g. Tighten main body to throttle body screws securely to prevent air leaks and cracked housings.

POOR ACCELERATION
Possible Causes:

    a. Corroded or bad seat on accelerator pump by-pass jet.

    b. Accelerator pump piston (or plunger) leather too hard, worn or loose on stem.

    c. Faulty accelerator pump discharge.

    d. Accelerator pump inlet check valve faulty.

    e. Incorrect fuel level.

    f. Worn or corroded needle valve and seat.

    g. Worn accelerator pump and throttle linkage.

    h. Automatic choke not operating properly.

Remedies:

    a. Disassemble carburetor. Clean and inspect accelerator pump by-pass jet. Replace by-pass jet if in questionable condition.

    b. Disassemble carburetor. Replace accelerator pump assembly if leather is hard, cracked or worn. Test follow up spring for compression.

    c. Disassemble carburetor. Use compressed air to clean the discharge nozzle and channels, after soaking main body in a suitable solvent. Check the pump capacity.

    d. Disassemble carburetor. Check accelerator pump inlet check valve for poor seat or release. If necessary, replace faulty part with a new unit.

    e. Check fuel level in carburetor. Adjust vertical lip of float lever to obtain correct level of ⅝ inch from top of fuel to top edge of fuel chamber.

CARBURETOR AND FUEL-INJECTION

f. Clean and inspect needle valve and seat. If found to be in questionable condition, replace assembly and then check fuel pump pressure. Pressure should be from 3½ to 5½ pounds.

g. Disassemble carburetor. Replace worn accelerator pump and throttle linkage, then check for correct position.

h. Check adjustment and operation of automatic choke. If necessary, replace choke to correct this condition.

CARBURETOR FLOODS OR LEAKS
Possible Causes:
- *a.* Cracked body.
- *b.* Defective body gaskets.
- *c.* High float level.
- *d.* Worn needle valve and seat.
- *e.* Excessive fuel pump pressure.

Remedies:

a. Disassemble carburetor. Replace cracked body, being sure main to throttle body screws are tight.

b. Disassemble carburetor. Replace defective gaskets, then check for leaks. Be sure the screws are tightened securely.

c. Check fuel level in carburetor. Adjust vertical lip of float lever to obtain correct level of ⅝ inch from top of fuel to top edge of fuel chamber.

d. Clean and inspect needle valve and seat. If found to be in questionable condition, replace complete assembly and then check fuel pump pressure. Pressure should be from 3½ to 5½ pounds.

e. Test fuel pump pressure as described in paragraph *e*, of this section. If pressure is in excess of 5½ pounds, replace fuel pump.

POOR PERFORMANCE—MIXTURE TOO RICH
Possible Causes:
- *a.* Restricted air cleaner.
- *b.* Excess of oil in air cleaner.
- *c.* Leaking float.
- *d.* High float level.
- *e.* Excessive fuel pump pressure.
- *f.* Worn main metering jet.

Remedies:

a. Remove and clean air cleaner.

b. Remove and clean air cleaner. Refill oil chamber to required level mark or one pint of SAE 50 engine oil.

Gas Engine Manual

    c. Disassemble carburetor. Replace leaking float with a new unit. Check float level, then if necessary, bend the vertical lip of float lever to obtain a ⅝ inch reading from top of fuel to top of fuel chamber.

    d. Adjust vertical lip of float as described in *c* above to secure correct float level.

    e. Check fuel pump pressure. Pressure should be from 3½ to 5½ pounds. If pressure is in excess of 5½ pounds, replace fuel pump assembly.

    f. Disassemble carburetor. Replace worn metering jet with a new one of the correct size and type.

FUEL-INJECTION SYSTEM SERVICE

    Since these systems are electronically controlled by preprogrammed, unserviceable devices, service operations are, for the most part, limited to unit malfunction tests and replacements. Well-equipped service departments utilize special testing devices to determine the proper functions of the system components. However, there are several tests (based on engine operation) that can help to isolate a system's malfunction. When removing and replacing a component, extra precautions must be taken to keep all parts physically and electrically clean.

    NOTE: Before suspecting the fuel injection system for engine malfunction, be sure the ignition system is in proper working order.

Engine Will Not Start

    If the starter rotates the engine but the engine will not start, check for a blown fuel-pump fuse, an open circuit between the battery and ECU, or a faulty jumper harness for the high-pressure fuel pump. Any of these situations could cause the fuel pump not to operate. In addition, the fuel pump itself could be inoperative.

    To check and see if any of the above situations are present, place the ignition switch to "ON" position and listen for the whine of the fuel pump (which should last for about one second). If no whine is heard, then check out the above trouble areas. Remove and replace faulty items.

    In addition to the aforementioned, other system components can prevent the engine from starting. Check for an open circuit between the ECU and the starter solenoid, a faulty ECU jumper harness or a faulty connection at the speed sensor. The speed sensor could be stuck closed. Also check for a defective throttle-position switch or hampered fuel flow. If the trouble is a defective throttle-position switch, and all other system

Carburetor and Fuel-Injection

components are operable, the engine should start if the switch is disconnected.

If a cold engine will not start, check for open wiring or a bad connection to the engine-coolant temperature sensor. This sensor can be checked with an ohmmeter connected across its terminals. If resistance is greater than specified, the sensor must be removed and replaced.

Engine Is Hard To Start

If a cool or cold engine has trouble starting, this could be caused by a defective engine coolant sensor. Check for specified resistance between the sensor's terminals with an ohmmeter. If resistance is greater than specified, remove and replace the sensor.

Malfunction of the high-pressure fuel pump, a defective pressure regulator or a faulty throttle-position switch will also cause cool or cold engine starting problems. If engine starts with the throttle-position switch disconnected, this switch must be replaced.

High Fuel Consumption

This can be caused by a manifold absolute pressure sensor (MAP) hose which has become disconnected or is leaking. The same holds true for the vacuum hoses for either the pressure regulator or the throttle body.

A faulty coolant sensor or air temperature sensor can also cause poor fuel economy. Check for specified resistance between the terminals of each sensor. If resistance is greater than specified, remove and replace the faulty sensor(s).

Engine Stalls After Statring

Check for an open circuit in the ignition wire between the fuse blocks and electronic control unit (ECU) and also for a faulty connector connection (located near the ECU jumper harness).

If the engine is either cold or warm and stalls after starting, check the condition of the engine coolant sensor or its wiring. Check for specified resistance between the terminals of this sensor with an ohmmeter. If resistance is greater than specified, replace the sensor.

Rough Idle

Check the manifold absolute pressure sensor (MAP) hose to see if it has been disconnected, if it is leaking or is restricted. Also check the system's harness line to the MAP and, if this line is defective, replace the entire harness.

GAS ENGINE MANUAL

A poor electrical connection at any of the injector valves, or a shorted coolant sensor, will also cause rough idling. To check the condition of the coolant sensor, check for specified resistance across its terminals. If sensor resistance is less than specified, it must be replaced.

When the engine is cold, poor electrical connections or "open" wiring to the air temperature sensor or the coolant sensor will cause idle difficulties. To check the condition of either sensor, check for specified resistance across the sensor terminals. If resistance (for either sensor) is greater than specified, replace the defective unit.

Prolonged Fast Idle

Check for a poor electrical connection to the fast-idle valve, or an inoperative heating element. Also, the throttle-position switch could be improperly adjusted and/or a vacuum leak could be causing the problem.

Engine Hesitates During Acceleration

Check the manifold absolute pressure sensor (MAP) hose to see if it has been disconnected, if it is leaking, or is restricted. Also check the system's harness line to the MAP for condition. If line is defective, replace the entire system harness.

Other causes can be an improperly adjusted or defective throttle-position switch, intermittent speed sensor operation or a faulty electronic control unit connector or jumper harness.

During cold engine operation, acceleration hesitation can be caused by the exhaust gas recirculation (EGR) solenoid. The solenoid could be stuck open or have a faulty electrical connection.

High-Speed Performance Inadequate

This can be caused by an improperly adjusted throttle-position switch (in its high-speed position only) or a malfunctioning throttle-position switch. Other units which could cause this condition are a defective high-pressure fuel pump, intermittent operation of the speed sensor, a blocked or restricted fuel filter, or an "open" wire between the electronic control unit and the starter solenoid.

CHAPTER 29

Electrical System Service

(Battery Ignition Service)

The breaker contacts not only serve to open the primary circuit and cause a high voltage spark, but they regulate the length of time that the current flows in the coil. This has a direct effect upon the value of the spark at the spark plugs and with the higher speeds and compression pressures of modern engines this effects the power and speed.

The manufacturer's specifications should always be followed when adjusting breaker contacts, to insure that the proper amount of separation is provided. If contact points are set too close they will tend to burn and pit rapidly, while points with too much separation will cause ignition failure at high speed.

After considerable use, contact points may not appear smooth and bright, but this is not necessarily an indication that they are not functioning properly and giving good ignition, and they should not be disturbed as long as proper operation is obtained. See Fig. 1.

Should the points become pitted or burned in operation, rub lightly with a dry oilstone. Points can also be dressed with a clean ignition file without removing them from the distributor.

Oxidized contact points condition may be caused by high resistance or loose connections in the condenser circuit, oil or foreign materials on the contact surfaces, or most commonly, high voltages. Check for these conditions where burned contacts are experienced.

Contact point pressure should be within the limits given. Weak tension will cause point chatter and ignition miss at high speed, while excessive tension will cause undue wear of the contact points, cam and rubbing block. Most data tables specifies breaker arm spring tension 17 to 20 ozs. See Fig. 2.

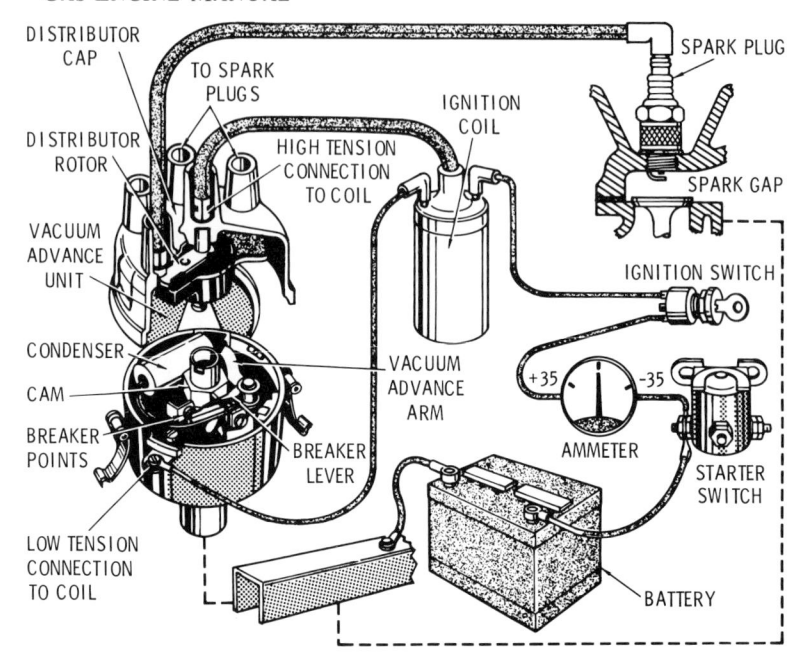

Fig. 1. Typical battery ignition system.

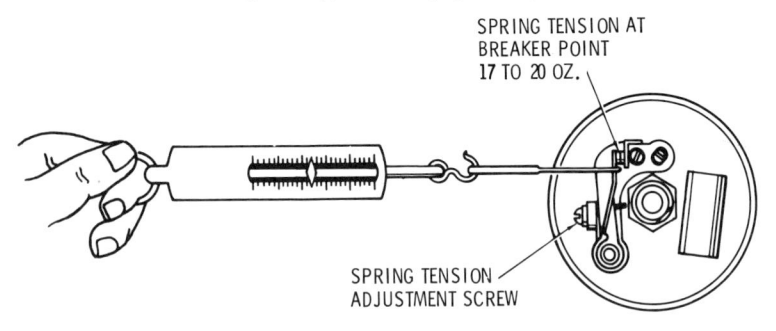

Fig. 2. Method of making breaker spring test. Spring tension as measured at back of breaker point should be adjusted to correspond to specifications.

DWELL

The contact points in a modern distributor must be adjusted under actual running conditions with a dwell angle meter. By definition the dwell angle is: *The angle of cam rotation through which the distributor points remain closed.* The points must remain closed long enough to

ELECTRICAL SYSTEM SERVICE

insure saturation and build up of the ignition coil. Eccentricity and bearing wear will cause variation of dwell angle. See Fig. 3.

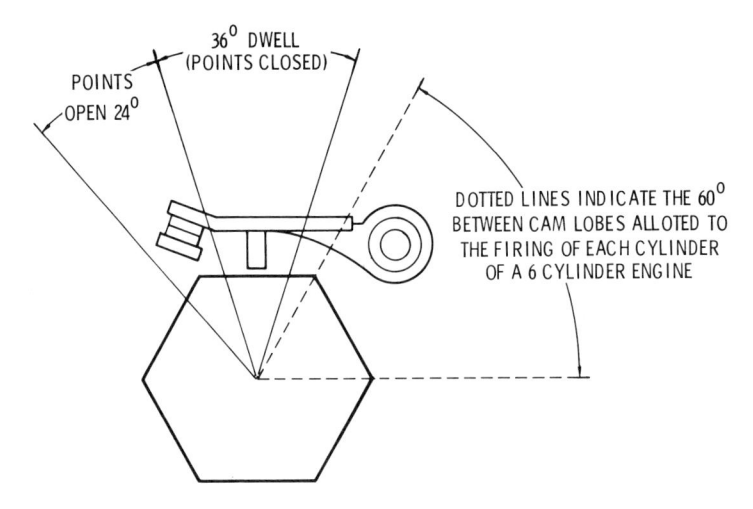

Fig. 3. Illustrating meaning of dwell angle.

CONDENSER

There are four factors which affect condenser performance and each must be considered in making condenser tests. They are:
1. Breakdown,
2. Low insulation resistance,
3. High series resistance,
4. Low series resistance.

Low insulation resistance or leakage prevents the condenser holding a charge.

A condenser with low insulation resistance is said to be weak. All condensers are subject to leakage, which up to a certain limit is not objectionable. When it is considered that the ignition condenser performs its function in approximately 1/12,000 of a second, it can be seen that leakage can be large without detrimental effects, but must be considered however, in making tests.

RELATIONSHIP OF COIL TO CONDENSER

The condenser controls the action or output of the coil. Its purpose is to induce high voltage current in the secondary winding of the coil which is

GAS ENGINE MANUAL

essential to proper ignition. The condenser may therefore, be considered an integral part of the coil as without it the coil cannot function.

Obviously, therefore, a condenser with improper capacity or one which leaks is failing to perform its proper function in the operation of the coil, and consequently the coil with an inefficient condenser cannot possibly deliver a spark to the plugs that represent its maximum efficiency.

It is important, therefore, that whenever a new coil is installed that the condenser is checked to insure that it is in good condition. As previously noted, the voltage in the primary part of the coil circuit is very low, the primary having a comparatively small number of turns of heavy wire.

The voltage on the secondary coil circuit, on the other hand, is very high, the secondary having a comparatively great number of turns of fine wire. The voltages in the coil sides vary approximately as the turn ratios, being between 10,000 and 25,000 volts in the secondary and usually 6 or 12 volts in the primary. See Fig. 4.

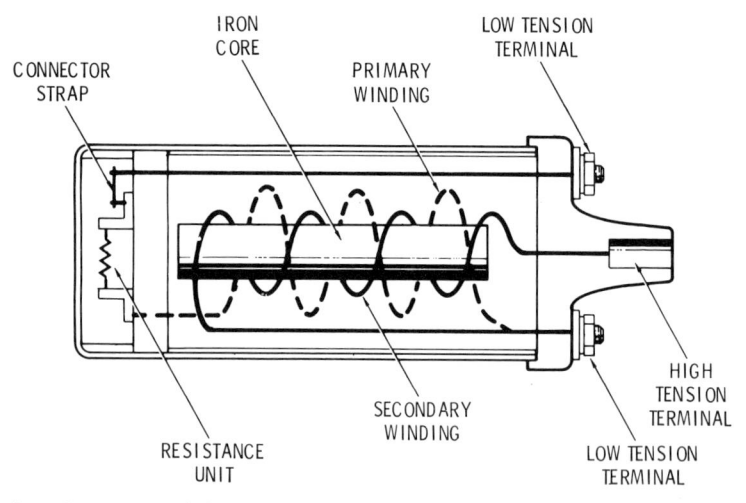

Fig. 4. Ignition coil details, illustrating internal connections of primary and secondary windings.

Without the condenser in the primary circuit, the collapse of the core could be so comparatively slow that the cutting of the lines of force would not be rapid enough to produce the high voltage necessary in the secondary circuit to adequately fire the plugs under high compression. The condenser functions, therefore, are (1) to prevent arcing, and (2) to cause

ELECTRICAL SYSTEM SERVICE

a rapid collapse of the magnetic field necessary to induce high secondary voltage.

SPARK TIMING

The term "timing" is used to describe the work of setting the ignition distributor to ignite the air fuel mixture within the cylinders at the correct time. See Fig. 5.

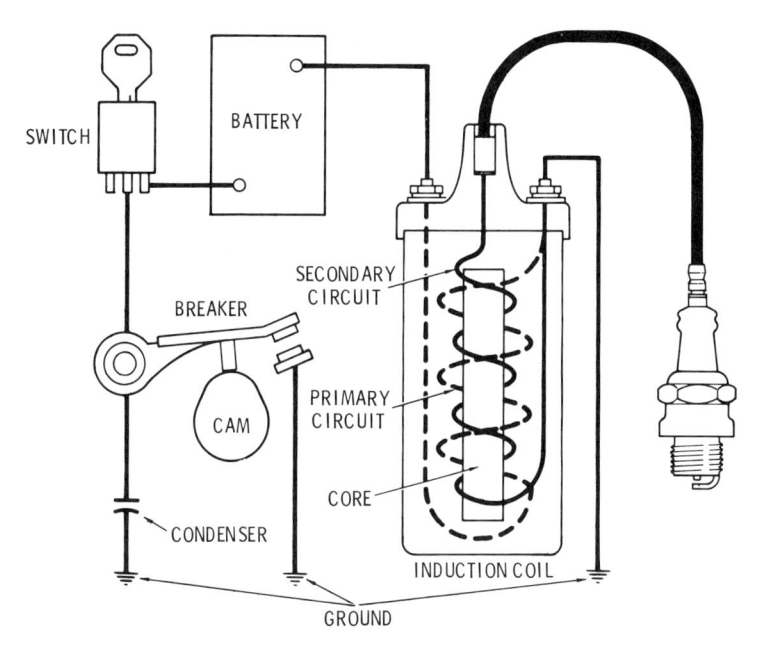

Fig. 5. Schematic diagram showing wiring methods for typical battery ignition system.

When the piston in the cylinder is at the top dead center, the air fuel charge must be ignited. In order to establish this, the ignition distributor must be set (timed correctly) according to the procedure outlined by the manufacturer. Late or retarded timing will cause: (1) Loss of engine power; (2) Increased fuel consumption. (3) Overheated engine and (4) Hard starting.

Early spark timing may cause; (1) Spark knock or detonation; (2) Early failure of engine bearings; (3) Excessive cylinder and piston wear; (4) Broken piston and piston pins. It is therefore important to set or adjust the spark timing as specified by the engine manufacturer.

GAS ENGINE MANUAL

Setting Spark Timing

Consult factory specifications for information and instructions as to the markings on the flywheel or balance wheel that are to be used for correct spark timing. The procedure after obtaining the foregoing data is generally as follows:
1. Connect test leads as shown in Fig. 6, and hold power light in line with flywheel or balance wheel markings.
2. Set engine idling at about 400 rpm or the speed specified by manufacturer. If the engine timing is correct the ball or degree line as specified by the manufacturer will be opposite the pointer. If timing is not correct, adjust the distributor as directed by engine manufacturer.

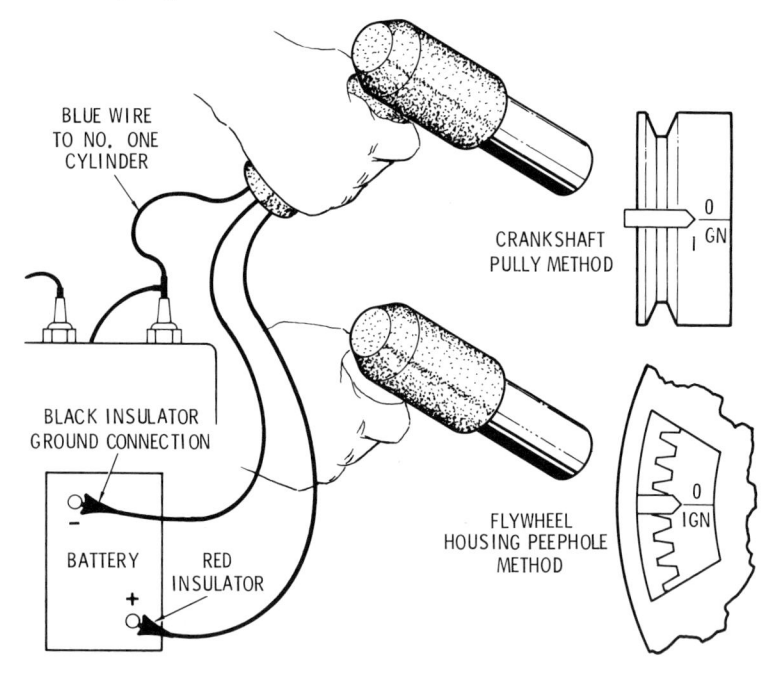

Fig. 6. Illustrating typical arrangement for distributor timing test.

ELECTRONIC IGNITION SERVICE

If the fuel system is known to be properly operational and the engine will not start, the system wiring harness and electric terminals should be checked for tight connections and cleanliness. These could be covered

ELECTRICAL SYSTEM SERVICE

with grease or loose. Also check for a faulty ballast resistor or ignition coil, deteriorated wiring, or a faulty control unit.

The electronic distributor (in most systems) is factory preset and requires little or no maintenance. However, the pick-up coil in the distributor could be defective and cause the system to be inoperative and prevent the engine from starting.

If the carburetor is properly set (not too lean), the trouble could be faulty system wiring (as above), loose pick-up leads (from the distributor), or a malfunctioning ignition coil.

Engine misses could be caused by dirty or faulty spark plugs, dirty or loosely connected ignition-coil secondary cables, a malfunctioning ignition coil, faulty primary wiring, or an inoperative electronic control unit.

Since the electronic control unit and the ballast resistor cannot be repaired or adjusted, determining whether they are operable (or not) can be accomplished by substituting units known to be in good condition in order to test the system. A malfunctioning ignition switch will also cause the system to become inoperative.

MAGNETO IGNITION SERVICE

Normally magnetos are not difficult to service nor are magneto troubles difficult to diagnose. When a magneto equipped engine ignition system does not operate properly, a visual inspection will (in a great many instances) reveal the source of trouble. Particular attention should be given to the spark plugs, distributor, cables, etc. since trouble of this sort may easily be remedied. See Figs. 7 and 8.

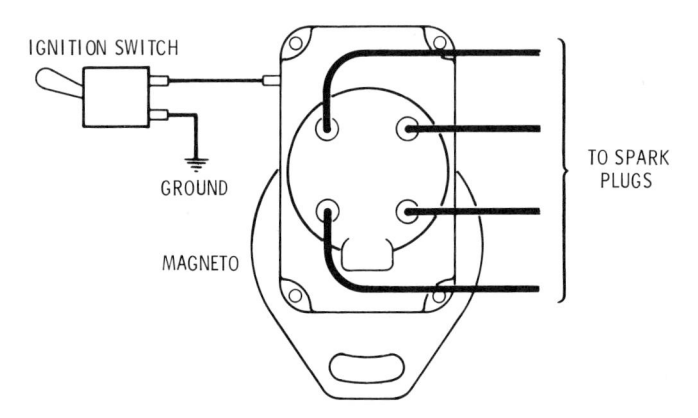

Fig. 7. Wiring diagram of typical magneto ignition system for four cylinder engine.

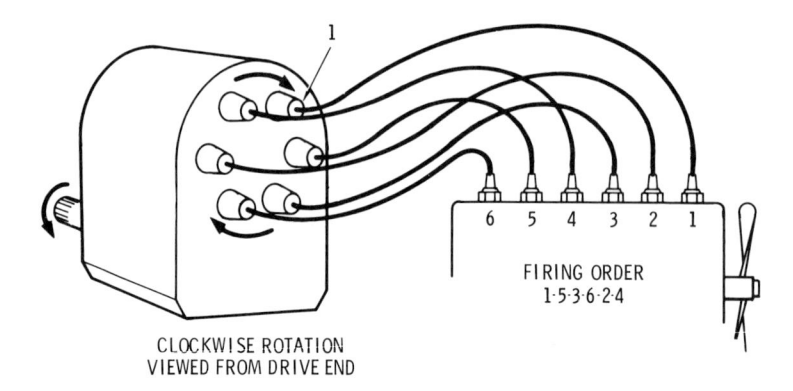

Fig. 8. Magneto spark plug connection for clockwise rotation of magneto.

If a satisfactory spark is not obtained and it is reasonable to believe that the ignition failure originates in the magneto, a series of simple tests may be made to determine whether the fault lies in the magneto or in other parts of the circuit. In general, when only one cylinder misfires, the fault is in the spark plug. The most common spark plug difficulties are as follows:

Spark Plug Gap Too Wide —Difficulty in starting an engine and missing at low speed are very often due to spark plug gaps being too wide, and as the spark will have a tendency to burn the electrode and thereby gradually increase the gap, it is especially important that the plugs be examined occasionally to see that the gap is not too great. Any difficulty due to this cause may readily be overcome by re-adjusting the electrodes.

The proper distance between the electrodes of the spark plug varies in different engines, but normally this distance should not be more than .025 in. Too wide a gap increases the electrical resistance and interferes with the operation of the engine at low speed.

Spark Plug Short-Circuited —This is usually caused by a cracked insulator, or by fouling of the electrodes or insulator. Any of these conditions will cause misfiring by permitting the current to stray away from its intended path.

Cable Troubles —Misfiring of one cylinder, either continuous or intermittently, may be due also to a chafed or broken cable, or a loose cable connection.

The metal terminals of the cables must not come in contact with any metal parts of the engine or the magneto, except those designated as being correct according to the instructions given.

ELECTRICAL SYSTEM SERVICE

Irregular Firing —If the cables and spark plugs are in good condition and the firing is still irregular, the trouble is probably in the magneto or the breaker assembly, which should be carefully examined. It is important that the breaker lever move freely and that the contacts are clean and in correct alignment.

Damaged Insulating Parts —If the distributor plate of the magneto is damaged, it should be carefully examined for possible carbonized tracks or leakage as a result of high-tension flashover.

Testing Method —With the magneto mounted on the engine, the first step is to ascertain whether the magneto is giving a spark. In this test, the spark-plug leads should be detached and the terminal supported one-eighth of an inch from the metal of the engine while the engine is cranked slowly. If a spark appears, the magneto and the breaker may be regarded as being in good condition.

If no sparks are produced at the spark gaps, inspect the breaker contacts for condition and spacing. Check the magneto switch and primary circuit for high resistance or damaged insulation.

When spare parts are available, the coil or condenser may be checked by the comparison method, provided parts can be readily removed. Install parts which are known to be in good condition in place of those parts suspected to be defective. If no improvement is found by the checks as outlined, and no sparks are produced, the magneto must be removed from the engine for shop service.

In this connection it should be noted that a very common cause of magneto failure is excessive lubrication. A film of oil on the breaker contacts results in arcing which shortens the life of the contacts. Use clean light bodied cylinder oil only sparingly or as recommended by the magneto manufacturer.

Inspection and Repairs —A convenient method of testing a magneto under operating conditions consists in removing the entire unit from the engine and operating it at various speeds on a suitable test stand. The following procedures are for the more common types of magnets for industrial engines:

1. Clamp the magneto on the test stand table, fastening securely so that it cannot slip when driven by the test stand motor.
2. Connect magneto to test stand motor and run it at approximately 1000 rpm and check for arcing at breaker contacts. Pin point arcing at breaker contacts indicates that primary circuit is in good condition. Check condenser capacity if breaker arcing is bright and spitting. If no secondary spark is produced, remove

distributor block and note if sparks jump the safety gap when magneto is turned at operating speed. If spark jumps gap when distributor block is off and does not jump when block is on, renew distributor block.
3. If magneto is still inoperative after previous tests, check the coil using a suitable test instrument.

Coil Testing — Fundamentally the coil tester provides a source of primary current intercepted by a built in breaker to induce high voltage current in the secondary winding of the coil to be tested. Primary current is controlled manually by a rheostat in the test unit, giving a definite amperage for each coil unit.

When testing any coil mounted on the armature plate, disconnect the condenser and separate the breaker by a strip of paper. One primary lead from the test unit is connected to the armature plate, with the other connected to the breaker racket. This completes the primary circuit for testing purposes.

If the coil is in good condition and suitable for use, the induced high voltage current to spark plugs should be of sufficient strength to consistently spark across the gap on the test unit, with primary current adjusted to amperage specified for the particular coil.

An irregular, seemingly weak or hesitating spark across the gap indicates a weak coil or damp and partially broken down secondary. Under no condition should an attempt be made to improve this spark by increasing the primary current. The coil is inoperative if it cannot be made to spark properly on the specified amperage. A completely dead coil is indicated by no visible spark.

GENERATOR SERVICING

As a general rule, the generator should be inspected and tested at frequent intervals to determine its condition. High speed operation, excessive dust or dirt, high temperature and operation of generator at or near full output are all factors which increase bearing, commutator and brush wear. See Fig. 9.

When a generator fails to operate properly and it has definitely been determined that the trouble is in the generator and not in some other part of the circuit, the generator should be removed from the engine and taken to a suitable test stand for examination.

ELECTRICAL SYSTEM SERVICE

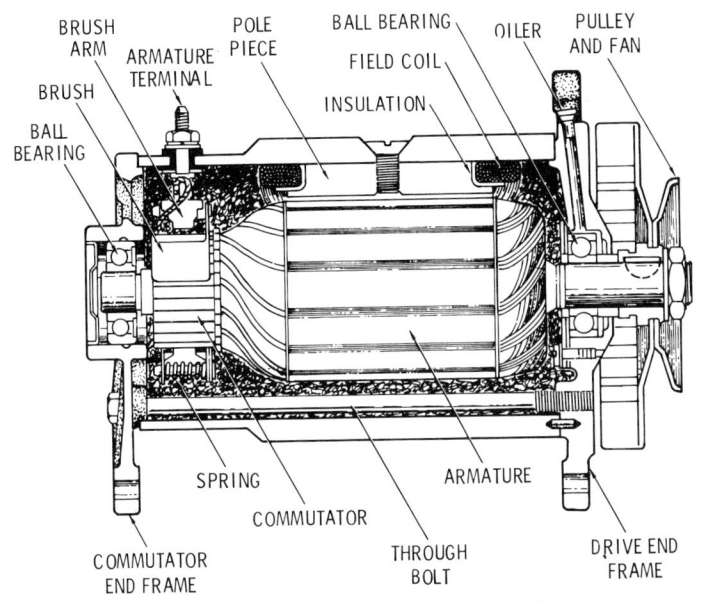

Fig. 9. Sectional view of typical belt-driven automotive type direct current generator.

Inspection

The following inspection will disclose whether the generator is in proper condition for service or in need of removal for repairs. Proceed as follows:

1. Using a good light and a mirror, inspect the commutator through the openings in the commutator end frame. Low or irregular output may result if the commutator being coated with grease or dirt, or is rough, out of round or has high mica between the bars. If commutator bars are burned, an open circuit is indicated. Check for proper air circulation, see Fig. 10.
2. Inspect commutator end of generator for thrown solder, indicating the generator has been overheated due to excessive output. Excessive output usually results when the generator field is grounded, either internally or at the regulator. If this is indicated, disconnect the "field" terminal of the generator or regulator and run engine at medium speed. If generator output drops off the regulator is at fault, but if output remains high the field is grounded internally in generator. If the field is found to be grounded, the regulator probably will have to be replaced.

Gas Engine Manual

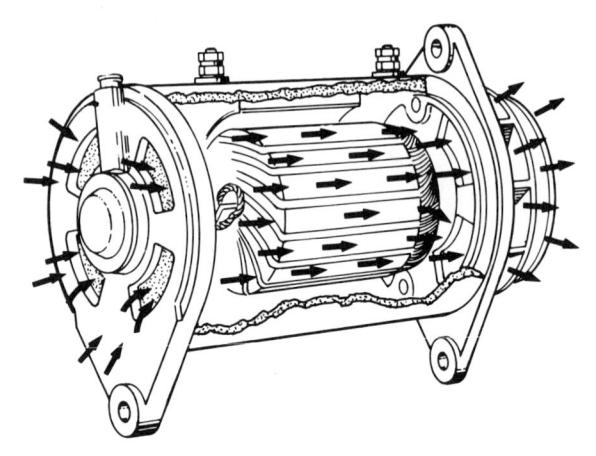

Fig. 10. Illustrating generator ventilation. Air cooling the generator allows it to carry a heavier load without danger of overheating.

3. Check conditions of brushes; make sure that they are not binding in holders and that they are resting on the commutator with sufficient tension to give good, firm contact. Brush leads and screws must be tight. If the brushes are worn down to one-half their original length compared with new brushes, the generator must be removed for installation of new brushes.
4. If the commutator or brushes are in bad condition, other than being dirty, the generator should be removed for repairs. If these parts are only dirty, however, they may be cleaned without removal of generator.
5. Check fan belt for condition and proper tension; make sure that all generator mounting bracket and brace bolts are tight. A loose fan belt will permit belt slippage, resulting in rapid belt wear and low or erratic generator output. An excessively tight belt will cause rapid belt wear and rapid wear of generator and water pump bearings. If belt requires adjustment, first loosen belt so that generator pulley is free, then check pulley for tightness and check generator bearings for freeness of rotation and excessive side play. Rough or excessively worn bearings should be replaced. See Fig. 11.
6. Inspect and manually check all wiring connections at generator, regulator, charge indicator, junction block and battery to make certain that connections are clean and tight.

ELECTRICAL SYSTEM SERVICE

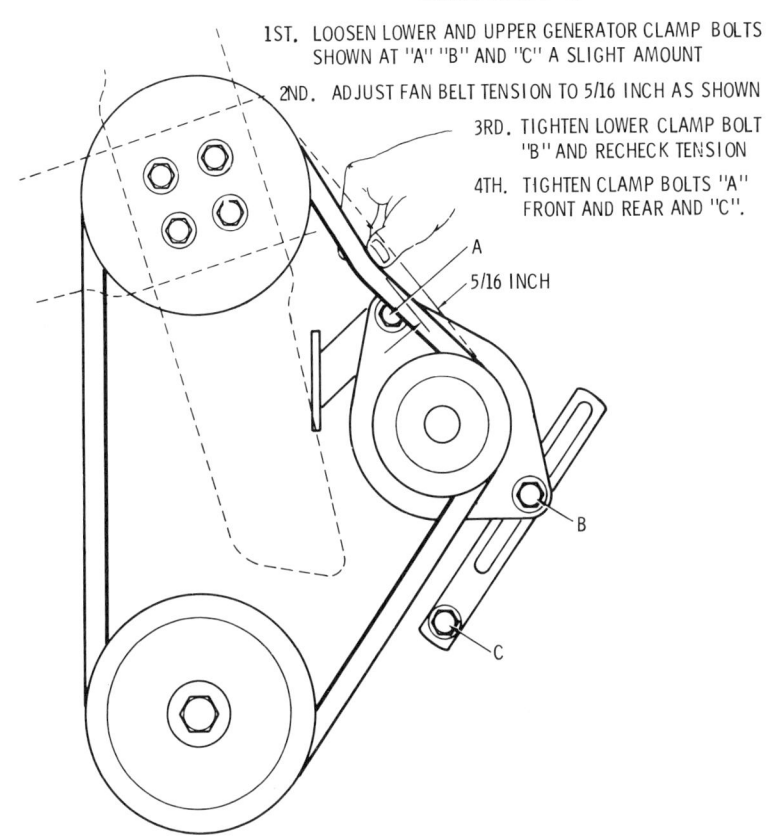

1ST. LOOSEN LOWER AND UPPER GENERATOR CLAMP BOLTS SHOWN AT "A" "B" AND "C" A SLIGHT AMOUNT

2ND. ADJUST FAN BELT TENSION TO 5/16 INCH AS SHOWN

3RD. TIGHTEN LOWER CLAMP BOLT "B" AND RECHECK TENSION

4TH. TIGHTEN CLAMP BOLTS "A" FRONT AND REAR AND "C".

Fig. 11. Method of fan belt adjustment.

GENERATOR REGULATOR SERVICE

Generators being attached and driven by the engine runs at variable speed with resulting variable output in the absence of output control. Accordingly, regulators are necessary to keep the current and voltage within proper limits, as well as to prevent reversal of the current.

Before testing and adjusting the generator regulator, it is advisable to first test the generator output and the charging circuit wiring. If generator output or charging circuit voltage drops are not within normal limits, repairs should be made before testing the regulator.

The following is the general procedure when making tests and adjustments of the *cutout relay, voltage regulator* and *current regulator* in the order named:

Gas Engine Manual

Cutout Relay

Air Gap: Disconnect regulator. Measure air gap between armature and center of winding core with the contact points held closed. Bend the spring fingers until both sets of points meet at the same time. If the air gap does not agree with specification, adjust by loosening the two adjusting screws. Raise or lower armature as required.

Point Opening: Measure point opening. If it does not agree with specifications, bend the upper armature stop until it does.

Closing Voltage Check: Connect the regulator to the proper generator and battery. (Fig. 12.) Connect a voltmeter between the regulator "GENERATOR" terminal and the regulator base. Connect an ammeter between the battery and the regulator "BATTERY" terminal. Slowly increase the generator speed until the points close. If the closing voltage is not according to specifications, bend the spring post until it is. Decrease the generator speed and note the reverse current necessary to open points. If not according to specifications, adjust by changing the air gap. Increasing the air gap lowers reverse current setting.

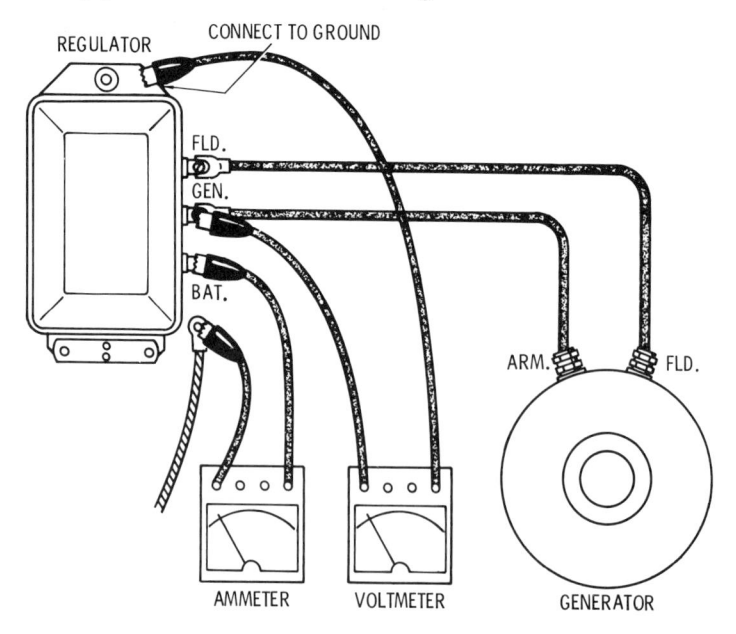

Fig. 12. Wiring diagram illustrating connections for cutout relay adjustment.

ELECTRICAL SYSTEM SERVICE

Voltage Regulator

Voltage Setting, Fixed-Resistance Method: Disconnect the lead from the battery terminal. Connect a test voltmeter and a fixed resistor from the battery terminal to the regulator base. With regulator at operating temperature, run generator at charging speed and note the voltage setting. If the setting does not agree with specifications, adjust by bending down one spring hanger until it does. Bending hanger down increases setting, bending hanger up lowers the setting. Confine the adjustment to one spring unless regulator is badly out of adjustment.

For complete adjustment, remove the second spring. Connect a voltmeter from the "GENERATOR" terminal to the regulator base. Open the voltage regulator points by hand, increase the generator speed until voltmeter reads approximately one-half the specified operating voltage, (open circuit). This establishes the approximate generator speed at which adjustment should be made. Let the points close and adjust the first spring hanger to one-half the total voltage setting. Install the second spring. Connect the voltmeter and resistance as illustrated. Complete the adjustment of the second spring hanger (without changing the first spring hanger) until the correct voltage setting is obtained. After each change of setting, check the adjustment by replacing cover and reduce generator speed until the points open, then increase the speed until the points close.

Voltage Setting, Variable-Resistance Method: Connect an ammeter and a ¼ ohm variable resistor between the battery and the "BATTERY" terminal, as illustrated in Fig. 13. Connect a voltmeter between the "BATTERY" terminal and the regulator base. Operate the generator at medium speed. If less than 8 amperes are obtained, turn on all lights and accessories to permit higher output. Increase the variable resistor setting until the current output is reduced from 8 to 10 amperes. Cycle the generator and note the setting. Adjust as in the fixed-resistance method.

Air Gap: Disconnect regulator. Open points by hand. Release armature slowly until points touch, then measure air gap. If not according to specifications, adjust by loosening the two mounting screws and raise or lower the contact brackets as required.

Current Regulator

Current Setting: Connect regulator to generator and battery. Remove regulator cover and connect jumper across voltage regulator

375

Gas Engine Manual

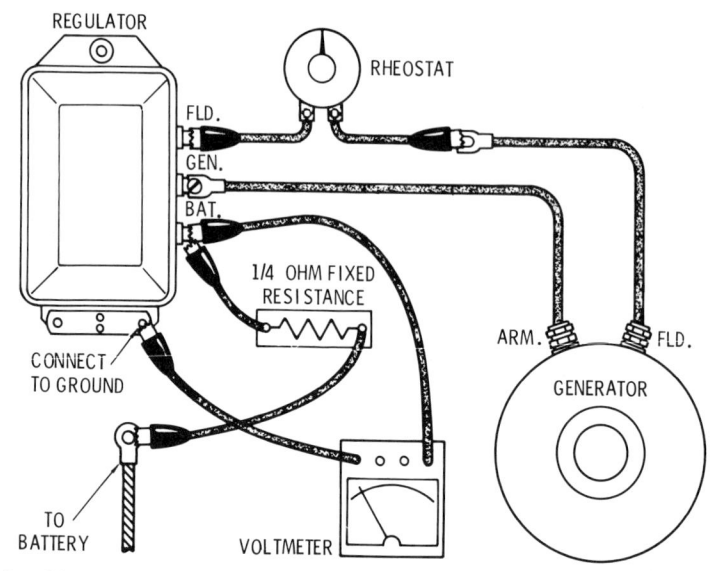

Fig. 13. Wiring diagram illustrating connection for voltage regulator adjustment.

points. Connect ammeter as indicated. (Fig. 14.) With regulator at operating temperature, turn on all lights and accessories. Run generator at medium speed and note the current setting. If not

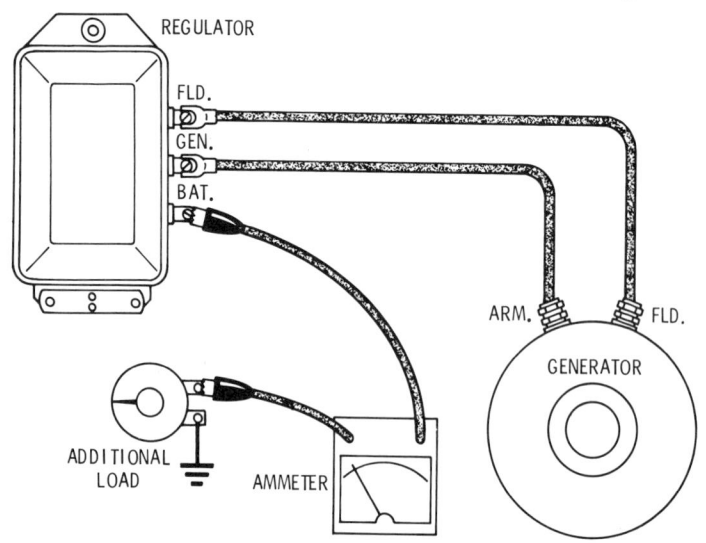

Fig. 14. Wiring diagram illustrating connection for current regulator adjustment.

according to specifications, adjust by bending the spring hanger. Bending spring hanger up decreases current setting. Bending hanger down increases current setting. Confine the adjustment to one spring unless regulator is completely out of adjustment.

For complete adjustment, remove one spring and adjust remaining spring hanger to one-half the specified setting. Reinstall the first spring and adjust its spring hanger to the full current setting. *Air Gap:* Disconnect regulator, open points by hand. Release armature slowly until points just touch, then measure air gap. If not according to specifications, adjust by loosening the two contact mounting screws and raise or lower contact bracket as required.

ALTERNATOR SERVICING

Like a generator (preceding), an alternator should be frequently inspected and tested. However, the checks that can be made without special equipment generally are limited to those that will determine whether or not the alternator, as a unit, must be replaced.

A battery and charging-circuit tester (there are various make commercial units) is used for testing (in accordance with instructions supplied with the particular unit). If alternator output is low or erratic, check fan belt and external wiring as previously explained for a generator. If this fails to locate the trouble, the alternator must be replaced.

Because of the many different alternator constructions, it is not practical to detail the service requirements here. Each manufacturer furnishes such instructions. Generally, an alternator must be disassembled for component checking, and diodes, resistors, windings, etc. must be separately checked—and replaced, if malfunctioning.

STARTING MOTOR SERVICE

The starting system consists of the *battery, starting switch* and *starting motor*. The battery supplies the energy, the switch completes the circuit allowing this energy to flow to the starting motor. The motor then delivers mechanical energy and does the actual work of cranking the engine. Because of its action in cranking the engine, the starting motor is also called the "cranking motor."

The starting motor assembly consists of motor, drive assembly, shift lever and solenoid switch. When the solenoid is energized, the starter armature spins, feeding the pinion rearwards on the threaded sleeve until it meshes with the flywheel gear.

Gas Engine Manual

The sudden shock of meshing is absorbed by the drive spring. After the engine starts the flywheel ring gear turns the starter pinion faster than the armature. At a pre-determined speed the gear is released and forced back along the sleeve threads to its normal position.

Locating Troubles in Starting Motor

In many respects, a starting motor is similar to a generator and the inspection for location of troubles are similar for both. Starting motor action is indicative to some extent of the starting motor condition. A starting motor that responds readily and cranks the engine at normal speed when the control circuit is closed, is usually in good condition.

If the motor does not develop rated torque and crank the engine slowly, or not at all, check the battery, battery terminals and connections, the ground cable and battery to cranking motor cable. Corroded, frayed or broken cables should be replaced and loose or dirty connections corrected. The magnetic switch should be checked for burned contacts and the contacts replaced if necessary.

If there are burned bars on the commutator, it may indicated open circuited armature coils which prevent proper cranking. Inspect the soldered connections at the commutator riser bars, resolder these connections and turn down the commutator as necessary. See Fig. 15 for typical starting motor circuit.

An open armature will show excessive arcing at the commutator bar which is open, on the no load test.

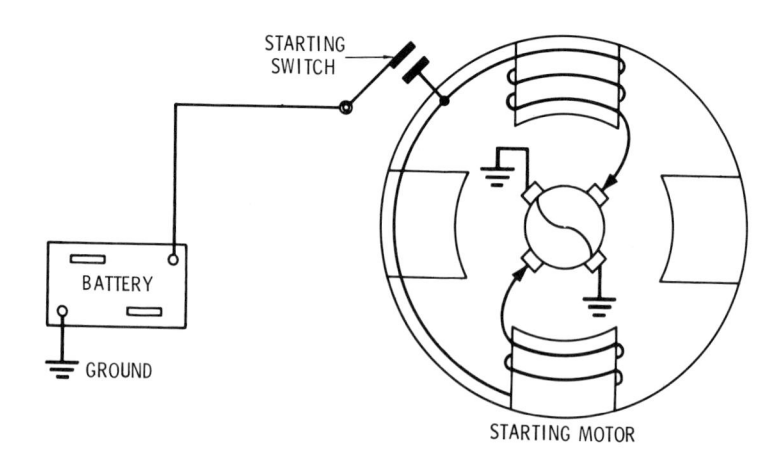

Fig. 15. Typical starting motor circuit.

ELECTRICAL SYSTEM SERVICE

Tight or dirty bearings will reduce armature speed or prevent the armature turning. A worn bearing, bent shaft or loose pole shoe will allow the armature to drag, causing slow speed or failure of the armature to rotate. Check for these conditons.

If the brushes, bearings, commutators, etc., appear in good condition, the battery and external circuit also in good condition, and the cranking motor still do not operate correctly, remove the cranking motor and submit it to the no load and torque test.

Interpretation of no load and torque tests results — The following indications apply:
1. Rated torque, current draw and no load speed indicates normal condition of cranking motor.
2. Low free speed and high current draw with low developed torque may result from:

a. Tight, dirty or worn bearings, bent armature shaft or loose field pole screws which would allow the armature to drag.

b. Shorted armature. Check armature further on growler.

c. A grounded armature or field. Check by raising the grounded brushes and insulating them from the commutator with cardboard, and then checking with a test lamp between the insulated terminal and the frame. If test lamp lights, raise other brushes from commutator and check fields and commutator separately to determine whether it is the fields or armature that is grounded.

3. Failure to operate with high current draw:

a. A direct ground in the switch, terminal or fields.

b. Frozen shaft bearings which prevent the armature turning.

4. Failure to operate with no current draw:

a. Open field circuit. Inspect internal connections and trace circuit with a lamp test.

b. Open armature coils. Inspect the commutator for badly burned bars. Running free speed, an open armature will show excessive arcing at the commutator bar which is open.

c. Broken or weakened brush springs, worn brushes, high mica on the commutator, or other causes which would prevent good contact between the brushes and commutator. Any of these conditions will cause burned commutator bars.

5. Low no-load speed, with low torque and low current draw indicates:

a. An open field winding. Raise and insulate ungrounded brushes from commutator and check fields with test lamp.

379

Gas Engine Manual

b. High internal resistance due to poor connections, defective leads, dirty commutator and causes listed under 4c.

6. High free speed with low developed torque and high current draw indicates shorted fields. There is no easy way to detect shorted fields, since the field resistance is already low. If shorted fields is suspected, replace the fields and check for improvement in performance.

TESTING STARTING MOTOR PARTS

Field coil test for continuous circuit — Place the test prod leads on the field coil leads.

If the test lamp lights, the field coils are all right.

If the test lamp does not light, there is an open circuit in one or both of the field coils.

Field coil test for ground — Place one test prod lead to frame and the other to the field coil lead.

If the test lamp does not light, the field coils are O.K. If the test lamp lights, one or both field coils are grounded.

Individual field coil test for ground — Break soldered connection between the two field coils and test each one separately, replacing the field coil found to be grounded.

Field coil leads inspection — Inspect the field coil leads where they are soldered at the starting switch terminal to be sure that they are tight.

Armature test for ground. — Place one test prod on the armature and the other on the commutator.

If the test lamp lights, the armature is grounded and should be replaced. If the test lamp does not light, the armature is O.K.

Armature test for short circuit — Place the armature on the growler, and with a saw blade over the armature core, rotate the armature and test.

If the saw blade does not vibrate, the armature is O.K. If the saw blade vibrates, the armature is short circuited and should be replaced.

Commutator — Inspect the commutator for roughness.

If it is rough, turn down on a lathe until it is thoroughly cleaned, then sand off the commutator with 00 sand paper.

Insulated brush holder test for ground. — Place one test prod lead to the cover and the other on the brush holder.

If the test lamp light, brush holder is grounded and should be replaced. If the test lamp does not light, the brush holder is O.K.

Brushes — Check condition of the brushes and if they are pitted or worn, they should be replaced.

Check the tension of the brush holder springs; they should have enough tension to hold the brushes snugly against the commutator.

Brush ground leads — Disconnect the brush ground leads from the end frame and clean all terminals and replace.

Check the insulation of the brush to field coil leads. The insulation should not be broken.

Drive housing bushing — Check the condition of the drive housing bushing.

The armature shaft should fit snugly in this bushing, if it is worn it should be replaced.

To facilitate the service diagnosis the following basic trouble pointers together with their possible causes and remedies are given:

CONDITIONS — POSSIBLE CAUSES — REMEDIES

STARTER FAILS TO OPERATE
Possible Causes:
  a. Corrosion at battery posts.
  b. Loose battery cables.
  c. Dead battery cell.
  d. Defective starter or solenoid switch.
  e. Defective starter.
  f. Weak battery.

Remedies:
  a. Remove battery cable and clean terminals and clamps. Check clamps for erosion and replace if necessary.
  b. Clean battery posts and cable clamps. Tighten securely for good contact.
  c. Replace defective battery. Check voltage regulator and generator output.
  d. Replace starter and ignition switch and check starting motor solenoid for operation. Replace if necessary.
  e. Remove and test starting motor. Replace parts as required or the complete unit.
  f. Test specific gravity of battery and check for dead cell.

STARTER FAILS AND LIGHTS DIM
Possible Causes:
    a. Weak battery.
    b. Loose connections.
    c. Dead battery cell.
    d. Battery terminals corroded.
    e. Internal ground in windings.

Remedies:
    a. Test specific gravity of battery and check for dead cell.
    b. Tighten loose connections as required, being sure to check terminals for corrosion.
    c. Replace defective battery. Check voltage regulator and generator output, which may have contributed to the battery failure.
    d. Remove battery cables and clean terminals and clamps. Check clamps for erosion and replace if necessary.
    e. Remove starting motor and test.

STARTER TURNS BUT DOES NOT ENGAGE
Possible Causes:
    a. Broken drive spring.
    b. Broken teeth on flywheel ring gear.
    c. Grease or dirt on screw shaft.

Remedies:
    a. Remove starting motor and install new drive spring. Check screw shaft for excessive wear or burring, replace if necessary.
    b. Replace flywheel ring gear; be sure and check the teeth on mating pinion for wear and replace if necessary.
    c. Remove starting motor and clean screw shaft in clean kerosene.

CHAPTER 30

Emission Controls Services

Electric-Assist Choke — Each unit is specifically calibrated for its application and must be checked against the manufacturer's specifications for performance. If either the choke unit or the heater control switch is malfunctioning, the faulty unit must be replaced.

Choke Hot-Air Modulator — With the air cleaner removed, tape a thermometer near the modulator. If the modulator does not open or close within specified temperatures, remove and replace the unit. Also check condition of tubing (or hose) which connects the modulator to the heater coil, and replace it if it is damaged. Clean the interior of the hose, if dirty.

Crankcase Ventilation (PCV) System — If the engine operates erratically, remove the PC valve. A dirty or stuck valve must be replaced. Also, at this time, check the condition of all the connecting hoses or tubing. All hoses and tubing must be in good (undamaged) condition and have airtight connections.

Air-Pump Systems — System components can be checked separately, as follows:

Air Pump: Remove the outlet hose from the pump and, with engine running, check to see if air is coming from the outlet where hose was removed. Air volume should increase as engine speed is increased. Reinstall hose and check to see if any air is leaking from the pump relief valve (if so equipped). The relief valve can be replaced (as a unit) if defective. If pump is excessively noisy or is seized, or if any of the above checks prove unsatisfactory, remove and replace the pump.

Check for proper drive-belt tension and pump mounting, and be sure all connecting hoses or tubes and the air manifold (on the engine) are free from air leakage. Also be sure carburetor air cleaner is properly installed.

Gas Engine Manual

Air By-Pass Valve: First determine whether valve is receiving a vacuum signal by removing the hose from the signal port on the valve. With engine running, place finger over end of hose to feel whether vacuum is present. If there is none, replace hose.

With valve signal hose reinstalled and with engine warmed up and idling, no air should be flowing out of valve outlet. Now, quickly open and close the engine throttle. This should cause a short blast of air to be expelled from the valve outlet. If the valve does not operate as described, it is defective and must be replaced.

Check Valve: Disconnect the hose from the check valve, but do not remove valve from air manifold. With engine cool and not running, check for air flow into the manifold only. This can be done orally. If air can be sucked out of the air manifold, the check valve is defective and must be replaced.

Mixture-Control or Backfire Valve—All hoses to the valve should be carefully checked as any defective hose will cause the valve to malfunction. Replace defective hoses.

With all hoses connected and engine running at idle, disconnect the vacuum hose from the valve and place a finger over the end of the hose to feel whether vacuum is present. If there is no vacuum, replace the vacuum hose.

Again, with all hoses connected and engine running at idle, check the condition of the valve by first removing the hose which leads from the pump to the valve (disconnect hose at pump). If an air noise is heard at the hose, this indicates a leaking valve. Continue by placing a hand over the hose end. If a strong vacuum is felt or no air noise is now heard, this also indicates a defective valve. A leaking or defective valve must be replaced.

Another check will determine valve condition. Open and close the engine throttle quickly. The result should be a distinct but slowly decreasing air noise. Either a total lack of noise or an exceptional amount of noise also indicates a defective valve, which must be replaced.

Pulse Air-Injection Reactor (PAIR) — During engine idle, all the check valves and the air shut-off valve should be taking in air. Check all hoses and valves (starting with the air cleaner hose) for this condition. If any hose or valve does not function in this way, replace the faulty component.

NOTE: Each check valve should take in air in a pulsating manner.

Thermostatic-Controlled Air Cleaner — Any defective components of this system must be removed and replaced. Particular attention should be

EMISSION CONTROLS SERVICES

given to the air door or air damper, actuating thermostats, heat sensors, vacuum controls, and all connecting hoses or tubing.

Transmission-Controlled Spark (TCS) — If engine is acting erratic or tends to overheat, the ignition, carburetion and cooling systems should be checked *first* to determine their condition. If everything checks out satisfactorily, then check out all components of the transmission controlled spark system. Remove and replace faulty components — these units must meet manufacturer's specifications and cannot be repaired.

NOTE: The Combined Emission Control (CEC) and the Speed-Controlled Spark (SCS) systems are both checked as above.

Exhaust-Gas Recirculation (EGR) — Remove the *EGR valve* from its manifold, apply a specified auxiliary source of vacuum to the vacuum port of the valve and observe whether the valve shaft rises and remains in this position. If it does not rise, or lowers again (after rising), there is a *malfunctioning valve diaphragm,* and the EGR valve assembly will have to be replaced.

Early Fuel Evaporation (EFE) — Begin with a cold engine. Grasp the hose (or tubing) that connects the EFE valve to the engine intake manifold. Start the engine. The hose should heat up rapidly, due to the passage of exhaust gases through it. If it does not heat up within a reasonable time, the EFE valve should be removed and replaced.

After the engine has warmed up to the manufacturer's specified temperature, the *thermal vacuum switch* should close the vacuum control to the EFE valve, and stop exhaust gas flow to the intake manifold. The thermal vacuum switch should be replaced if this does not occur.

Cleaner Air Package (CAP) — If the engine is operating erratically or tends to overheat, the ignition, carburetion and cooling systems should be checked first to determine their conditions. If everything checks out satisfactorily, then check-out all components of the cleaner air package system in accordance with the manufacturer's operating specifications. Replace malfunctioning units.

Computer-Controlled Timing — Malfunctioning could be caused by dirty, damaged or loosely connected wiring. Check this first. To check system components, replace them one at a time with units known to be in operating condition, until the faulty one(s) has been detected. Faulty components must be replaced.

The distributor breaker points (there are two sets) should be checked for specified gap. Also check condition of spark plugs for condition and proper gaps.

Gas Engine Manual

Spark-Delay Valve — This valve is a one-way valve. To check its condition, remove the valve and orally check for free air flow in one direction and completely blocked flow in the other. If these conditions are not met, replace the valve (be certain to "face" it correctly).

Deceleration Valve — First check, and replace or tighten as necessary, all hoses and hose connections. Remove the valve from the engine and apply a specified auxilliary source of vacuum to the valve. Observe whether the valve opens (by viewing through the valve's intake-manifold port). If valve does not open, it is defective and must be replaced.

Spark-Delay Valve: This is a two-way valve and should be checked for immediate free-flow of air in one direction and restricted flow in the other. Remove the valve from the system and check it orally. Replace it if defective ("facing" it in the right direction).

Check Valve: This valve is also a two-way valve, but allows free flow of air in one direction and no flow in the other. Remove the valve from the system and check it orally. Replace it if defective ("facing" it correctly).

Delay-Vacuum By-Pass System (DVB) — First check all hoses (or tubing) and wiring for condition. Tighten any loose connections and replace any dirty or damaged parts. Also, make certain that the ignition and carburetion systems are operating properly. Other system components are checked as follows:

Spark-Delay Valve: This is a two-way valve and should be checked for immediate free-flow of air in one direction and restricted flow in the other. Remove the valve from the system and check it orally. Replace it if defective ("facing" it in the right direction).

Check Valve: This valve is also a two-way valve, but allows free flow of air in one direction and no flow in the other. Remove the valve from the system and check it orally. Replace it if defective ("facing" it correctly).

Temperature Switch: If it does not open or close at specified temperatures, it must be replaced.

Solenoid Vacuum Valve (SV): If the temperature switch is operating properly, the solenoid valve should open and close at specified temperatures. If it does not operate properly, replace it.

Temperature-Activated Vacuum (TAV) — If unsatisfactory engine performance becomes apparent it could be caused by defective components of this system, provided the ignition and carburetion systems are in good working order.

Check all hoses (or tubing) for apparent condition. If they are clogged, collapsed or show signs of leakage, they should be replaced.

If the *ambient temperature switch* or the *solenoid valve* does not operate at specified temperatures, the faulty component should be removed and replaced.

Since the *in-line bleed* clears the system of excessive gasoline vapors, *any* indication that this unit is defective is reason enough to replace it.

Cold-Temperature-Activated Vacuum (CTAV) — This system is essentially the same as the temperature activated vacuum (TAV) system, previously described, and all like components should be checked for operation in the same way. However, one additional component, a latching relay, has been added.

If the temperature switch or latching relay does not operate within specified temperatures, the faulty component should be replaced.

Catalytic Converters — If a catalytic converter looses its effectiveness, it should be removed and replaced. Use of leaded fuel and/or excessive overheating of the unit are the principal causes of unit failure. Special emission-level checking equipment is required.

FUEL-EVAPORATION EMISSION CONTROLS

CAUTIONS:
1. Since this is a system which contains gasoline vapors. be extremely careful not to use a torch or any other type of open flame near any of its components. A fire could occur, and lead to an explosion.
2. If a new gas tank cap is needed, be sure to replace it with the proper kind. If a vented cap is mistakenly used, it will render the system inoperative. If a cap without a double check valve is installed (again, mistakenly), engine fuel pump suction of the fuel from the storage tank will collapse the tank.

Check all system hoses to see that they have not been pinched shut or broken. If any lines need replacement, be sure they are marked "EVAP". Standard fuel hoses *cannot* be used on these systems as they may clog or deteriorate.

Be sure the *charcoal canister* is replaced at specified intervals.

INDEX

A

Accelerating pump, carburetor, 151
Admission stroke, 15, 16
Air
   and oil circulating systems, 116
   bleed principle, 149
   cleaner, 126
   cooling, 22, 116
   -fuel ratio, 137
   -injector system, 188
Alternator, 375
Aluminum-alloy
   connecting rods, 63
   piston, 39
Ammeter, 230
Antifreeze solution, 114

B

Balance, crankshaft, 65
Bendix drive, starting, 274
Breakerless distributor, 244
Built-up and single piece crankshaft, 72

C

Cam grinding pistons, 42
Camshaft, 78, 344
Capacitor-discharge, 241
Carburetion, 352
Carburetor
   accelerating pump, 157
   and fuel-injection components, 137, 160, 347
   choke
      circuit, 158
      valve, 142
   downdraft, 152
   economize, 150
   float, 143
      circuit, 155
   heating method, 150
   low speed circuit, 155
   main metering circuit, 156
   metering rod, 150
   operating principles, 138
   pump circuit, 156
   venturi effect, 145
Cast-iron pistons, 38
Catalytic converters, 199
Cells, battery, 224

circuits, 225
Centrifugal governor, 166
Chemical action, storage battery, 259
Choke, 55
   circuit, carburetor, 158
   valve, 142
   hot-air modulator, 177
Clearance, piston, 44
Clutch
   elements, 202
   heavy-duty, 203
   operation, 202
Coil, ignition, 232
Combination lubricating, 94
Combustion
   control system, 174
   of gasoline, 134
Comparison of engines, 20
Compression
   rings, 47
   stroke, 15, 17
Condensers, ignition, 235, 361
Connecting rods, 59, 333
   construction of, 61
Coolant
   -sensing switch, 187
   temperature switch, 197
Cooling system, 105
   air, 22, 116
   antifreeze solution, 114
      precautions, 115
   liquid, 23
   temperature control, 111
   variable
      -speed fan, 108
   water
      and oil circulating system, 115
      circulating pump, 110
      circulation system, 106
      jacket, 108
Cracking process of gasoline, 134
Crankcase
   casting, 31
   -vapor venting, 173
Crankshaft, 65, 335
   balance, 65
   bearing clearance, 335, 337

construction of, 65
throw arrangement, 69
Cushioning devices, clutch, 206
Cylinder
  arrangement, 26
    in-line, 26
    horizontally opposed, 27
    radial, 27
    V-type, 26
  block service, 317
  boring, 318
  honing, 320
  head, 31

D

Deceleration valve, 181
Delay time, 197
Dilution, oil, 102
Distribution, 227, 242
  breakerless, 244
  breaker points, 248
Diverter valve, 190
Downdraft carburetor, 152
Dual diaphragm distributor, 183
Dwell, distributor, 360
Dynamometer, horsepower, 211

E

Economizer, carburetor, 150
Efficiency
  horsepower, 213
  mechanical, 215
  thermal, 214
  volumetric, 215
EGR system, 193
  valve, 195
Electrically-operated fuel pump, 123
Electrical system, 252, 359
  alternator, 375
  dwell, 360
  generator, 262, 368
    a-c, 268
    d-c, 263
  regulator, 265, 371
  starter, 271, 375
  storage battery, 252
    charge, 259
    construction of, 254
    discharge, 259
    electrolyte, 256

    rating, 261
    specific gravity, 256
    timing, 363
Electrolyte, battery, 256
Electromagnetic induction, 223
Electronic
  and computer-controlled timing, 187
  battery ignition, 240
  control unit, fuel injector, 162
Emission control systems, 169, 381
Engine
  flywheel, 73
  torsional vibration, 73
  vibration damper, 74
  speed governor, 166
  tune-up, 303
Exhaust
  back-pressure transducer, 195
  manifold, 127
  stroke, 19
Expansion of piston, 323

F

Fast-idle valve, fuel injector, 165
Filter, oil, 96
Final-exhaust controls,
  emission controls, 171
Firing order, 28
Float, carburetor, 143, 347
  circuit, 155
Flywheel, 73
  torsional vibration, 73
  vibration dampers, 74
Forced lubrication, 93
Four-stroke engine, 14
  admission stroke, 15, 16
  compression stroke, 15, 17
  exhaust stroke, 19
  power stroke, 18
Frequency of oil changer, 103
Friction clutches, 201
  principles of, 201
Fuel
  evaporation control, emission, 171
  filter, 124
  flow circuit, 154
  gauge, 128
  -injector system, 130, 356
    electronic control unit, 162

fast-idle valve, 165
manifold absolute pressure sensor, 162
oil-pressure sensor, 165
pressure
  sensor and switch, 162
  regulator, 161
  pump, 161, 166
  temperature sensor, 165
  throttle-position switch, 164
systems, 119
  air cleaner, 126
  gasoline, 134
    combustion of, 134
    octane rating, 135
  manifold,
    exhaust, 127
    intake, 127
  muffler, 127
  tank, 128
Fundamental electricity, 219
  current, 221
  electromagnetic induction, 223
  magnetism, 222
  Ohms law, 219
  primary induction coil, 225
  secondary induction coils, 226

G

Gas engine
  fundamentals, 9
    admission stroke, 15, 16
    comparison of engines, 20
    compression stroke, 15, 17
    exhaust stroke, 19
    four-stroke, 14
    parts, 30
      moving, 34
      stationary, 31
        casting,
          crankcase, 31
          cylinder head, 31
          oil pan, 31
    power
      stroke, 18
      output, 20
    two-stroke engine, 10
    valve overlap, 20
Gap adjustment, spark plug, 278
Gasoline, 134
  combustion of, 134
  cracking process, 134
  octane rating, 135
Gauge
  fuel, 128
  oil, 97
Generator, 368
Governors,
  centrifugal, 166
  vacuum, 167
Gravity-flow carburetor, 120
Grinding, cylinder block, 319
Guides, valve, 78, 340

H

Heat
  control valve, 181
  range, spark plug, 279
Heating method, carburetor, 150
Heavy duty clutches, 203
High
  oil consumption, 293
  pumping at rings, 294
  -speed modulator, 198
Honing, cylinder block, 320
Horizontally opposed cylinder
  arrangement, 27
Horsepower measurements, 207
  indicated horsepower, 212
  prony brake, 207
  rope brake, 209
  SAE horsepower, 212
Hydrometer, storage battery, 256

I

Idle-stop solenoid, 181
Ignition systems, 229
  capacitor-discharge, 241
  coil, 232
  distributor, 233, 248
    breakerless, 244
  electro-mechanical battery, 229
  magneto, 244
  plugs, 235
  resistor, 232
  spark action, 233
  switch, 231
I-head valve arrangement, 23
Indicated horsepower, 212
In-line cylinder arrangement, 26
Intake manifold, 127

L

Lapping, cylinder block, 320
Liquid cooling, 23
L-head valve arrangement, 23
Loss of power—engine, 287
Low
   oil pressure, 301
   speed circuit, carburetor, 155
Lubricating system, 93
   combination, 94
   forced, 93

M

Magnetism, 222
Magneto ignition, 244
   flywheel, 250
Main
   bearing, crankshaft, 72
   noise, 298
   metering circuit, carburetor, 156
Manifold absolute pressure sensor,
   fuel-injector, 162
Major engine tune-up, 311
Mechanical efficiency, horsepower, 215
Mechanically-actuated, fuel pump, 121
Metering rod, carburetor, 150
Minor engine tune-up, 303
Miscellaneous rings, piston, 56
Muffler, 127
Multicylinder engines, 25

N

Noise, engine, 296
Noisy valves, 297

O

Octane rating, gasoline, 135
Ohms law, 219
Oil
   filter, 96
   gauge, 97
   pan, 31
   -pressure sensor, fuel injector, 165
   pump, 94
   rings, piston, 52
   strainer, 95
   viscosity, 93
Operating
   mechanism, valves, 81
   principles, carburetor, 138

Overrunning clutch drive, starting, 272

P

Piston, 37, 323
   clearance, 44
   constant clearance, 41
   displacement, 215
   expansion of, 323
   fitting, 324
   material, 37
      aluminum-alloy, 39
      cast-iron, 38
      semisteel, 39
   removing from cylinder, 324
   requirements, 37
   rings, 47, 326
      fitting, 327
   slap, 40
   temperature, 45
Prony brake, horsepower, 207
Power
   stroke, 18
   output, 20
Precautions, antifreeze, 115
Pressure
   cap, radiator, 113
   -feed carburetor, 120
   sensor and switch, fuel-injector, 162
Primary induction coils, 225
Pulse air-injection reactor system, 192
Pump,
   accelerating, 151
   circuit, carburetor, 156
Purpose and type of fuel pump, 121
   electrically-operated, 123
   mechanically-actuated, 121

R

Radial cylinder arrangement, 27
Radiator, 108
   pressure cap, 113
   tubular, 110
Reaming
   cylinder block, 320
   valve guides, 342
Reconditioning valves and seats, 339
Refacing valves, 340
   seats, 341
Regulator, voltage, 371

Resistors, ignition, 232
Rings, piston, 47
  compression, 47
  miscellaneous, 56
  oil, 52
Rope brake, horsepower, 209
Rust inhibitors, antifreeze, 115

S

SAE horsepower, 212
Seats, valve, 77
Secondary induction coils, 226
Semisteel pistons, 39
Skips or misses—engine, 289
Slap, piston, 40
Solenoid valve, 192
Spark
  action, 233, 247
  control, 236
    centrifugal force, 237
    engine vacuum, 238
  -delay valve, 186
  plugs, 235, 277
    cleaning, 280
    gap, 278
    heat range, 279
    thread size, 279
Specific gravity, battery, 256
Spray nozzle, 147
Springs, valves, 342
Stalls—engine, 286
Starting motor, 271, 375
  Bendix drive, 274
  overruning clutch drive, 272
Storage battery, 252
  charge, 253
  chemical action, 259
  construction of, 254
  discharge, 259
  hydrometer, 256
  rating, 261
  specific gravity, 256
  temperature correction, 257
Super-chargers, 216
Switch, ignition, 231

T

T-
  head valve arrangement, 23
  slot piston, 42

Tank
  fuel, 128
  vapor venting, 172
Temperature
  control, 111
  correction,
    battery, 257
  piston, 147
  sensors,
    fuel-injector, 165
Thermal efficiency, horsepower, 214
Thread size, spark plug, 279
Throttle
  idling, carburetor, 348
  -position switch, fuel-injector, 164
Timing, 363
  finding dead center, 90
  ignition, 249
Torque, horsepower, 210
Torsional vibration, flywheel, 73
Troubleshooting, 283
Tune-up, engine, 303
Two-stroke engines, 10

V

Vacuum
  -advance control valve, 184
  differential valve, 191
  signal modulator, 197
Valve
  arrangement
    I-head, 23
    L-head, 23
    T-head, 23
  guide, 75, 340
    remaining, 342
  operating mechanism, 81
  seats, 74
  springs, 342
Valves and valve gears, 77
Venturi effect, 145
Vibration dampers, flywheel, 74
Voltage regulator, 266, 371

W

Water
  and oil circulating system, 115
  circulation system, 106
  jacket, 108
Wrist pins, 56

The Audel® Mail Order Bookstore

Here's an opportunity to order the valuable books you may have missed before and to build your own personal, comprehensive library of Audel books. You can choose from an extensive selection of technical guides and reference books. They will provide access to the same sources the experts use, put all the answers at your fingertips, and give you the know-how to complete even the most complicated building or repairing job, in the same professional way.

Each volume:
- **Fully illustrated**
- **Packed with up-to-date facts and figures**
- **Completely indexed for easy reference**

APPLIANCES

REFRIGERATION: HOME AND COMMERCIAL
Covers the whole realm of refrigeration equipment from fractional-horsepower water coolers, through domestic refrigerators to multi-ton commercial installations. 656 pages; 5½ x 8¼; hardbound. **Cat. No. 23286 Price: $11.95**

AIR CONDITIONING: HOME AND COMMERCIAL
A concise collection of basic information, tables, and charts for those interested in understanding, troubleshooting, and repairing home air conditioners and commercial installations. 464 pages; 5½ x 8¼; hardbound. **Cat. No. 23288 Price: $8.95**

HOME APPLIANCE SERVICING, 3rd Edition
A practical book for electric & gas servicemen, mechanics & dealers. Covers the principles, servicing, and repairing of home appliances. 592 pages; 5¼ x 8¼; hardbound. **Cat. No 23214 Price: $16.95**

REFRIGERATION AND AIR CONDITIONING LIBRARY—2 Vols.
Cat. No. 23305 Price: $16.95

OIL BURNERS, 3rd Edition
Provides complete information on all types of oil burners and associated equipment. Discusses burners—blowers—ignition transformers--electrodes—nozzles—fuel pumps—filters—Controls. Installation and maintenance are stressed. 320 pages; 5½ x 8¼; hardbound. **Cat. No. 23277 Price: $8.95**

Use the order coupon on the back page of this book.
All prices are subject to change without notice.

AUTOMOTIVE

AUTO BODY REPAIR FOR THE DO-IT-YOURSELFER

Shows how to use touch-up paint; repair chips, scratches, and dents; remove and prevent rust; care for glass, doors, locks, lids, and vinyl tops; and clean and repair upholstery. 96 pages; 8½ x 11; softcover. **Cat. No. 23238 Price: $5.95**

AUTOMOBILE REPAIR GUIDE, 4th Edition

A practical reference for auto mechanics, servicemen, trainees, and owners Explains theory, construction, and servicing of modern domestic motorcars. 800 pages; 5½ x 8¼; hardbound. **Cat. No. 23291 Price: $13.95**

CAN-DO TUNE-UP™ SERIES

Each book in this series comes with an audio tape cassette. Together they provide an organized set of instructions that will show you and talk you through the maintenance and tune-up procedures designed for your particular car. All books are softcover.

AMERICAN MOTORS CORPORATION CARS

(The 1964 thru 1974 cars covered include: Matador, Rambler, Gremlin, and AMC Jeep (Willys).). 112 pages; 5½ x 8½; softcover. **Cat. No. 23843 Price: $7.95**
Cat. No. 23851 Without Cassette **Price: $4.95**

CHRYSLER CORPORATION CARS

(The 1964 thru 1974 cars covered include: Chrysler, Dodge, and Plymouth.) 112 pages; 5½ x 8½; softcover. **Cat. No. 23825 Price $7.95**
Cat. No. 23846 Without Cassette **Price: $4.95**

FORD MOTOR COMPANY CARS

(The 1954 thru 1974 cars covered include: Ford, Lincoln, and Mercury.) 112 pages; 5½ x 8½; softcover. **Cat. No. 23827 Price: $7.95**
Cat. No. 23848 Without Cassette **Price: $4.95**

GENERAL MOTORS CORPORATION CARS

(The 1964 thru 1974 cars covered include: Buick, Cadillac, Chevrolet, Oldsmobile and Pontiac.) 112 pages; 5½ x 8½; softcover. **Cat. No. 23824 Price: $7.95**
Cat. No. 23845 Without Cassette **Price: $4.95**

PINTO AND VEGA CARS,

1971 thru 1974. 112 pages· 5½ x 8½; softcover. **Cat. No. 23831 Price: $7.95**
Cat. No. 23849 Without Cassette **Price: $4.95**

TOYOTA AND DATSUN CARS,

1964 thru 1974. 112 pages; 5½ x 8½; softcover. **Cat. No. 23835 Price: $7.95**
Cat. No. 23850 Without Cassette **Price: $4.95**

VOLKSWAGEN CARS

(The 1964 thru 1974 cars covered include: Beetle, Super Beetle, and Karmann Ghia.) 96 pages; 5½ x 8½; softcover. **Cat. No. 23826 Price: $7.95**
Cat. No. 23847 Without Cassette **Price: $4.95**

AUTOMOTIVE AIR CONDITIONING

You can easily perform most all service procedures you've been paying for in the past. This book covers the systems built by the major manufacturers, even after-market installations. Contents: introduction—refrigerant—tools—air conditioning circuit—general service procedures—electrical systems—the cooling system—system diagnosis—electrical diagnosis—troubleshooting. 232 pages; 5½ x 8½; softcover. **Cat. No. 23318 Price: $6.95**

Use the order coupon on the back page of this book.
All prices are subject to change without notice.

DIESEL ENGINE MANUAL, 3rd Edition

A practical guide covering the theory, operation, and maintenance of modern diesel engines. Explains diesel principles—valves—timing—fuel pumps—pistons and rings—cylinders—lubrication—cooling system—fuel oil and more. 480 pages; 5½ x 8¼; hardbound. **Cat. No. 23199 Price: $9.95**

GAS ENGINE MANUAL, 2nd Edition

A completely practical book covering the construction, operation, and repair of all types of modern gas engines. 400 pages; 5½ x 8¼; hardbound. **Cat. No. 23245 Price: $8.95**

BUILDING AND MAINTENANCE

ANSWERS ON BLUEPRINT READING, 3rd Edition

Covers all types of blueprint reading for mechanics and builders. This book reveals the secret language of blueprints, step-by-step in easy stages. 312 pages; 5½ x 8¼; hardbound. **Cat. No. 23283 Price: $7.95**

BUILDING MAINTENANCE, 2nd Edition

Covers all the practical aspects of building maintenance. Painting and decorating; plumbing and pipe fitting; carpentry; heating maintenance; custodial practices and more. (A book for building owners, managers, and maintenance personnel.) 384 pages; 5½ x 8¼; hardbound. **Cat. No. 23278 Price: $8.95**

COMPLETE BUILDING CONSTRUCTION

At last—a *one-volume* instruction manual to show you how to construct a frame or brick building from the footings to the ridge. Build your own garage, tool shed, other outbuilding—even your own house or place of business. Building construction tells you how to lay out the building and excavation lines on the lot; how to make concrete forms and pour the footings and foundation; how to make concrete slabs, walks, and driveways; how to lay concrete block. brick and tile; how to build your own fireplace and chimney: It's one of the newest Audel books, clearly written by experts in each field and ready to help you every step of the way. 800 pages; 5½ x 8¼; hardbound. **Cat. No. 23323 Price: $19.95**

GARDENING & LANDSCAPING

A comprehensive guide for homeowners and for industrial, municipal, and estate groundskeepers. Gives information on proper care of annual and perennial flowers; various house plants; greenhouse design and construction; insect and rodent controls; and more. 384 pages; 5½ x 8¼; hardbound. **Cat. No. 23229 Price: $8.95**

CARPENTERS & BUILDERS LIBRARY, 4th Edition (4 Vols.)

A practical, illustrated trade assistant on modern construction for carpenters, builders. and all woodworkers. Explains in practical, concise language and illustrations all the principles, advances, and shortcuts based on modern practice. How to calculate various jobs. **Cat. No. 23244 Price: $27.95**

Vol. 1—Tools, steel square. saw filing, joinery cabinets. 384 pages; 5½ x 8¼; hardbound. **Cat. No. 23240 Price: $7.95**
Vol. 2—Mathematics, plans, specifications, estimates 304 pages; 5½ x 8¼; hardbound. **Cat. No. 23241 Price: $7.95**
Vol. 3—House and roof framing, laying out foundations. 304 pages; 5½ x 8¼; hardbound. **Cat. No. 23242 Price: $7.95**
Vol. 4—Doors, windows, stairs, millwork, painting. 368 pages; 5½ x 8¼; hardbound. **Cat. No. 23243 Price: $7.95**

Use the order coupon on the back page of this book.
All prices are subject to change without notice.

PLUMBERS AND PIPE FITTERS LIBRARY—3 Vols.

A practical, illustrated trade assistant and reference for master plumbers, journeymen and apprentice pipe fitters, gas fitters and helpers, builders, contractors, and engineers. Explains in simple language, illustrations, diagrams, charts, graphs, and pictures, the principles of modern plumbing and pipe-fitting practices. **Cat. No. 23255 Price $20.95**

  Vol. 1—Materials, tools, roughing-in. 320 pages; 5½ x 8¼; hardbound. **Cat. No. 23256 Price: $7.95**

  Vol. 2—Welding, heating, air-conditioning. 384 pages; 5½ x 8¼; hardbound. **Cat. No. 23257 Price: $7.95**

  Vol. 3—Water supply, drainage, calculations. 272 pages; 5½ x 8¼; hardbound. **Cat. No. 23258 Price: $7.95**

PLUMBERS HANDBOOK

A pocket manual providing reference material for plumbers and/or pipe fitters. General information sections contain data on cast-iron fittings, copper drainage fittings, plastic pipe, and repair of fixtures. 288 pages; 4 x 6; softcover. **Cat. No. 23339 Price: $5.95**

QUESTIONS AND ANSWERS FOR PLUMBERS EXAMINATIONS, 2nd Edition

Answers plumbers' questions about types of fixtures to use, size of pipe to install, design of systems, size and location of septic tank systems, and procedures used in installing material. 256 pages; 5½ x 8¼; softcover. **Cat. No. 23285 Price: $5.95**

TREE CARE MANUAL

The conscientious gardener's guide to healthy, beautiful trees. Covers planting, grafting, fertilizing, pruning, and spraying. Tells how to cope with insects, plant diseases, and environmental damage. 224 pages; 8½ x 11; softcover. **Cat. No. 23280 Price: $8.95**

UPHOLSTERING

Upholstering is explained for the average householder and apprentice upholsterer. From repairing and regluing of the bare frame, to the final sewing or tacking, for antiques and most modern pieces, this book covers it all. 400 pages; 5½ x 8¼; hardbound. **Cat. No. 23189 Price: $7.95**

WOOD FURNITURE: Finishing, Refinishing, Repairing

Presents the fundamentals of furniture repair for both veneer and solid wood. Gives complete instructions on refinishing procedures, which includes stripping the old finish, sanding, selecting the finish and using wood fillers. 352 pages; 5½ x 8¼; hardbound. **Cat. No. 23216 Price: $8.95**

ELECTRICITY/ELECTRONICS

ELECTRICAL LIBRARY

If you are a student of electricity or a practicing electrician, here is a very important and helpful library you should consider owning. You can learn the basics of electricity, study electric motors and wiring diagrams, learn how to interpret the NEC, and prepare for the electrician's examination by using these books. **Cat. No. 23359 Price: $38.95**

  Electric Motors, 3rd Edition. 528 pages; 5½ x 8¼; hardbound. **Cat. No. 23264 Price: $8.95**

  Guide to the 1978 National Electrical Code. 672 pages; 5½ x 8¼; hardbound. **Cat. No. 23308 Price: $9.95**

  House Wiring, 4th Edition. 256 pages; 5½ x 8¼; hardbound. **Cat. No. 23315 Price: $7.95**

  Practical Electricity, 3rd Edition. 496 pages; 5½ x 8¼; hardbound. **Cat. No. 23218 Price: $8.95**

  Questions and Answers for Electricians Examinations, 6th Edition. 288 pages; 5½ x 8¼; hardbound. **Cat. No. 23307 Price: $7.95**

ELECTRICAL COURSE FOR APPRENTICES AND JOURNEYMEN

A study course for apprentice or journeymen electricians. Covers electrical theory and its applications. 448 pages; 5½ x 8¼; hardbound. **Cat. No. 23209 Price: $8.95**

Use the order coupon on the back page of this book.

All prices are subject to change without notice.

CARPENTRY AND BUILDING

Answers to the problems encountered in today's building trades. The actual questions asked of an architect by carpenters and builders are answered in this book. 448 pages; 5½ x 8¼; hardbound: **Cat. No. 23142 Price: $8.95**

WOOD STOVE HANDBOOK

The wood stove handbook shows how wood burned in a modern wood stove offers an immediate, practical, low-cost method of full-time or part-time home heating. The book points out that wood is plentiful, low in cost (sometimes free), and nonpolluting, especially when burned in one of the newer and more efficient stoves. In this book, you will learn about the nature of heat and its control, what happens inside and outside a stove, how to have a safe and efficient chimney, and how to install a modern wood burning stove. You will also learn about the different types of firewood and how to get it, cut it, split it, and store it. 128 pages; 8½ x 11; softcover. **Cat. No. 23319 Price: $6.95**

HEATING, VENTILATING, AND AIR CONDITIONING LIBRARY (3 Vols.)

This three-volume set covers all types of furnaces, ductwork, air conditioners, heat pumps, radiant heaters, and water heaters, including swimming-pool heating systems. **Cat. No. 23227 Price: $32.95**

Volume 1

Partial Contents: Heating Fundamentals . . . Insulation Principles . . . Heating Fuels . . . Electric Heating System . . . Furnace Fundamentals . . . Gas-Fired Furnaces . . . Oil-Fired Furnaces . . . Coal-Fired Furnaces . . . Electric Furnaces. **Cat. No. 23248 Price: $11.95**

Volume 2

Partial Contents: Oil Burners . . . Gas Burners . . . Thermostats and Humidistats . . . Gas and Oil Controls . . . Pipes, Pipe Fitting, and Piping Details . . . Valves and Valve Installations. 560 pages; 5½ x 8¼; hardbound. **Cat. No. 23249 Price: $11.95**

Volume 3

Partial Contents: Radiant Heating . . . Radiators, Convectors, and Unit Heaters . . . Stoves, Fireplaces, and Chimneys . . . Water Heaters and Other Appliances . . . Central Air Conditioning Systems . . . Humidifiers and Dehumidifiers. 544 pages; 5½ x 8¼; hardbound. **Cat. No. 23250 Price: $11.95**

HOME MAINTENANCE AND REPAIR: Walls, Ceilings, and Floors

Easy-to-follow instructions for sprucing up and repairing the walls, ceiling, and floors of your home. Covers nail pops, plaster repair, painting, paneling, ceiling and bathroom tile, and sound control. 80 pages; 8½ x 11; softcover. **Cat. No. 23281 Price: $5.95**

HOME PLUMBING HANDBOOK, 2nd Edition

A complete guide to home plumbing repair and installation. 200 pages; 8½ x 11; softcover. **Cat. No. 23321 Price: $7.95**

MASONS AND BUILDERS LIBRARY—2 Vols.

A practical, illustrated trade assistant on modern construction for bricklayers, stonemasons, cement workers, plasterers, and tile setters. Explains all the principles, advances, and shortcuts based on modern practice—including how to figure and calculate various jobs. **Cat. No. 23185 Price: $13.95**

Vol. 1—Concrete, Block, Tile, Terrazzo. 368 pages; 5½ x 8¼; hardbound. **Cat. No. 23182 Price: $7.95**

Vol. 2—Bricklaying, Plastering, Rock Masonry, Clay Tile. 384 pages; 5½ x 8¼; hardbound. **Cat. No. 23183 Price: $7.95**

Use the order coupon on the back page of this book.

All prices are subject to change without notice.

RADIOMANS GUIDE, 4th Edition

Contains the latest information on radio and electronics from the basics through transistors. 480 pages; 5½ x 8¼; hardbound. **Cat. No. 23259 Price: $10.95**

TELEVISION SERVICE MANUAL, 4th Edition

Provides the practical information necessary for accurate diagnosis and repair of both black-and-white and color television receivers. 512 pages; 5½ x 8¼; hardbound. **Cat. No. 23247 Price: $8.95**

ENGINEERS/MECHANICS/ MACHINISTS

MACHINISTS LIBRARY, 2nd Edition

Covers modern machine-shop practice. Tells how to set up and operate lathes, screw and milling machines, shapers, drill presses, and all other machine tools. A complete reference library. **Cat. No. 23300 Price: $23.95**

Vol. 1—Basic Machine Shop. 352 pages; 5½ x 8¼; hardbound. **Cat. No. 23301 Price: $8.95**

Vol. 2—Machine Shop. 480 pages; 5½ x 8¼; hardbound. **Cat. No. 23302 Price: $8.95**

Vol. 3—Toolmakers Handy Book. 400 pages; 5½ x 8¼; hardbound. **Cat. No. 23303 Price: $8.95**

MECHANICAL TRADES POCKET MANUAL

Provides practical reference material for mechanical tradesmen. This handbook covers methods, tools, equipment, procedures, and much more. 256 pages; 4 x 6; softcover. **Cat. No. 23215 Price: $5.95**

MILLWRIGHTS AND MECHANICS GUIDE, 2nd Edition

Practical information on plant installation, operation, and maintenance for millwrights, mechanics, maintenance men, erectors, riggers, foremen, inspectors, and superintendents. 960 pages; 5½ x 8¼; hardbound. **Cat. No. 23201 Price: $12.95**

POWER PLANT ENGINEERS GUIDE, 2nd Edition

The complete steam or diesel power-plant engineer's library. 816 pages; 5½ x 8¼; hardbound. **Cat. No. 23220 Price: $15.95**

QUESTIONS AND ANSWERS FOR ENGINEERS AND FIREMANS EXAMINATIONS, 3RD EDITION

Presents both legitimate and "catch" questions with answers that may appear on examinations for engineers and firemans licenses for stationary, marine, and combustion engines. 496 pages; 5½ x 8¼; hardbound. **Cat. No. 23327 Price $8.95**

WELDERS GUIDE, 2nd Edition

This new edition is a practical and concise manual on the theory, practical operation, and maintenance of all welding machines. Fully covers both electric and oxy-gas welding. 928 pages; 5½ x 8¼; hardbound. **Cat. No. 23202 Price: $12.95**

WELDER/FITTERS GUIDE

Provides basic training and instruction for those wishing to become welder/fitters. Step-by-step learning sequences are presented from learning about basic tools and aids used in weldment assembly, through simple work practices, to actual fabrication of weldments. 160 pages · 8½ x 11; softcover; **Cat. No. 23325 Price: $7.95**

Use the order coupon on the back page of this book.

All prices are subject to change without notice.

FLUID POWER

PNEUMATICS AND HYDRAULICS, 3rd Edition
Fully discusses installation, operation, and maintenance of both HYDRAULIC AND PNEUMATIC (air) devices. 496 pages; 5½ x 8¼; hardbound. **Cat. No. 23237 Price: $9.95**

PUMPS, 3rd Edition
A detailed book on all types of pumps from the old-fashioned kitchen variety to the most modern types. Covers construction, application, installation, and troubleshooting. 480 pages; 5½ x 8¼; hardbound. **Cat. No. 23292 Price: $9.95**

HYDRAULICS FOR OFF-THE-ROAD EQUIPMENT
Everything you need to know from basic hydraulics to troubleshooting hydraulic systems on off-the-road equipment. Heavy-equipment operators, farmers, fork-lift owners and operators, mechanics—all need this practical, fully illustrated manual. 272 pages; 5½ x 8¼; hardbound. **Cat. No. 23306 Price: $6.95**

HOBBY

COMPLETE COURSE IN STAINED GLASS
Written by an outstanding artist in the field of stained glass, this book is dedicated to all who love the beauty of the art. Ten complete lessons describe the required materials, how to obtain them, and explicit directions for making several stained glass projects. 80 pages; 8½ x 11; softbound. **Cat. No. 23287 Price: $5.95**

BUILD YOUR OWN AUDEL DO-IT-YOURSELF LIBRARY AT HOME!

Use the handy order coupon today to gain the valuable information you need in all the areas that once required a repairman. Save money and have fun while you learn to service your own air conditioner, automobile, and plumbing. Do your own professional carpentry, masonry, and wood furniture refinishing and repair. Build your own security systems. Find out how to repair your TV or Hi-Fi. Learn landscaping, upholstery, electronics and much, much more.

All prices are subject to change without notice.

HERE'S HOW TO ORDER

Select the Audel book(s) you want, fill in the order card below, detach and mail today. Send no money now. You'll have 15 days to examine the books in the comfort of your own home. If not completely satisfied, simply return your order and owe nothing.

If you decide to keep the books, we will bill you for the total amount, plus a small charge for shipping and handling.

1. Enter the correct catalog number(s) of the book(s) you want in the space(s) provided.

2. Print your name, address, city, state and zip code, clearly.

3. Detach the order card below and mail today. No postage is required.

Detach postage-free order card on perforated line

FREE TRIAL ORDER CARD

☐ Please rush the following book(s) for my free trial. I understand if I'm not completely satisfied, I may return my order within 15 days and owe nothing. Otherwise, you will bill me for the total amount plus a small postage & handling charge.

Write book catalog numbers at right.
(Numbers are listed with titles)

NAME_____

ADDRESS_____

CITY_____STATE_____ZIP_____

☐ Save postage & handling costs. Full payment enclosed (Plus sales tax, if any.)

Cash must accompany orders under $5.00.
Money-Back guarantee still applies.

DETACH POSTAGE-PAID REPLY CARD BELOW AND MAIL TODAY!

Just select your books, enter the code numbers on the order card, fill out your name and address, and mail. There's no need to send money.

15-Day Free Trial On All Books . . .

BUSINESS REPLY CARD
FIRST CLASS   PERMIT NO. 1076   INDIANAPOLIS, IND.

POSTAGE WILL BE PAID BY ADDRESSEE

ATTENTION: ORDER DEPT.

Theodore Audel & Company
4300 West 62nd Street
P.O. Box 7092
Indianapolis, Indiana 46206

NO POSTAGE NECESSARY IF MAILED IN THE UNITED STATES